Ergebnisse der Mathematik und ihrer Grenzgebiete 34

A Series of Modern Surveys in Mathematics

D. Mumford J. Fogarty F. Kirwan

Geometric Invariant Theory

Third Enlarged Edition

Springer-Verlag
Berlin Heidelberg New York
London Paris Tokyo
Hong Kong Barcelona
Budapest

David Mumford

Department of Mathematics
Harvard University
1 Oxford Street
Cambridge, MA 02138, USA

John Fogarty

Department of Mathematics and Statistics
University of Massachusetts
Amherst, MA 01003, USA

Frances Kirwan

Balliol College
Oxford OX1 3BJ, England

Mathematics Subject Classification (1991): 14D20, 14D25, 14Jxx, 14Kxx, 14L30, 20Gxx, 58F05

ISBN 3-540-56963-4 Springer-Verlag Berlin Heidelberg New York
ISBN 0-387-56963-4 Springer-Verlag New York Berlin Heidelberg

ISBN 3-540-11290-1 2. Aufl. Springer-Verlag Berlin Heidelberg New York
ISBN 0-387-11290-1 2nd edn. Springer-Verlag New York Berlin Heidelberg

Library of Congress Cataloging-in-Publication Data.
Mumford, David. Geometric invariant theory/D. Mumford, J. Fogarty, F. Kirwan –
3rd enlarged ed. p. cm.
(Ergebnisse der Mathematik und ihrer Grenzgebiete; 34)
Includes bibliographical references and index.
ISBN 0-387-56963-4
1. Geometry, Algebraic. 2. Invariants. 3. Moduli theory.
I. Fogarty, John, 1934– . II. Kirwan, Frances Clare, 1959– . III. Title.
IV. Series: Ergebnisse der Mathematik und ihrer Grenzgebiete; 2. Folge, Bd. 34.
QA564.M85 1994 516.3'5–dc20 93-33772 CIP

Printed in the United States of America

41/3140 - 5 4 3 2 1 — Printed on acid-free paper

Preface to third edition

Just over ten years have passed since the publication of the second edition of this book. In these ten years there have been many developments relating to geometric invariant theory. So many, in fact, that although the new references in this edition outnumber the references in the second edition by a very large margin they form only a selection of the work done in this area in the last decade[1].

This edition of the book has been extended to take account of one of these developments, one which was just hinted at in the second edition[2]. A close and very fruitful relationship has been discovered between geometric invariant theory for quasi-projective complex varieties and the moment map in symplectic geometry, and a chapter has been added describing this relationship and some of its applications. In an infinite-dimensional setting the moment map links geometric invariant theory and Yang-Mills theory, which has of course been the focus of much attention among mathematicians over the last fifteen years.

In style this extra chapter is closer to the appendices added in the second edition than to the original text. In particular no proofs are given where satisfactory references exist.

On the many other exciting developments related to geometric invariant theory since the publication of the second edition I regret that I lack sufficient expertise to do more than add some relevant publications to the list of references. Among these are several survey articles and books on invariant theory, moduli problems and related areas published in the last ten years, including [119], [349], [486], [489], [494], [508], [529], [545], [565], [579], [598], [602], [608], [613], [648], [686–688], [694], [698], [743], [769], [792], [800], [806], [809], [837], [845], [859], [868], [873], [879], [895], [899], and [908].

The reader should be warned that this edition is something of a collage. Chapters 1–7 were written twenty eight years ago. Chapter 8 is new. The Appendices were written about twelve years ago, but references to recent work in each area have been added in footnotes at the

[1] Indeed they form a rather haphazard selection, I fear, and my apologies are due to the many authors whose works have been omitted.

[2] See Appendix 2C.

beginning of each subsection. The other authors and I hope that it will still be useful.

Finally I would like to express my thanks to all those who have given me comments and advice, and in particular to David Mumford and Simon Donaldson for their help.

Oxford, December, 1992 FRANCES KIRWAN

Preface to second edition

In the 16 years since this book was published, there has been an explosion of activity in algebraic geometry. In particular, there has been great progress in invariant theory and the theory of moduli. Reprinting this monograph gave us the opportunity of making various revisions. The first edition was written primilarly as a research monograph on a geometric way to formulate invariant theory and its applications to the theory of moduli. In this edition we have left the original text essentially intact, but have added appendices, which sketch the progress on the topics treated in the text, mostly without proofs, but with discussions and references to all the original papers.

In the preface to the 1st edition, it was explained that most of the invariant theoretic results were proven only in char. 0 and that therefore the applications to the construction of moduli spaces were either valid only in char. 0, or else depended on the particular invariant theoretic results in Ch. 4 which could be established by elementary methods in all characteristics. However, a conjecture was made which would extend all the invariant theory to char. p. Fortunately, W. HABOUSH [128] has proved this conjecture, hence the book is now more straightforward and instead of giving one construction of the moduli space $\mathcal{M}_g$ of curves of genus g over $\boldsymbol{Q}$ by means of invariant theory, and one construction of $\mathcal{M}_g$ and $\mathcal{A}_{g,1}$ (the moduli space of principally polarized abelian varieties) over $\boldsymbol{Z}$ by means of the covariant of points of finite order and Torelli's theorem, we actually give or sketch the following constructions of these spaces:

1) $\mathcal{M}_g$ and $\mathcal{A}_g$ are constructed by proving the stability of the Chow forms of both curves and abelian varieties (Ch. 4, § 6; Appendix 3B)

2) $\mathcal{M}_g$ and $\mathcal{A}_g$ are independently constructed by covariants of finite sets of points (Appendix 7C; Ch. 7, § 3)

3) $\mathcal{A}_g$ is constructed by an explicit embedding by theta constants (Appendix 7B).

All of this is valid over $\boldsymbol{Z}$ except that the covariant approach to $\mathcal{M}_g$ uses higher Weierstrass points and is valid only over $\boldsymbol{Q}$ (unless one can prove their finiteness for high enough multiples $|nK|$, in char. p: see Appendix 7C).

This preface gives us the opportunity to draw attention to some basic open questions in invariant theory and moduli theory. 3 questions raised in the 1st edition have been answered:

a) the geometric reductivity of reductive groups has been proven by HABOUSH, op. cit.,

b) the existence of canonical destabilizing flags for unstable points has been proven by KEMPF [171] and ROUSSEAU [285],

c) the stability of Chow forms and Hilbert points of pluricanonically embedded surfaces of general type has been proven by GIESEKER [116]. Pursuing the ideas in c), leads one to ask:

d) which polarized elliptic and $K3$-surfaces have stable Chow forms? and what is more difficult probably:

e) can one compactify the moduli spaces of smooth surfaces by allowing suitable singular polarized surfaces which are still „asymptotically stable", i.e., have stable Chow forms when embedded by any complete linear system which is a sufficiently large multiple of the polarization? In a more classical direction, now that the reasons for the existence of $\mathcal{M}_g$ and $\mathcal{A}_g$ are so well understood, the time seems ripe to try to understand their geometry more deeply, e.g.

f) Can one calculate, or bound, some birational invariants* of $\mathcal{M}_g$ or $\mathcal{A}_g$? Investigate the cohomology ring and chow ring of $\mathcal{M}_g$ or $\mathcal{A}_g$.
and

g) Find explicit Siegel modular forms vanishing on the Jacobian locus or cutting out this locus, and relate the various known special properties of Jacobians.

For those who want to learn something of the subject of moduli, we want to say what they will *not* find here and where more background on these topics may be found. The subject of moduli divides at present into 3 broad areas: *deformation theory, geometric invariant theory*, and *the theory of period maps*. Deformation theory deals with local questions: infinitesimal deformations of a variety, or analytic germs of deformations. Period maps deal with the construction of moduli of Hodge structures and the construction of moduli spaces of vareties by attaching a family of Hodge structures to a family of vareties. Both of these subjects are discussed only briefly in this monograph (see Appendix 5B). Unfortunately, deformation theory has not received a systematic treatment by anyone: a general introduction to the theory of moduli as a whole including deformation theory is given in SESHADRI [304], deformations of *singularities* are treated in ARTIN [45]. The origins of the algebraic

* For large g, $\mathcal{M}_g$ and $\mathcal{A}_g$ are varieties of general type: FREITAG [108], [109], HARRIS-MUMFORD [337], HARRIS [603], MUMFORD [755]. See new references to Appendix 5D.

treatment of the subject are in GROTHENDIECK [13], exp. 195 and SCHLESSINGER, LICHTENBAUM [289], [185]. The theory of period maps is largely the creation of GRIFFITHS [121], and a survey of the theory can be found in GRIFFITHS-SCHMID [124]. Expository or part expository/part research articles on geometric invariant theory proper and related questions of moduli have been written by DIEUDONNÉ-CARREL [85], GIESEKER [119], NEWSTEAD [247], and MUMFORD [213], [218], [220]. We hope this monograph will help to make the subject accessible.

Writing these appendices has brought home to us very clearly how many people have been involved in invariant theory and moduli problems. It has been exciting to try to express coherently all their results and their interconnections. We are sure, however, that some people have been overlooked. For this we can only offer the hackneyed excuse that even together we had only four hands and two heads.

D. MUMFORD
J. FOGARTY
Cambridge, Mass.
November, 1981

Preface to first edition

The purpose of this book is to study two related problems: when does an orbit space of an algebraic scheme acted on by an algebraic group exist? And to construct moduli schemes for various types of algebraic objects. The second problem appears to be, in essence, a special and highly non-trivial case of the first problem. From an Italian point of view, the crux of both problems is in passing from a birational to a biregular point of view. To construct both orbit spaces and moduli "generically" are simple exercises. The problem is whether, within the set of all models of the resulting birational class, there is one model whose geometric points classify the set of orbits in some action, or the set of algebraic objects in some moduli problem. In both cases, it is quite possible that some orbits, or some objects are so exceptional, or, as we shall say, are *not stable*, so that they must be left out of the model. The difficulty is to pin down the meaning of stability in a given case. One of the most intriguing unsolved problems, in this regard, is that of the moduli of non-singular polarized surfaces. Which such surfaces are not stable, in the sense that there is no moduli scheme for them and their deformations? This property is very delicate.

One of my principles has been not to worry too much about the difference between characteristic 0, and finite characteristics. A large part of this book is, therefore, devoted to a theory developed only in characteristic 0.* I am convinced, however, that it is almost entirely valid in all characteristics. What is necessary is to find some property of semi-simple algebraic groups in all characteristics which takes the place of the full reducibility of representations which is valid only in characteristic 0. I conjecture, in fact, that if a semi-simple algebraic group G is represented in a vector space V, and if V_0 is an invariant subspace of codimension 1, then *for some* ν the invariant subspace of codimension 1

$$V_0 \cdot S^{p^\nu - 1}(V) \subset S^{p^\nu}(V)$$

* The hypothesis of characteristic 0 is disguised in the assumption that the group which is acting is reductive (by reductive we always mean that all its representations are completely reducible). But, in characteristic p, only relatively unimportant groups are reductive (cf. [28]), so the theory is uninteresting.

is complemented by an invariant 1-dimensional subspace.** Nonetheless, this is unknown, and one consequence is that this book is divided fairly sharply into two halves. Although both parts are closely analogous, they are logically independent. One half consists in Chapters 1, 2, 4 and 5 which deal essentially only with characteristic zero, and yield a construction of the moduli scheme for curves over $\boldsymbol{Q}$. The, other half consists in Chapters 3, 6, and 7, which deal with the ,,arithmetic case", i.e., over Spec ($\boldsymbol{Z}$), and yield a construction of the moduli schemes for curves over $\boldsymbol{Z}$. From a standpoint of content, however, Chapters 1, 2, 3 and 4 deal with orbit space problems, while Chapters 5, 6 and 7 deal with moduli.

This book is written entirely in the language of schemes. Of course, the results, for most purposes, could have been stated and proven in a classical language. However, it seems to me that algebraic geometry fulfills only in the language of schemes that essential requirement of all contemporary mathematics: to state its definitions and theorems in their natural abstract and formal setting in which they can be considered independently of geometric intuition. Moreover, it seems to me incorrect to assume that any geometric intuition is lost thereby: for example, the underlying variety in an algebraic scheme is rediscovered, and perhaps better understood through the concept of geometric points. As another example, the theory of schemes has made it possible, in a very intuitive way, to finally dispose of that famous embarrassment to the Italian school: the lack of an algebraic proof of the completeness of the characteristic linear system of suitable complete continuous systems on a surface in characteristic O (cf. [40, 18, 33 and 36]).

It is my pleasure to acknowledge at this point the great encouragement and stimulation which *I* have received from OSCAR ZARISKI, JOHN TATE, and ALEXANDER GROTHENDIECK. In addition, *I* want to give credit to the many mathematicians from whom *I* have taken a great deal. This book is primarily an original monograph, but secondarily an exposition of a whole topic, so *I* have taken the liberty of including anybody else's results when relevant. *I* am particularly conscious of my indebtedness to GROTHENDIECK, HILBERT, and NAGATA. It is impossible to enumerate all the sources from which *I* have borrowed, but this is a partial list:

Ch. 1 and 2 owe a great deal to D. HILBERT [14],

§ 1.2. was developed independently by C. CHEVALLEY, N. IWAHORI, and M. NAGATA,

§ 2.2. is largely a theory of J. TITS,

Ch. 3 was worked out by J. TATE and myself,

** Here $S^k V$ stands for the kth symmetric power of V.

§ 4.3 includes an example of M. NAGATA,

§ 4.5 is a theorem of B. KOSTANT.

In Ch. 5 and 7, the whole approach to moduli via functors is due to A. GROTHENDIECK,

§ 5.3 and 5.4 follow suggestions of A. GROTHENDIECK,

Ch. 6 is almost entirely the work of A. GROTHENDIECK.

Finally a word about references: the tremendous contributions made by GROTHENDIECK to both the technique and the substance of algebraic geometry have not always been paralleled by their publication in permanently available form. In particular, for many of his results, we have only the barest outlines of proofs, as presented in the Bourbaki Seminar (reprinted in [13]). Nonetheless, since all the results which *I* want to use have been presented in detail in seminars at Harvard and will be published before too long by GROTHENDIECK, there seems no harm in making full use of them. For the convenience of the reader, the results which are only to be found in [13], and some others for which no good reference is available, are reproduced in Ch. 0, § 5. We have not reproduced, however, the results which we need from the semi-published Seminar Notes "Séminaire géométrie algébrique, IHES, 1960—61" since full proofs appear there. The results in exposés 3 and 8 of these notes are among the most vital tools which we use, and a familiarity with them is essential in order to read Chapters 6 and 7.

Harvard University, March, 1965

DAVID MUMFORD

Contents

Chapter 0

Preliminaries

We list first some notations and conventions which we will follow:

(1) A "pre-scheme X/S" means a morphism from the pre-scheme X to the pre-scheme S. If $S = \mathrm{Spec}\,(R)$, we shall abbreviate this to "a pre-scheme X/R".

(2) An S-valued point of a pre-scheme X means a morphism from S to X. If $S = \mathrm{Spec}\,(R)$, we shall abbreviate this to "an R-valued point" of X. If, moreover, R is an algebraically closed field, such a point will be referred to as a "geometric point" of X.

(3) Given pre-schemes X/k, Y/k, where k is a field fixed in some discussion, then all morphisms $f: X \to Y$ will be understood to be k-morphisms; and $X \underset{\mathrm{Spec}(k)}{\times} Y$ will be be abbreviated to $X \times Y$. Moreover, $\bar{k}$ will stand for an algebraic closure of k, and we shall abbreviate $X \times \mathrm{Spec}\,(\bar{k})$ to $\bar{X}$. In this case, an algebraic pre-scheme* X/k will be called a pre-*variety* if $\bar{X}$ is irreducible and reduced. Finally, given a k-rational point $x \,\varepsilon\, X$, the image point, as reduced subscheme of X, will be denoted $\{x\}$.

(4) If Z is a closed subscheme of X, or a cycle on X, then supp (Z) will denote the closed subset of X which is the support of Z.

(5) The symbols $\boldsymbol{A}^n$, and $\boldsymbol{P}_n$ will denote affine n-space and projective n-space *over* $\boldsymbol{Z}$, (i.e. Proj $\boldsymbol{Z}\,[X_0, \ldots, X_n]$) unless, in a particular chapter, all considerations are over a ground field k, in which case they will denote affine n-space and projective n-space over k. $PGL\,(n)$ will be the projective group acting on $\boldsymbol{P}_n$, $GL\,(n)$ the general linear group acting on $\boldsymbol{A}^n$, and $\boldsymbol{G}_m = GL\,(1)$. The same conventions on the base scheme hold for these group schemes.

(6) 1_X is the identity morphism from X to X. p_1 and p_2 are the projections from $X \underset{S}{\times} Y$ to X and Y; p_{12}, etc. are the projections from $X \underset{S}{\times} Y \underset{S}{\times} Z$ to $X \underset{S}{\times} Y$, etc. If $f: X_1 \to X_2$ and $g: Y_1 \to Y_2$ are S-morphisms, then $f \times g: X_1 \underset{S}{\times} Y_1 \to X_2 \underset{S}{\times} Y_2$ is the product. If $f: X \to Y_1$ and

---

* i.e. a scheme of finite type over Spec(k).

$g: X \to Y_2$ are S-morphisms, then $(f, g): X \to Y_1 \times_S Y_2$ is the induced S-morphism.

(7) If X is a pre-scheme, and $x \in X$, then $\varkappa(x)$ will denote the residue field of $\mathfrak{o}_{x,X}$. If X is a scheme over a field k, X_k will denote the set of points $x \in X$ such that $k \cong \varkappa(x)$.

§ 1. Definitions

Definition 0.1. A group pre-scheme G/S is a morphism $\pi: G \to S$ of pre-schemes, plus S-morphisms $\mu: G \times_S G \to G$, $\beta: G \to G$, $e: S \to G$ satisfying the usual identities:

(a) *Associativity:*

$$\begin{array}{ccc} G \times_S G \times_S G & \xrightarrow{1_G \times \mu} & G \times_S G \\ \downarrow{\scriptstyle \mu \times 1_G} & & \downarrow{\scriptstyle \mu} \\ G \times_S G & \xrightarrow{\mu} & G \end{array}$$ commutes.

(b) *Law of inverse:* The compositions

$$G \xrightarrow{\Delta} G \times_S G \underset{\beta \times 1_G}{\overset{1_G \times \beta}{\rightrightarrows}} G \times_S G \xrightarrow{\mu} G$$

both equal $e \circ \pi$ (here Δ is the diagonal).

(c) *Law of identity:* The compositions

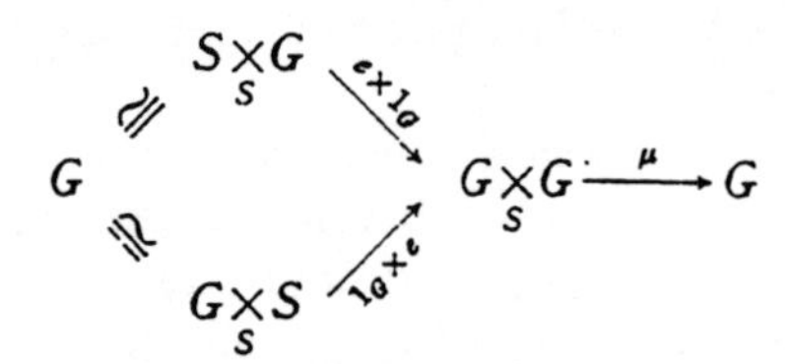

both equal 1_G.

Definition 0.2. An *algebraic group* G over a field k, is a group pre-scheme G/k which is an algebraic *scheme, smooth* over k.

Definition 0.3. A group pre-scheme G/S *acts* or *operates* on a pre-scheme X/S if an S-morphism $\sigma: G \times_S X \to X$ is given, such that:

(a)

$$\begin{array}{ccc} G \times_S G \times_S X & \xrightarrow{1_G \times \sigma} & G \times_S X \\ \downarrow{\scriptstyle \mu \times 1_X} & & \downarrow{\scriptstyle \sigma} \\ G \times_S X & \xrightarrow{\sigma} & X \end{array}$$

commutes (where μ is the group law for G).

(b) The composition:

$$X \cong S \underset{S}{\times} X \xrightarrow{e \times 1_X} G \underset{S}{\times} X \xrightarrow{\sigma} X$$

equals 1_X (where e is the identity morphism for G).

Definition 0.4. Let $f: T \to X$ be a T-valued point of X. Then $\sigma \circ (1_G \times f)$ is a morphism from $G \underset{S}{\times} T$ to X. Define the morphism

$$\psi_f^G : G \underset{S}{\times} T \to X \underset{S}{\times} T$$

as $(\sigma \circ (1_G \times f), p_2)$. If no confusion arises, ψ_f^G will be shortened to ψ_f. If $f = 1_X$, ψ_f will be denoted Ψ. Ψ is simply:

$$(\sigma, p_2) : G \underset{S}{\times} X \to X \underset{S}{\times} X.$$

The image of ψ_f will be denoted $0(f)$ and called the *orbit* of f. Now $X \underset{S}{\times} T$, as a scheme over T, has a canonical section, namely $(f, 1_T)$. Via this, we set up a fibre product defining $S(f)$:

$$\begin{array}{ccc} S(f) & \longrightarrow & T \\ \downarrow & & \downarrow {\scriptstyle (f, 1_T)} \\ G \underset{S}{\times} T & \xrightarrow{\psi_f} & X \underset{S}{\times} T \end{array}$$

Now $G \underset{S}{\times} T$ is, of course, a group pre-scheme over T, and it is not hard to show that $S(f)$ is a subgroup pre-scheme over T. It is called the *stabilizer* of f.

In case $T = \text{Spec}\,(k)$, and f is a closed immersion of a point $x \in X$ into X, we shall also write $0(x)$ for $0(f)$, and $S(x)$ for $S(f)$. In case $T = \text{Spec}\,(\Omega)$ and Ω is an algebraically closed field, then f is a geometric point of X, and it will usually be denoted by a letter x, y, etc. Then note that $0(x)$ is a subset of $\bar{X}$ and $S(x)$ is a subgroup pre-scheme of $\bar{G}$, where $\bar{X}, \bar{G}$ are the geometric fibres of X, G respectively over the geometric point of S which is under f.

Definition 0.5. Given an action σ of G/S on X/S, a pair (Y, ϕ) consisting of a pre-scheme Y/S and an S-morphism $\phi: X \to Y$ will be called a *categorical quotient* (of X by G) if

i) the diagram:

$$\begin{array}{ccc} G \underset{S}{\times} X & \xrightarrow{\sigma} & X \\ \downarrow {\scriptstyle p_2} & & \downarrow {\scriptstyle \phi} \\ X & \xrightarrow{\phi} & Y \end{array}$$ commutes,

ii) given any pair (Z, ψ) consisting of a pre-scheme Z over S, and an S-morphism $\psi: X \to Z$ such that $\psi \circ \sigma = \psi \circ p_2$, i.e. (i) holds for Z and ψ, then there is a unique S-morphism $\chi: Y \to Z$ such that $\psi = \chi \circ \phi$.

Definition 0.6. Given an action σ of G/S on X/S, a pair (Y, ϕ) consisting of a pre-scheme Y over S and an S-morphism $\phi: X \to Y$ will be called a *geometric quotient* (of X by G) if

i) $\phi \circ \sigma = \phi \circ p_2$ (as in definition 0.5),

ii) ϕ is surjective, and the image of Ψ is $X \underset{Y}{\times} X$ (cf. definition 0.4); [equivalently, the geometric fibres of ϕ are precisely the orbits of the geometric points of X, for geometric points over an algebraically closed field of sufficiently high transcendence degree].*

iii) ϕ is submersive, i.e. a subset $U \subset Y$ is open if and only if $\phi^{-1}(U)$ is open in X.

$U' \subset Y'$ is open if and only if $\phi'^{-1}(U')$ is open in X'.

iv) the fundamental sheaf ϱ_Y is the subsheaf of $\phi_*(\varrho_X)$ consisting of invariant functions, i.e. if $f \varepsilon \Gamma(U, \phi_*(\varrho_X)) = \Gamma(\phi^{-1}(U), \varrho_X)$, then $f \varepsilon \Gamma(U, \varrho_Y)$ if and only if:

$$\begin{array}{ccc} G \times \phi^{-1}(U) & \xrightarrow{\sigma} & \phi^{-1}(U) \\ \downarrow{\scriptstyle p_2} & & \downarrow{\scriptstyle F} \\ \phi^{-1}(U) & \xrightarrow{F} & A^1 \end{array} \quad \text{commutes,}$$

(where F is the morphism defined by f).

Definition 0.7. Given an action σ of G/S on X/S, a pair (Y, ϕ) as above will be called a *universal categorical quotient* (resp. *universal geometric quotient*) if, for all morphisms $Y' \to Y$, we put $X' = X \underset{Y}{\times} Y'$ and let $\phi': X' \to Y'$ denote p_2, then (Y', ϕ') is a categorical quotient (resp. geometric quotient) of X' by. G. If this holds only for *flat* morphisms $Y' \to Y$, then (Y, ϕ) will be called a *uniform categorical quotient* (resp. *uniform geometric quotient*).

§ 2. First Properties

The above definitions are the basic concepts for everything that follows. Their first properties will be given in this section.

Proposition 0.1. Let σ be an action of G/S on X/S and suppose (Y, ϕ) is a geometric quotient of X by G. Then (Y, ϕ) is a categorical

* If G and X are of finite type over S and Y respectively, this is true for any algebraically closed field.

quotient of X by G, hence it is unique up to isomorphism. Moreover if (Y, ϕ) is a universal geometric quotient, then it is also a universal categorical quotient.

Proof. Suppose $\psi: X \to Z$ is any S-morphism such that $\psi \circ \sigma = \psi \circ p_2$, as morphisms from $G \underset{S}{\times} X$ to Z. To construct a morphism $\chi: Y \to Z$, let $\{V_i\}$ be any affine open covering of Z. Then for each i, $\psi^{-1}(V_i)$ is an invariant open subset of X, hence by condition (ii) of definition 0.6, $\psi^{-1}(V_i) = \phi^{-1}(U_i)$ for *some subset* U_i of Y*. But then, by condition (iii) of definition 0.6, U_i is necessarily open.

Now since ϕ is surjective, $\{U_i\}$ is an open covering of Y, and any morphism $\chi: Y \to Z$ such that $\psi = \chi \circ \phi$ must satisfy $\chi(U_i) \subset V_i$. Therefore it must be defined by a set of homomorphisms h_i such that the diagram:

$$\begin{array}{ccc} \Gamma(V_i, \varrho_Z) & \xrightarrow{h_i} & \Gamma(U_i, \varrho_Y) \\ \downarrow{\scriptstyle \psi^*} & & \downarrow{\scriptstyle \phi^*} \\ \Gamma(\psi^{-1}(V_i), \varrho_X) & = & \Gamma(\phi^{-1}(U_i), \varrho_X) \end{array}$$

commutes. Since ϕ^* is injective by condition (iv) of definition 0.6, h_i is also uniquely determined — if it exists — and hence at most one χ exists. But for any $g \,\varepsilon\, \Gamma(V_i, \varrho_Z)$, one checks that $\psi^*(g)$ is an *invariant* element of $\Gamma(\phi^{-1}(U_i), \varrho_X)$ in the sense of condition (iv): hence it is in the sub-ring $\phi^*[\Gamma(U_i, \varrho_Y)]$. Therefore such an h_i does exist.

This h_i defines $\chi_i: U_i \to V_i$. It remains only to check that $\chi_i = \chi_j$ on $U_i \cap U_j$ and this is immediate. This constructs χ. *QED.*

In the rest of this section, we shall analyze informally the various concepts of quotients, in a series of remarks. Therefore, we fix the notations S, G, X, Y, σ, ϕ as in Definitions 0.5, 0.6, and 0.7.

(1) S plays no essential role in any of these definitions. That is to say, we can replace S by Y, and G by the Y-group pre-scheme $G \underset{S}{\times} Y$ if we wish.

(2) Suppose (Y, ϕ) is a categorical quotient of X by G. Then, by means of the universal mapping property, it is easy to check the following implications:

X reduced $\Rightarrow Y$ reduced

X connected $\Rightarrow Y$ connected

X irreducible $\Rightarrow Y$ irreducible

X locally integral $\Rightarrow Y$ locally integral

X locally integral and normal $\Rightarrow Y$ locally integral and normal.

---

* In fact, by condition (ii), one proves easily that, if x, y are two points of X, then $\phi x = \phi y$ implies $\psi x = \psi y$.

On the other hand, it does not seem likely that if X is noetherian, then Y is noetherian, or even locally noetherian in all cases (this is an analog of Hilbert's 14th problem). I only know one fact in this direction: if $S = \text{Spec}\,(k)$, k a field, X is a normal algebraic scheme over k, and (Y, ϕ) is a *geometric* quotient, then Y is an algebraic scheme over k. Since we have no use for this, I omit the proof.

(3) Suppose (Y, ϕ) is a universal categorical quotient. Applying this assumption to the base extensions given by the inclusion of open subsets U in Y, one deduces condition (iv) of definition 0.6. Moreover, applying it to the inclusion of a single point y in Y, and noting that the categorical quotient of an empty scheme by any group is an empty scheme, one deduces that ϕ is surjective. Therefore, for (Y, ϕ) to be a universal geometric quotient, it is necessary and sufficient that (1) ϕ is universally submersive, and (2) the image of Ψ is $X \times_Y X$.

(4) In this remark we shall assume that all schemes are noetherian, all morphisms of finite type, and that the base S is even **normal**. A very useful hypothesis is that G is universally open over S. By Chevalley's criterion (EGA Ch. 4,14.4)* this is equivalent to assuming either that G is open over S, or that all the group schemes which occur as fibres of G over S have the same dimension. This implies, for example, that if (Y, ϕ) is a geometric quotient of X by G, ϕ is also universally open:

To see that ϕ is open, let $U \subset X$ be any open set. But $p_2 : G \times_S X \rightarrow X$ is open since G is universally open over S. Now σ can be factored:

$$G \times_S X \xrightarrow{(p_1, \sigma)} G \times_S X \xrightarrow{p_2} X$$

and (p_1, σ) is an isomorphism. Therefore σ is open. In particular $\sigma(G \times_S U) = U'$ is open. But $\phi(U) = \phi(U')$, and since $\phi(U')$ is a subset of X invariant under G, $U' = \phi^{-1}(\phi(U'))$. Since ϕ is submersive, this proves that $\phi(U')$, hence $\phi(U)$ is open. The same argument proves that ϕ is even universally open.

Another consequence concerns dimensions. Suppose ϕ: $X \rightarrow Y$ is a dominating S-morphism such that

i) $\phi \circ \sigma = \phi \circ p_2$

ii) for every algebraically closed field k, the geometric fibres of ϕ over k contain at most one orbit under $\overline{G} = G \times_S \text{Spec}\,(k)$.

---

* The abreviation SGA will always refer to [12], and the number will always be the exposé referred to. Similarly the abreviation EGA will always refer to [11], and the number will then be the chapter referred to.

For all $x \in X$, let

$\sigma(x) =$ dimension of stabilizer of x,

$\tau(x) =$ dimension of fibre $\phi^{-1}(\phi(x))$ at x.

By standard theorems on upper semi-continuity of dimensions (applied to the stabilizer $S(1_x)$ over S, and to ϕ) σ and τ are upper semi-continuous (cf. EGA 4, 13.1). On the other hand, if g is the dimension of all the group schemes occurring as fibres in G over S, then by (ii)

$$\sigma(x) + \tau(x) = g$$

for all x. Therefore σ and τ are both constant on all topological components of X.

Using this remark, we obtain the important criterion:

Proposition 0.2. Suppose X and Y are irreducible, normal noetherian pre-schemes over S, and suppose $\phi: X \rightarrow Y$ is a dominating S-morphism of finite type. Suppose that the residue field of the generic point of Y has characteristic 0. Suppose a group pre-scheme G, of finite type and universally open over S, acts on X via $\sigma: G \underset{S}{\times} X \rightarrow X$. Then if

i) $\phi \circ \sigma = \phi \circ p_2$,

ii) for every algebraically closed field k, the geometric fibres of ϕ over k contain at most one orbit under $\overline{G} = G \underset{S}{\times} \mathrm{Spec}\,(k)$,

it follows that ϕ is a universally open morphism, and $(\phi X, \phi)$ is a geometric quotient of X by G.

Proof. First of all, since G is universally open over S, all the group schemes $\overline{G}$ have the same dimension. Therefore, by ii) all components of all geometric fibres of ϕ have the same dimension; hence ϕ is universally open by Chevalley's criterion. To prove the second statement, we need only verify condition (iv) of Definition 0.6. Therefore let $U \subset \phi(X)$ be an open set. Since ϕ is dominating and Y is reduced, it is clear that:

$$\Gamma(U, \mathfrak{o}_Y) \hookrightarrow \Gamma(\phi^{-1}(U), \mathfrak{o}_X).$$

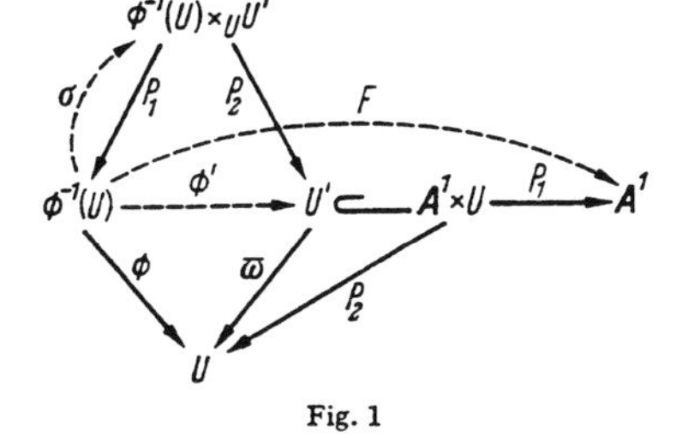

Fig. 1

Now let f be an invariant section of $\mathfrak{o}_X$ over $\phi^{-1}(U)$: we must show that f is a section of $\mathfrak{o}_Y$. Let f define the morphism:

$$F: \phi^{-1}(U) \rightarrow A^1.$$

Let U' be the reduced and irreducible subscheme of $A^1 \times U$ whose support is the closure of $(F, \phi)\,[\phi^{-1}(U)]$, (cf. Fig. 1). I claim that the projection ϖ from U' to U is an isomorphism: it then follows that F factors through U, as required. But first of all, ϖ is geometrically injective when restricted to the image of $\phi^{-1}(U)$ in U' (which is a dense constructible subset): for the geometric fibres of ϕ already contain only one orbit each, and an invariant function F cannot split up these fibres any further. In particular, since the generic characteristic is 0, $\bar{\omega}$ is birational. But secondly, suppose we form the fibre product $\phi^{-1}(U) \times_U U'$ (cf. Fig. 1). Since $\bar{\omega}$ is a separated morphism, $p_1 : \phi^{-1}(U) \times_U U \to \phi^{-1}(U)$ is separated. The morphism ϕ' from $\phi^{-1}(U)$ to U' induces a section σ of p_1: by EGA 1, 5.4.6, its image, $\sigma[\phi^{-1}(U)]$, is closed. But $p_2 : \phi^{-1}(U) \times_U U' \to U'$ is obtained from ϕ by base extension. Therefore it is open; in particular

$$p_2 \{\phi^{-1}(U) \times_U U' - \sigma[\phi^{-1}(U)]\}$$

is *open* in U'. But the geometric points y' of this set are those such that there is a geometric point x of $\phi^{-1}(U)$ and (a) $\phi'(x) \neq y'$, (b) $\phi(x) = \bar{\omega}(y')$. By our first remark, all geometric points $\tilde{x}$ of $\phi^{-1}(U)$ for which $\phi(\tilde{x}) = \bar{\omega}(y')$ are mapped by ϕ' to the same geometric point $\phi'(x)$ of U': therefore such a y' is not in the image of ϕ'. Since this image is dense, there are no such y's. Therefore $\sigma[\phi^{-1}(U)] = \phi^{-1}(U) \times_U U'$, and ϕ' is surjective. Therefore $\bar{\omega}$ is geometrically injective without restriction, and by Zariski's Main Theorem, $\bar{\omega}$ is an isomorphism. *QED.*

Remark (5). The following are equivalent:

(a) (Y, ϕ) is a universal categorical quotient of X by G,

(b) for all affine schemes Y', and morphisms $Y' \to Y$, if $\phi' : X' \to Y'$ is the base extension, then (Y', ϕ') is a categorical quotient of X' by G,

(c) there is an affine open covering $\{U_i\}$ of Y such that if $\phi_i : \phi^{-1}(U_i) \to U_i$ is the restriction of ϕ, then (U_i, ϕ_i) is a universal categorical quotient of $\phi^{-1}(U_i)$ by G, for all i.

(6) In Chapter 1, we will need a criterion for (Y, ϕ) to be a categorical quotient which does not imply that (Y, ϕ) is a geometric quotient. In fact, the following 3 conditions suffice:

i) $\phi \circ \sigma = \phi \circ p_2$,

ii) ϱ_Y is the subsheaf of invariants of $\phi_*(\varrho_X)$,

iii) if W is an invariant closed subset of X, then $\phi(W)$ is closed in Y; if W_i, $i \in I$, is a set of invariant closed subsets of X, then:

$$\phi\Big(\bigcap_{i \in I} W_i\Big) = \bigcap_{i \in I} \phi(W_i).$$

Moreover, if these conditions hold, ϕ is submersive.

Proof. Note first of all that ϕ is dominating by (ii); hence by (iii), ϕ is actually surjective. Now suppose $\psi : X \to Z$ is any S-morphism such that $\psi \circ \sigma = \psi \circ p_2$. We proceed exactly as in the proof of Proposition 0.1: choose an affine open covering $\{V_i\}$ of Z. Once we show that there is an open covering $\{U_i\}$ of Y such that $\psi^{-1}(V_i) \supset \phi^{-1}(U_i)$, we can conclude the proof as in Proposition 0.1. To show this, let $W_i = X - \psi^{-1}(V_i)$. These are closed invariant subsets of X. Therefore, by (iii), $U_i = Y - \phi W_i$ is open; and $\phi^{-1}(U_i) \subset \psi^{-1}(V_i)$ for all i. Finally, $\{\psi^{-1}(V_i)\}$ cover X, hence $\bigcap_i W_i = \Phi$, hence by (iii) $\bigcap_i \phi W_i = \Phi$, hence $\{U_i\}$ cover Y. Finally, let $Z \subset Y$ be any subset such that $\phi^{-1}(Z)$ is closed in X. Then $\phi^{-1}(Z)$ is invariant by (i), hence by (iii) $\phi(\phi^{-1}(Z))$ is closed in Y. But ϕ is surjective, so $Z = \phi(\phi^{-1}(Z))$. This proves that ϕ is submersive. *QED.*

(7). Suppose (Y, ϕ) is a geometric quotient of X by G. Notice that conditions (i) and (ii) of Definition 0.6 are preserved by any base extension $Y' \to Y$. But the more delicate condition (iv) might (and indeed sometimes does) break down. However, if $Y' \to Y$ is *flat*, then condition (iv) is preserved. To see this, let $\psi = \phi \circ \sigma = \phi \circ p_2$. Then (iv) asserts the exactness of:

$$(*) \qquad 0 \longrightarrow \mathcal{O}_Y \longrightarrow \phi_*(\mathcal{O}_X) \xrightarrow{\sigma^* - p_2^*} \psi_*\left(\mathcal{O}_{G \times_S X}\right).$$

But taking direct image sheaves commutes with *flat* base extension, i.e.

$$\phi_*(\mathcal{O}_X) \otimes \mathcal{O}_{Y'} \cong \phi'_*(\mathcal{O}_{X'})$$

$$\psi_*\left(\mathcal{O}_{G \times_S X}\right) \otimes \mathcal{O}_{Y'} \cong \psi'_*\left(\mathcal{O}_{G \times_S X'}\right)$$

(where $X' = X \times_Y Y'$, etc.), and the exactness of a sequence is preserved under flat base extension — hence the analogous sequence $(*)'$ on Y' is exact.

(8). On the other hand, consider the converse. I claim that if $Y' \to Y$ is faithfully flat and quasi-compact, and if the extended morphism $\phi' : X' \to Y'$ makes (Y', ϕ') a universal categorical quotient or geometric quotient or universal geometric quotient of X' by G, then (Y, ϕ) is the same type of quotient of X by G. These assertions are all simple corollaries of the general theory of descent of SGA 8.

§ 3. Good and bad actions

We shall now look somewhat more closely at the structure of an action of a group. As above, let G/S act via σ on X/S.

Definition 0.8. The action σ is said to be

i) *closed* if for all geometric points x of X, the orbit $0(x) \subset \overline{X}$ is closed, (i.e. $\overline{X} = X \times_S \operatorname{Spec} \Omega$ if x is an Ω-valued point),

ii) *separated* if the image of

$$\Psi = (\sigma, p_2) : G \times_S X \to X \times_S X$$

is closed,

iii) *proper* if Ψ is proper,

iv) *free* if Ψ is a closed immersion.

Moreover, let $\mathcal{S} = S(1_X)$ be the stabilizer of $1_X : X \to X$ (cf. Definition 0.4). It is a sub-group pre-scheme of the group pre-scheme $G \underset{S}{\times} X/X$. Let $\omega : \mathcal{S} \to X$ be the projection and let $e_X : X \to \mathcal{S}$ be the identity section. Let $\sigma(x)$ be the dimension at $e_X(x)$ of $\omega^{-1}(x)$, as a scheme over $\operatorname{Spec} \varkappa(x)$, for all points $x \varepsilon X$. If X is noetherian and G is of finite type over S, this is finite and upper semi-continuous (cf. EGA 4, 13.1; compare remark 4, § 2). Assuming this, we make:

Definition 0.9. $S_r(X) = \{x \varepsilon X \mid \sigma(x) \geq r\}$. By the above, this is closed. Moreover, say x is *regular* for the action of G if σ is constant in some neighborhood of x.

The set of regular points of X for the action of G forms an open set X^{reg} in X. According to remark 4 of § 2, if G is of finite type and universally open over S and if a geometric quotient (Y, ϕ) of X by G exists such that Y is noetherian and ϕ is of finite type, then every point of X must be regular for the action: $X^{\text{reg}} = X$.

A useful observation is that if G is of finite type over S and $X^{\text{reg}} = X$, then the action of G is closed — even if X is not of finite type over S. To prove this, it suffices to take $S = \operatorname{Spec}(k)$, k algebraically closed; we may assume that G is irreducible too. Note that if $x \varepsilon X_k$, then $0(x)$ is irreducible, and if y is its generic point, then $\varkappa(y)$ is a subfield of $k(G)$, the function field of G. Moreover, it is not hard to verify that:

$$\text{tr.d.}\, \varkappa(y)/k + \sigma(y) = \dim G.$$

Now if the action were not closed, at least over some k there would be two points $x_1, x_2 \varepsilon X_k$ such that $0(x_2) \subset \overline{0(x_1)} - 0(x_1)$. Let y_i be the generic point of $0(x_i)$. Since x_2 is a regular point of the action, one finds that $\sigma(x_1) = \sigma(x_2)$, hence

$$\text{tr.d.}\, \varkappa(y_1)/k = \text{tr.d.}\, \varkappa(y_2)/k < \infty.$$

But y_2 is a specialization of y_1, hence this is absurd.

Concerning restrictive conditions on the action of G, a notion intermediate between the closedness of one orbit and the properness of the whole action is sometimes convenient:

Lemma 0.3. Let $S = \operatorname{Spec}(k)$, k an algebraically closed field, and assume X and G are algebraic pre-schemes over k. Let σ be an action of G on X, and let $x \varepsilon X_k$. Then ψ_x is proper if and only if $0(x)$ is closed in X and $S(x)$ is proper over k.

Proof. If ψ_x is proper, then its image $0(x)$ must be closed, and its fibre over x — which is $S(x)$ — must be proper over k. On the other hand, assume that $0(x)$ is closed and $S(x)$ is proper over k. To prove that ψ_x is proper, we may as well replace G by G_{red}; then let Z be the reduced closed subscheme of X with support $0(x)$. Then G and Z are homogeneous reduced pre-schemes, hence they are non-singular. If we define ψ'_x to be ψ_x as a morphism from G to Z, then it suffices to prove that ψ'_x is proper. But ψ'_x is flat — since G and Z are non-singular and all components of all fibres of ψ_x have the same dimension, i.e. $\dim S(x)/k$ (cf. EGA 4, 15.4). Therefore, to prove ψ'_x proper we may make a base extension by ψ_x itself; i.e. it suffices to prove that

$$\psi''_x = p_2\colon G \underset{Z}{\times} G \to G$$

is proper (cf. SGA 8, Cor. 4.7). But let μ' be the restriction of the group law in G to a morphism from $S(x) \times G$ to G. Then $(\mu', p_2)_Z$ is a morphism from $S(x) \times G$ to $G \underset{Z}{\times} G$. One checks formally that it is an isomorphism. Therefore it suffices to prove that:

$$\psi''_x \circ (\mu', p_2)_Z\colon S(x) \times G \to G$$

is proper. But $\psi''_x \circ (\mu', p_2)_Z$ is the second projection $p_2\colon S(x) \times G \to G$. This is proper since $S(x)$ is proper over k. *QED.*

From the proof of this lemma, it is clear that one can play various tricks with group actions to obtain implications that are not entirely obvious. However, the (global) properness of an action, i.e. the properness of Ψ, is subtler than the properness of ψ_x, for one x. To illustrate this, we give an example and a lemma which suggest opposite conclusions (the example shattered over-optimistic conjectures that the author had entertained!).

Example 0.4. Of a semi-simple group G over $\mathbf{C}$, the complex numbers, acting on a non-singular quasi-affine scheme of finite type over $\mathbf{C}$, such that:

i) all stabilizers are reduced to e itself, i.e. the action is set-theoretically free,

ii) a geometric quotient exists, which is also a scheme of finite type over $\mathbf{C}$; hence (cf. § 4) the action is separated,

iii) the action is not proper, and, in particular, not free (algebro-geometrically).

Proof. Take $G = SL(2)$. Let V_n stand for the $(n+1)$-dimensional affine space whose closed points are homogeneous forms in X and Y of degree n over $\mathbf{C}$: then $SL(2)$ acts on V_n by means of substitutions in X and Y. Define a reduced subscheme $X \subset V_1 \times V_4$ by means of

$$(F_1, F_4) \in X_{\mathbf{C}} \Leftrightarrow \begin{array}{l}\text{(a) } F_1 \neq 0 \\ \text{(b) } F_4 \text{ is the square of a homogeneous} \\ \quad\text{quadratic form of discriminant 1.}\end{array}$$

One checks that X is non-singular, invariant under $SL(2)$, and that no non-trivial $\alpha \in SL(2)_C$ leaves fixed any point of X. Define $\phi: X \to A^1$ by

$$\phi(\alpha X + \beta Y, F_4(X, Y)) = F_4(-\beta, \alpha).$$

Then it is easy to check that (A^1, ϕ) is a geometric quotient of X by $SL(2)$. Finally, the morphism Ψ is not even closed. Let Z be the closed subscheme of $G \times X$ whose closed points are the pairs:

$$\begin{pmatrix} 0 & -\lambda^{-1} \\ \lambda & 0 \end{pmatrix} \times (\lambda X + Y, X^2 Y^2)$$

for $\lambda \in C - (0)$. The image under Ψ of the above point is:

$$(-\lambda X + Y, X^2 Y^2) \times (\lambda X + Y, X^2 Y^2).$$

In the closure of this set is the extra point:

$$(Y, X^2 Y^2) \times (Y, X^2 Y^2).$$

Lemma 0.5. Let σ be an action of G_m on an algebraic scheme X, all over a field k. Assume that X admits an immersion in projective space P_n for which the action σ extends to an action of G_m on P_n. Then:

$$\sigma \text{ is proper} \Leftrightarrow \text{(i) } \sigma \text{ is separated,}$$
$$\text{(ii) } S_1(X) = \Phi.$$

Proof. The implication $\Rightarrow$ is clear. Conversely, assume σ is separated and $S_1(X) = \Phi$. To prove that the morphism Ψ is proper, we use the valuative criterion (EGA 2, § 7). Let R be a valuation ring over k, and let K be its quotient field. We may assume that R contains an algebraically closed overfield of k, Ω, which is isomorphic to its residue field (Remark 7.3.9, EGA 2). Now suppose $\phi \times \xi$ is a K-valued but not R valued point of $G \times X$ such that $\Psi(\phi \times \xi)$ is an R-valued point of $X \times X$. But

$$\Psi(\phi \times \xi) = \sigma(\phi, \xi) \times \xi,$$

hence $\sigma(\phi, \xi)$ and ξ are R-valued, but ϕ is not. Let $\overline{\sigma(\phi, \xi)}$ and $\bar{\xi}$ be the induced Ω-valued points of X obtained via the inclusion: Spec $(\Omega) \to$ Spec (R). Since σ is separated, $\overline{\sigma(\phi, \xi)} \times \bar{\xi}$ is in the image of Ψ, hence there is an Ω-valued point $\bar{\phi}_0$ of G such that:

$$\overline{\sigma(\phi, \xi)} = \sigma(\bar{\phi}_0, \bar{\xi}).$$

Since R contains Ω, $\bar{\phi}_0$ can be lifted to an R-valued point ϕ_0 of G; replacing ϕ by $\phi_0^{-1} \cdot \phi$, we may assume from the beginning that $\sigma(\phi, \xi)$ and ξ induce the same Ω-valued point $\bar{\xi}$ of X (more classically: specialize to the same geometric point of X).

Now let $I: X \to P_n$ be a G_m-linear immersion for a suitable action of G_m on P_n. We require the following:

(*) If G_m acts on P_n over the field Ω, then for every point $x \in P_n$, there is an invariant affine subspace U containing x (i.e. of the form,

$\boldsymbol{P}_n$ minus a hyperplane), and there are coordinates $x_1, \ldots, x_n$ on U, such that the action $\sigma: \boldsymbol{G}_m \times U \to U$ is defined by

$$\sigma^*(x_i) = \alpha^{r_i} \cdot x_i, \; 1 \leq i \leq n$$

for suitable integers r_i, (identifying $\boldsymbol{G}_m$ with Spec $\Omega[\alpha, \alpha^{-1}]$). (For proof of (*), cf. [8]).

We apply (*) to obtain a neighborhood U of $I(\bar{\xi})$ where the action has this form. Since $I(\bar{\xi})$ is a point of U, so are $I(\xi)$ and $I(\sigma(\phi, \xi))$. Using the assumption $S_1(X) = \Phi$, it follows that $I(\bar{\xi})$ is not left fixed by $\boldsymbol{G}_m$; hence for some i, the ith coordinate $I(\bar{\xi})_i$ is not zero and $r_i \neq 0$. Now ϕ is a morphism:

$$\text{Spec } K \to \boldsymbol{G}_m = \text{Spec } k\,[\alpha, \alpha^{-1}].$$

Suppose the function α on $\boldsymbol{G}_m$ induces $A \,\varepsilon\, K$. Then the ith coordinates of $I(\xi)$ and $I(\sigma(\phi, \xi)) = \sigma(\phi, I(\xi))$ are related by:

$$I\big(\sigma(\phi, \xi)\big)_i = A^{r_i} \cdot I(\xi)_i.$$

Here $I(\xi)_i$ and $I(\sigma(\phi, \xi))_i$ are elements of R whose reductions in the residue field Ω of R are both $I(\bar{\xi})_i$. Since $r_i \neq 0$, it follows that A must be a unit in R. But then ϕ factors through Spec R, i.e. ϕ is actually an R-valued point of $\boldsymbol{G}_m$. *QED.*

§ 4. Further properties

In this section we wish to relate the properties of an action σ of G/S on X/S to the properties of the geometric quotient (Y, ϕ), assuming that it exists. We fix these notations for the whole of this section.

Lemma 0.6. If a geometric quotient (Y, ϕ) exists at all, the action σ is closed. In this case, Y is a scheme over S if and only if σ is separated.

Proof. Let x be a geometric point of X over an algebraically closed field Ω. If $\bar{X} = X \times_S \text{Spec}\,(\Omega)$, $\bar{Y} = Y \times_S \text{Spec}\,(\Omega)$ and $\bar{\phi}: \bar{X} \to \bar{Y}$ is induced by ϕ, then

$$0(x) = \bar{\phi}^{-1}\bar{\phi}\big((x)\big).$$

Therefore $0(x)$ is closed, since $\bar{\phi}$ is continuous, and an Ω-rational point (such as $\bar{\phi}(x)$) of any pre-scheme over Ω is a closed point.

Now since (Y, ϕ) is a geometric quotient, the image of $G \times_S X$ in $X \times_S X$ under Ψ is exactly $X \times_Y X$. But we have the diagram:

$$\begin{array}{ccc} X \times_Y X & \longrightarrow & X \times_S X \\ \downarrow{\scriptstyle \phi \circ p_i} & & \downarrow{\scriptstyle \phi \times \phi} \\ Y & \xrightarrow{\Delta} & Y \times_S Y \end{array}$$

where Δ is the diagonal morphism, and this diagram makes $X \underset{Y}{\times} X$ the fibre product of Y and $X \underset{S}{\times} X$ over $Y \underset{S}{\times} Y$. Moreover, $\phi \times \phi$ is submersive since ϕ is universally submersive. Therefore

$$\left[\Delta(Y) \text{ closed in } Y \underset{S}{\times} Y\right] \Leftrightarrow \left[(\phi \times \phi)^{-1}(\Delta(Y)) \text{ closed in } X \underset{S}{\times} X\right]$$
$$\Leftrightarrow \left[X \underset{Y}{\times} X \text{ closed in } X \underset{S}{\times} X\right].$$

This proves that Y is an S-scheme if and only if σ is separated. *QED.*

A more subtle point is:

Proposition 0.7. Let $S = \text{Spec}\,(k)$, k a field, and assume that G, X, and Y are algebraic schemes over k. Assume that G is affine, and that the action σ is proper. Then ϕ is affine.

Proof. To prove this, suppose first of all that ϕ has a section $s: Y \to X$. Then consider the following diagram, where the top triangle is obtained from the bottom by the base extension s:

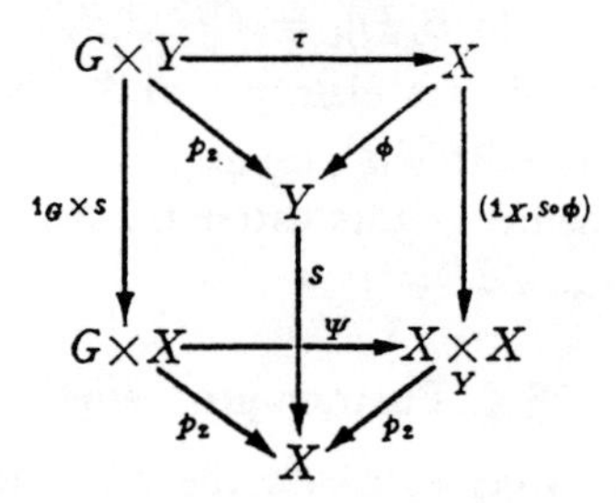

Since the action σ is proper, the morphism Ψ is proper. Therefore τ is proper. But p_2 is an affine morphism since G is affine. Therefore by Chevalley's Theorem (EGA 2, Theorem 6.7.1), ϕ is affine.

In general, we wish to reduce to the case where ϕ has a section by considering a base extension $\pi: Y' \to Y$. Suppose that after such an extension, the new morphism $\phi': X \underset{Y}{\times} Y' \to Y'$ is affine. In some cases, this allows us to conclude that ϕ was affine. This is so if π is finite and surjective, again by Chevalley's Theorem. Since being affine is a local property on Y, it is also true if $\{U_i\}$ is an open covering of Y, and $\pi: \cup U_i \to Y$ is given by the inclusion morphisms. Therefore, by Remark 4, § 2, the Proposition will follow from:

Lemma. Let $\phi: X \to Y$ be a universally open surjective morphism of finite type of algebraic schemes. Then the extended morphism $\phi': X \underset{Y}{\times} Y' \to Y'$ will have a section after a suitable base extension $\pi: Y' \to Y$ where π is a composition of

i) finite surjective morphisms,

ii) the union of inclusion morphisms $\cup U_i \to Y$, where $\{U_i\}$ is an open covering of Y.

Proof. As a first base extension, let $\pi_1: Y_1 \to Y$ be the canonical morphism from the normalization of Y_{red} to Y. Let $X_1 = X \times_Y Y_1$ and let $\phi_1: X_1 \to Y_1$ be the extended morphism. Since ϕ is open, the image of every component of X_1 is a component of Y_1, and the fibres of ϕ_1 in this component all have the same dimension (cf. EGA 4, 14.2). Now for every closed point $y \in Y_1$, let x be a closed point of X_1 over y. First of all, there exists a reduced and irreducible closed subscheme $H \subset X_1$ such that

i) $\dim H = \dim Y_1$,

ii) x is an isolated point of $H \cap \phi_1^{-1}(y)$.

Such a subscheme can be constructed because the fibres of ϕ_1 have the same dimension, e.g., as the set of zeroes of $f_1, \ldots, f_n \in \mathfrak{o}_{x,X_1}$, where $n = \dim_x \phi_1^{-1}(y)$, and f_i, restricted to $\mathfrak{o}_{x,\phi^{-1}(y)}$ generate an ideal primary to the maximal ideal. Then the function field $k(H)$ is a finite algebraic extension of $k(Y_1)$ (or of $k(Y_y)$, where Y_y is the component of Y_1 containing y). Let $L \supset k(H)$ be a further finite algebraic extension which is purely inseparable over a Galois extension of $k(Y_1)$. Let Y' be the normalization of Y_1 in the field L (or of that component of Y_1 containing y). Then I claim that $X_1 \times_{Y_1} Y'$ has a section over Y' locally in a neighborhood of each $y' \in Y'$ which lies over $y \in Y$. But since all y' over y are conjugate, it suffices to prove this for *one* such y'. Now consider the *rational* map $Y' \to H$ induced by $L \supset k(H)$. I claim that it is a morphism in some neighborhood U' of some y' over y. When composed with the inclusion morphism of H in X_1, this will define a section of $X_1 \times_{Y_1} Y'$ over U'. But let H' be the normalization of H in L. Then since $H \to Y_1$ is a morphism, the birational map of the normalizations $H' \to Y'$ is a morphism. Let x' be a point of H' over $x \in H$. Then since x is isolated in its fibre for the morphism $H \to Y_1$ (assumption ii), it follows that x' is isolated in its fibre for the morphism $H' \to Y'$. Therefore, by Zariski's Main Theorem, an open set containing x' is isomorphic to its image $U' \subset Y'$. The rational map from Y' to H must be a morphism on this U'. This proves that $X_1 \times_{Y_1} Y'$ has sections over Y' in a neighborhood of each $y' \in Y'$ over $y \in Y_1$, (compare Fig. 2).

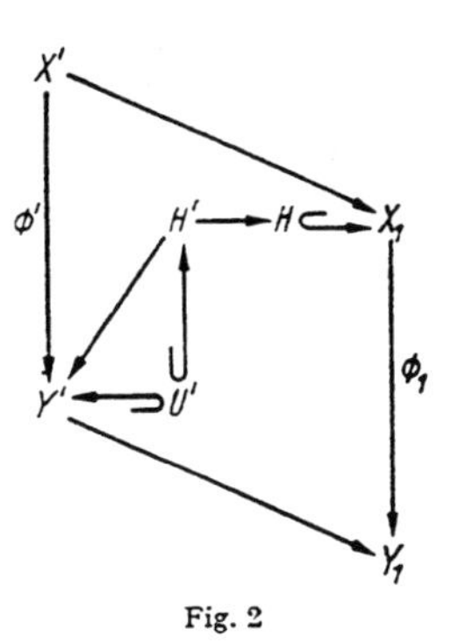

Fig. 2

Therefore there is an open set $U_y \subset Y_1$ containing y such that if V_y is the inverse image of U_y in Y', then $X_1 \times_{Y_1} V_y$ has sections over every point of V_y. Therefore, we first replace Y_1 by a finite covering by these U_y's. Call this Y_2. Then we pass to the corresponding collection of V_y's over these U_y's: this is a Y_3. finite over Y_2. Finally, we replace Y_3 by an open covering Y_4, such that in each component of Y_4, a global section of $X_1 \times_{Y_1} Y_4$ over Y_4 is defined. *QED.*

This result can be generalized with some more technique to more general base schemes S. We omit this. More interesting for the sequel is that, using the methods of chapters 1 and 2, we can prove that the converse of Proposition 0.7 is true in one case:

Proposition 0.8. Let $S = \text{Spec}(k)$, k a field, and assume that G, X, and Y are algebraic schemes over k. Assume that G is a reductive algebraic group (cf. § 1.1) and that the stabilizers of geometric points of X under the action of G are finite. Then if ϕ is affine, the action σ is proper.

Proof. Since $\Psi: G \times X \to X \times X$ is a Y-morphism, it suffices to prove that it is proper over every set U_i of an affine open covering $\{U_i\}$ of Y. Therefore assume that Y is affine. Then X is affine, and by Definition 1.7:

$$X = X^s_{(0)}(\varrho_X).$$

Therefore G acts properly on X by Corollary 2.5. *QED.*

Putting these two results together, it follows that if G is reductive and the stabilizers are *finite*, ϕ being affine is, in essence, a weak topological restiction, i.e. it is equivalent to the properness of σ. But if the stabilizers are positive dimensional, ϕ being affine is a much stronger restriction. At least in characteristic 0, it implies that these stabilizers are themselves reductive groups (cf. [37]).

The final "preliminary" result is aimed at further motivating the concept of a geometric quotient by showing that, for a free action, it implies all that one might hope for.

Definition 0.10. Let (Y, ϕ) be a geometric quotient of X by G (over S). Assume that G is flat and of finite type over S. Then X is a ***principal fibre bundle*** over Y, with group G, if

i) ϕ is a flat morphism of finite type,
ii) Ψ is an isomorphism of $G \times_S X$ and $X \times_Y X$.

Proposition 0.9. Let $S = \text{Spec}(k)$, k a field, assume that G is an algebraic group, and that X and Y are algebraic schemes over k. If the action σ is free, and (Y, ϕ) is a geometric quotient of X by G, then X is a principal fibre bundle over Y with group G.*

Proof. First of all, we may assume that k is algebraically closed: for by such a base extension, i) σ remains free, ii) (Y, ϕ) remains a geometric quotient (Remark 7, § 2), iii) the property of being a principal

---

* The assumption that Y is algebraic is not neccssary. If $\varphi: X \to Y$ is ***any*** geometric quotient, then X algebraic $\Rightarrow$ Y algebraic. (see [102]).

fibre bundle descends (SGA 8, 5.4). Secondly, we may replace Y by one of its local rings Spec $(\underline{o}_{y,Y})$, for some closed point $y \in Y$: for all conditions of Definition 0.10 are expressible locally. And thirdly, we may replace Spec $(\underline{o}_{y,Y})$ by Spec $(\hat{\underline{o}}_{y,Y})$ (its completion). For this is also a faithfully flat quasi-compact base extension and, as before, i) σ remains free, ii) (Y, ϕ) remains a geometric quotient, and iii) the property of being a principal fibre bundle descends.

Therefore we have $Y = \mathrm{Spec}(A)$, where A is a complete local ring with algebraically closed residue field. Let y be the closed point of Y, and let x be a closed, hence k-rational point of X over y. Let $I_x \subset \underline{o}_{x,X}$ be the ideal defining the orbit $0(x)$ of x. Since σ is free, the morphism ψ_x defines an isomorphism:

$$\underline{o}_{x,X}/I_x \xrightarrow{\sim} \underline{o}_{e,G}.$$

Let $g = \dim G$, and let $f_1, \ldots, f_g \in \underline{m}_{e,G}$ be a basis of this maximal ideal. Let $x_1, \ldots, x_g \in \underline{m}_{x,X}$ be elements whose reductions mod I_x are $f_1, \ldots, f_g$ respectively. Then the ideal $(x_1, \ldots, x_g) \subset \underline{o}_{x,X}$ defines a germ of a subscheme of X, at the point x. Since this germ intersects $\phi^{-1}(y)$ in the isolated point x, it follows from Proposition 5.5.1, EGA 3, that there is a unique connected closed subscheme H in X extending this germ, and that H is finite over Y.

Consider Ψ as an X-morphism from $G \times X$ to $X \underset{Y}{\times} X$. Then by the base extension $H \to X$, we define

$$\Psi' : G \times H \to X \underset{Y}{\times} H.$$

Since σ is free, Ψ is a closed immersion, hence Ψ' is a closed immersion. Therefore the morphism $\Psi'' = p_1 \circ \Psi'$ from $G \times H$ to X is a finite morphism. I claim that Ψ'' is an isomorphism. Note first of all that this will prove the Proposition. For, by the definition of Ψ'', if we let G act on $G \times H$ by the product of left multiplication on G, and the identity on H, then Ψ'' is a G-linear morphism. But (H, p_2) is a geometric quotient of $G \times H$ by G. By the uniqueness of geometric quotients, there is a commutative diagram:

$$\begin{array}{ccc} G \times H & \xrightarrow{\sim} & X \\ \downarrow p_2 & & \downarrow \phi \\ H & \xrightarrow{\sim} & Y \end{array}$$

Since p_2 is flat, ϕ is flat; and since $G \times (G \times H)$ is isomorphic to $(G \times H) \underset{H}{\times} (G \times H)$, it follows that $G \times X$ is isomorphic to $X \underset{Y}{\times} X$.

To prove that Ψ'' is an isomorphism, consider the kernel and cokernel of the homomorphism:

$$\underline{o}_X \to \Psi''_*(\underline{o}_{G \times H}).$$

The union of the supports of these two coherent sheaves is a closed subset Z in X. Since Ψ'' commutes with the action of G, it follows that Z is invariant under G. But Y is a topological quotient space of X under G, since ϕ is submersive. Therefore $Z = \phi^{-1}(Z_0)$ for some closed subset Z_0 in Y. Therefore, if $Z \neq \Phi$, it follows that $y \in Z_0$, hence $x \in Z$.

We now prove that $x \notin Z$. Note first that $(\Psi'')^{-1}(x)$ consists of the single point $(e, x) \in G \times H$. For if (α, x') is a geometric point in $(\Psi'')^{-1}(x)$ — possibly over a larger algebraically closed field $\Omega \supset k$ — then $\sigma(\alpha, x') = x$. Therefore $\phi(x') = \phi(x) = y$, and $x' \in H \cap \phi^{-1}(y)$. This implies that $x' = x$ by definition of H; and since the action σ is free, this implies that $\alpha = e$. Therefore the stalk of $\Psi''_*(\underline{o}_{G\times H})$ at x is the local ring of $\underline{o}_{G\times H}$ at (e, x). Now since this stalk is finite over $\underline{o}_{x,X}$, to show that it is isomorphic to $\underline{o}_{x,X}$ it suffices to show that the *complete* local rings:

$$\hat{\underline{o}}_{x,X} \quad \text{and} \quad \hat{\underline{o}}_{e,G} \hat{\otimes}_k \underline{o}_{x,H}$$

are isomorphic. We now rephrase this whole situation in a lemma on local rings as follows, where $\hat{\sigma}$ is dual to the group action σ, $\underline{o} = \hat{\underline{o}}_{x,X}$, $k[[f_1, \ldots, f_g]] = \hat{\underline{o}}_{e,G}$, and the f_i and x_i are the same as above:

Lemma. Let o be a complete local ring, $\underline{\mathfrak{m}} \subset \underline{o}$ its maximal ideal, and $k \subset \underline{o}$ a subfield such that $k \xrightarrow{\sim} \underline{o}/\underline{\mathfrak{m}}$. Let $\hat{\sigma}$ be a local homomorphism from $\underline{o}$ to the formal power series ring $\underline{o}[[f_1, \ldots, f_g]]$ such that:

i) The composition $\hat{\psi}$

$$\underline{o} \xrightarrow{\hat{\sigma}} \underline{o}[[f_1, \ldots, f_g]] \longrightarrow \underline{o}/\underline{\mathfrak{m}}[[f_1, \ldots, f_g]]$$

is surjective,

ii) the composition

$$\underline{o} \xrightarrow{\hat{\sigma}} \underline{o}[[f_1, \ldots, f_g]] \longrightarrow \frac{\underline{o}[[f_1, \ldots, f_g]]}{(f_1, \ldots, f_g)} \cong \underline{o}$$

is the identity.

Let x_i be elements of $\underline{o}$ such that $\hat{\psi}(x_i) = f_i$. Then the composition $\hat{\psi}''$

$$\underline{o} \to \underline{o}[[f_1, \ldots, f_g]] \to \frac{\underline{o}}{(x_1, \ldots, x_g)}[[f_1, \ldots, f_g]]$$

is an isomorphism.

Proof. It follows easily that $\hat{\psi}''$ is surjective since the local rings are complete. To prove that $\hat{\psi}''$ is injective, it suffices to prove by induction that if $\hat{\psi}''(\alpha) = 0$, then $\alpha \in (x_1, \ldots, x_g)^{n+1}$. Suppose $\hat{\psi}''(\alpha) = 0$ and $\alpha \in (x_1, \ldots, x_g)^n$. Write

$$\alpha = \sum_{|A|=n} c_A x^A$$

in symbolic notation $A = (a_1, \ldots, a_g)$, $|A| = \sum a_i$. But by (ii) and (i):

$$\hat{\sigma}(x_i) \in x_i + f_i + \sum_k f_k \underline{\mathfrak{m}} [[f_1, \ldots, f_g]]$$

hence

$$\hat{\psi}''(x_i) = \sum_{j=1}^{g} a_{ij} f_j, \quad \det(a_{ij}) \in 1 + \underline{m}.$$

Therefore $\frac{o}{(x_1, \ldots, x_g)} [[f_1, \ldots, f_g]]$ is isomorphic to

$$\frac{o}{[x_1, \ldots, x_g]} [[\hat{\psi}'' x_1, \ldots, \hat{\psi}'' x_g]].$$

Now since

$$0 = \hat{\psi}'' \alpha = \sum_{|A|=n} \hat{\psi}''(c_A) \cdot \hat{\psi}''(x)^A,$$

it follows that $\hat{\psi}''(c_A) \in (\hat{\psi}''(x_1), \ldots, \hat{\psi}''(x_g))$. Therefore by (ii), the original coefficients c_A are in $(x_1, \ldots, x_g)$. Therefore $\alpha \in (x_1, \ldots, x_g)^{n+1}$ *QED.*

§ 5. Resumé of some results of Grothendieck

In this section, we want to list some definitions and theorems which are not specifically related to group actions, but for which there is no satisfactory reference available.

a) The first topic is a theorem in EGA 3, § 7, which is extremely useful, but which is unfortunately buried there in a mass of generalizations*. Let

$$X \xrightarrow{f} Y$$

be a proper morphism of noetherian schemes. Let $\mathcal{F}$ be a coherent sheaf on X, flat over Y. If $y \in Y$, let X_y (resp. $\mathcal{F}_y$) denote the fibre of f over y (resp. the sheaf induced by $\mathcal{F}$ on the fibre). Assume that for all $y \in Y$,

$$H^1(X_y, \mathcal{F}_y) = (0).$$

Then $f_*(\mathcal{F})$ is a locally free sheaf on Y, and "the formation of f_* commutes with base extension", i.e. in all fibre product situations:

$$\begin{array}{ccc} X' & \xrightarrow{g'} & X \\ {\scriptstyle f'}\downarrow & & \downarrow{\scriptstyle f} \\ Y' & \xrightarrow[g]{} & Y \end{array}$$

the natural homomorphism:

$$g^*(f_* \mathcal{F}) \to f'_*(g'^*(\mathcal{F}))$$

is an isomorphism. In particular, if $Y' = \operatorname{Spec} \varkappa(y)$, for a point $y \in Y$, then:

$$f_* \mathcal{F} \otimes \varkappa(y) \xrightarrow{\sim} H^0(X_y, \mathcal{F}_y).$$

This result follows immediately from Theorem 7.7.5 and Proposition 7.7.10 applied when $p = 1$, and Proposition 7.8.4 applied when $p = 0$ (all in EGA 3).

* A self-contained presentation of GROTHENDIECK's theory is given in [11], § 5.

b) The second topic is the functorial interpretation of $PGL(n)$. As usual, $PGL(n+1)$ is the open subset of Proj $\mathbf{Z}[a_{00}, \dots, a_{nn}]$ where det $(a_{ij}) \neq 0$. It is a group scheme over Spec $(\mathbf{Z})$ and it acts on $\mathbf{P}_n$ over Spec $(\mathbf{Z})$ in the usual way. Now let $\mathcal{PGL}(n+1)$ denote the functor defined by $PGL(n+1)$ in the category of noetherian schemes, i.e.

$$\mathcal{PGL}(n+1)(S) = \text{Hom}\big(S, PGL(n+1)\big).$$

Let Aut $(\mathbf{P}_n)$ denote the functor which assigns to every S the group of automorphisms α of $\mathbf{P}_n \times S$ over S (i.e. such that $p_2 \circ \alpha = p_2$). The action of $PGL(n+1)$ on $\mathbf{P}_n$ defines a morphism of functors:

$$\mathcal{PGL}(n+1) \to \text{Aut}(\mathbf{P}_n).$$

(*) *This is an isomorphism.*

Proof. Most of the proof can be found in EGA 2, § 4.2. The key point which is not proven there is that when S is connected, every invertible sheaf on $\mathbf{P}_n \times S$ is isomorphic to

$$p_1^*(\underline{o}(k)) \otimes p_2^*(L)$$

for some integer k, and some invertible sheaf L on S. A proof of this (based on (a) above) can be found in [40], Lecture 13 (with trivial modifications). It follows that if

$$\begin{array}{ccc} \mathbf{P}_n \times S & \xrightarrow{\alpha} & \mathbf{P}_n \times S \\ & {\scriptstyle p_2}\searrow \quad \swarrow{\scriptstyle p_2} & \\ & S & \end{array}$$

is *any* S-morphism, then:

$$\alpha^*[p_1^*(\underline{o}(1))] \cong p_1^*(\underline{o}(k)) \otimes p_2^*(L).$$

If α is an isomorphism, then

$$p_{2,*}[p_1^*(\underline{o}(1))] \quad \text{and} \quad p_{2,*}[p_1^*(\underline{o}(k)) \otimes p_2^*(L)]$$

must be isomorphic sheaves on S. But

$$\begin{aligned} p_{2.*}[p_1^*(\underline{o}(k)) \otimes p_2^*(L)] &= (0), \quad \text{if } k < 0 \\ &\cong L, \quad \text{if } k = 0 \\ &\cong \bigoplus_{r_0+\cdots+r_n=k} (X_0^{r_0} \dots X_n^{r_n}) \cdot L, \quad \text{if } k > 0. \end{aligned}$$

Therefore, $k = 1$; and by EGA 2, § 4.2, the group of automorphisms α is isomorphic to the group of

a) invertible sheaves L on S, plus

b) isomorphisms of

$$\bigoplus_{i=0}^{n} X_i \cdot \underline{o}_S \quad \text{and} \quad \bigoplus_{i=0}^{n} X_i \cdot L.$$

But such an isomorphism is given by an $(n+1)\times(n+1)$ matrix of sections α_{ij} of L (if X_i is mapped to $\sum_{j=0}^{n} X_j \cdot \alpha_{ij}$), provided that $\det(a_{ij})$, as a section of L^{n+1}, is nowhere zero. But a unique S-valued point

$$S \xrightarrow{f} PGL(n+1)$$

of $PGL(n+1)$ is defined by the conditions:

$$f^*(\underline{o}(1)) = L$$

$$f^*(a_{ij}) = \alpha_{ij}.$$

and one checks that this and only this f induces the automorphism α.

c) The third topic is the definition and existence theorems connected with the Hilbert scheme. Various details can be found in [13], exposé 221, in SERRE's talk at the International Congress in Stockholm, and in my Notes [40].* For all locally noetherian schemes S, let

$$\mathcal{H}ilb_{\mathbf{P}_n}(S) = \{\text{Set of closed subschemes } Z \subset \mathbf{P}_n \times S, \text{ flat over } S\}$$

$$\underset{\text{intuitively}}{=} \{\text{Set of families of subschemes of } \mathbf{P}_n, \text{ parametrized by } S\}.$$

This is a contravariant functor from the category of such S to the category of sets. The fundamental existence theorem states that it is representable, i.e. there is a locally noetherian scheme $Hilb_{\mathbf{P}_n}$ and isomorphisms

$$\mathcal{H}ilb_{\mathbf{P}_n}(S) \cong \mathrm{Hom}(S, Hilb_{\mathbf{P}_n})$$

one for each S, which are functorial in S.

Equivalently, this means that there is a closed subscheme

$$W \subset \mathbf{P}_n \times Hilb_{\mathbf{P}_n}$$

flat over $Hilb_{\mathbf{P}_n}$, which is "universal", i.e. given any $Z \subset \mathbf{P}_n \times S$, flat over S, there is a unique morphism $f: S \to Hilb_{\mathbf{P}_n}$ such that $Z = (1_{\mathbf{P}_n} \times f)^*(W)$.

Now given $Z \subset \mathbf{P}_n \times S$, flat over S, for all $s \in S$, let $Z_s \subset \mathbf{P}_n \times \mathrm{Spec}\, \varkappa(s)$ be the induced subscheme over s. Then each Z_s has a Hilbert polynomial P_s, i.e.

$$P_s(n) = \chi(\underline{o}_{Z_s}(n)).$$

Then if S is connected, all the polynomials P_s are equal. Therefore the universal family breaks up via Hilbert polynomials: i.e.

$$Hilb_{\mathbf{P}_n} = \coprod_{\substack{\text{polynomials} \\ P(n)}} Hilb^P_{\mathbf{P}_n}$$

where $Hilb^P_{\mathbf{P}_n}$ is an open and closed subset of $Hilb_{\mathbf{P}_n}$, and if $W \subset \mathbf{P}_n \times Hilb_{\mathbf{P}_n}$ is the universal subscheme, then W_s has Hilbert polynomial P if and only if

$$s \in Hilb^P_{\mathbf{P}_n}.$$

* A complete treatment was published recently by ALTMAN and KLEIMAN [42].

The strong form of the fundamental existence theorem states that $Hilb^P_{\mathbf{P}_n}$ is projective over Spec $(\mathbf{Z})$.

As corollaries of this theorem, one checks that many other functors are representable:

i) $Hilb^P_{\mathbf{P}_n}$ represents

$S \to$ {Set of closed subschemes $Z \subset \mathbf{P}_n \times S$, flat over S, such that every Z_s has Hilbert polynomial P}.

ii) If $W \subset \mathbf{P}_n \times Hilb^P_{\mathbf{P}_n}$ is the universal subscheme then

$$\overbrace{W \underset{Hilb^P_{\mathbf{P}_n}}{\times} \cdots \underset{Hilb^P_{\mathbf{P}_n}}{\times} W}^{k\times} \text{ represents}$$

$S \to$ {Set of closed subscheme $Z \subset \mathbf{P}_n \times S$, as in (i), *plus* k sections of Z over S}.

iii) If X is a closed subscheme of $\mathbf{P}_n \times T$, then a suitable closed subscheme of $Hilb_{\mathbf{P}_n} \times T$, called $Hilb_{X/T}$, represents the functor:

$S \to$ {Set of closed subschemes $Z \subset X \underset{T}{\times} S$, flat over S}

(in the category of locally noetherian T-schemes).

iv) If X and Y are closed subschemes of $\mathbf{P}_n \times T$, and X is flat over T, then a suitable subscheme of $Hilb_{X \underset{T}{\times} Y/T}$, called $Hom_T(X, Y)$, represents the functor:

$$S \to \{\text{Set of } S\text{-morphisms from } X \underset{T}{\times} S \text{ to } Y \underset{T}{\times} S\}.$$

d) The fourth topic is the definition and existence theorems concerned with the Picard Scheme. Various details can also be found in [13], exposés 232 and 236, SERRE's talk, and [40].* Start with a locally noetherian base scheme T, and a flat, projective morphism

$$\pi : X \to T$$

whose geometric fibres are varieties. For all locally noetherian T-schemes $f: S \to T$, let

$$\mathcal{P}ic_{X/T}(S) = \frac{\{\text{group of invertible sheaves } L \text{ on } X \underset{T}{\times} S\}}{\{\text{subgroup of sheaves of the form } p_2^*(K), \text{ for } K \text{ on } S\}}$$

If X/T has a section $\sigma: T \to X$, then one checks immediately:

$$\mathcal{P}ic_{X/T}(S) \cong \left\{ \begin{matrix} \text{group of isomorphism classes of invertible} \\ \text{sheaves } L \text{ on } X \underset{T}{\times} S, \text{ plus isomorphisms} \\ (\sigma \circ f, 1_S)^*(L) \cong \mathcal{O}_S \end{matrix} \right\}$$

* Complete proofs, plus the construction of the bigger scheme $\mathrm{Pic}_{X/T}$ projective locally over T containing $\mathrm{Pic}^{\tau}_{X/T}$ as an open set, have been given by ALTMAN and KLEIMAN [42].

This is what GROTHENDIECK calls "trivializing" or "normalizing" a family of invertible sheaves along the section σ.

In any case, $\mathcal{P}ic_{X/T}$ is a functor from the category of locally noetherian T-schemes to the category of abelian groups. The fundamental existence theorem states that under the hypotheses above (including the existence of σ), there is a group scheme over T, called $Pic_{X/T}$, representing $\mathcal{P}ic_{X/T}$. In case X is still flat and projective over T with varieties as geometric fibres, *but* no section σ exists, the fundamental theorem states only: there is a unique locally noetherian group scheme over T, called $Pic_{X/T}$, and a homomorphism ϕ of functors in S:

$$\mathcal{P}ic_{X/T}(S) \xrightarrow{\Phi(S)} \mathrm{Hom}_T\,(S, Pic_{X/T})$$

such that

a) $\phi(S)$ is injective for all S,

b) $\phi(S)$ is surjective when $X \times_T S$ admits a section over S.

Just like *Hilb*, *Pic* can be broken up via Hilbert polynomials: fix an invertible sheaf $\underline{o}(1)$ on X, relatively ample for π. For all invertible sheaves L on $X \times_T S$, and all $s \in S$, let L_s be the sheaf induced on $X \times_T \mathrm{Spec}\,\varkappa(s)$, and let

$$P_s(n) = \chi\bigl(L_s(n)\bigr).$$

Then P_s is constant, for s in each component of S, and $Pic_{X/T}$ is the disjoint union of open and closed subsets $Pic^P_{X/T}$ where each piece represents sheaves L with a given Hilbert polynomial. And the strong form of the fundamental existence theorem, that $Pic^P_{X/T}$ is a quasi-projective scheme over T, appears to be true.

More important than $Pic^P_{X/T}$ is an open subgroup scheme $Pic^\tau_{X/T}$. More generally start with any group scheme G, locally of finite type over T. For all geometric points:

$$\mathrm{Spec}\,\Omega \xrightarrow{f} T,$$

let G_f be the induced group scheme over Ω. Then let G^0_f be the component of G_f containing the identity, and let G^τ_f be the union of the components of finite order, i.e. consisting of points x some multiple of which is in G^0_f. Then there is a (unique) open subgroup scheme $G^\tau \subset G$ whose intersection with each G_f is G^τ_f. The key results are:

(A) $Pic^\tau_{X/T}$ is quasi-projective over T.

(B) If X is smooth over T, then $Pic^\tau_{X/T}$ is projective over T.

Finally, when $T = \mathrm{Spec}\,(\Omega)$, what is the connection between $Pic^\tau_{X/\Omega}$ and the classical Picard variety of X?

(C) The Picard variety of X is the reduced part of the connected component of $Pic^\tau_{X/\Omega}$ containing the identity. And $Pic^\tau_{X/\Omega}$ is reduced if and only if

$$\dim Pic^\tau_{X/\Omega} = \dim_\Omega H^1(X, \underline{o}_X).$$

e) The last topic is the concept of a *relative* Cartier divisor. For details, cf. EGA IV, 21.15 and [40], Lecture 10. Suppose $f: X \to Y$ is a flat morphism and $D \subset X$ is an effective Cartier divisor. Then the following are equivalent:

i) D is flat over Y,

ii) for all $y \in Y$, the induced subscheme $D_y \subset X_y$ is a Cartier divisor.

Such a divisor is called a *relative* Cartier divisor, and can be thought of as a family of Cartier divisors in the fibres of f over Y. Moreover, given any fibre product:

$$\begin{array}{ccc} X' & \xrightarrow{g'} & X \\ {\scriptstyle f'}\downarrow & & \downarrow \\ Y' & \longrightarrow & Y \end{array}$$

then $g'^{-1}(D)$ is a relative Cartier divisor on X', over Y'.

Chapter 1

Fundamental theorems for the actions of reductive groups

To set the stage, we shall be concerned in this chapter exclusively with schemes X over a fixed (but not necessarily algebraically closed) ground field k. We also fix an algebraically closed over-field $\Omega \supset k$, and write $\overline{X}$ for $X \times_k \Omega$ whenever X is a scheme over k. The basic set up will be an action of a reductive algebraic group G on a scheme X. Our purpose is to investigate for which open sets $U \subset X$, invariant under the action of G, does a geometric quotient U/G exist? To this end, we will introduce the basic concept of a stable point of X. The rest of this book is devoted to exploiting this concept.

§ 1. Definitions

Due to the fact that we are working over a non-algebraically closed field, we require a preliminary analysis of the notions of an action of an algebraic group on a ring, of invariants, etc. We have grouped all these trivial facts together in this section.

Definition 1.1. Let G be an algebraic group. By a representation of G we mean a morphism $\varrho: G \to GL(n)$ such that, if $\alpha: G \times G \to G$ and $\beta: GL(n) \times GL(n) \to GL(n)$ denote multiplication in G and $GL(n)$ respectively, then $\beta \circ (\varrho \times \varrho) = \varrho \circ \alpha$ (as morphisms from $G \times G$ to $GL(n)$).

Such a representation induces *three* types of actions: (a) a morphism $\sigma: G \times \mathbf{A}^n \to \mathbf{A}^n$ which is an action on the *scheme* $\mathbf{A}^n$ in the sense of

Chapter 0; (b) a map of sets from $G_k \times V \to V$, where V is an n-dimensional vector space over k; (c) a dual action in the following sense, (if G is linear):

Definition 1.2. Let G be a linear algebraic group, let $S = \Gamma(G, \varrho_G)$, let $\hat{\alpha}: S \to S \otimes_k S$ be the homomorphism defining multiplication, and let $\hat{\beta}: S \to k$ be the homomorphism defining the inclusion of the identity in G. Let V be a vector space (resp. R a ring) over k. Then a *dual action* of G on V (resp. R) is a homomorphism of vector spaces (resp. of rings):

$$\hat{\sigma}: V \to S \otimes_k V \quad \left(\text{resp. } \hat{\sigma}: R \to S \otimes_k R\right)$$

such that (i)

$$\begin{array}{ccccc} & & S \otimes_k V & & \\ & \nearrow^{\hat{\sigma}} & & \searrow^{\hat{\alpha} \otimes 1_V} & \\ V & & & & S \otimes_k S \otimes_k V \\ & \searrow_{\hat{\sigma}} & & \nearrow_{1_S \otimes \hat{\sigma}} & \\ & & S \otimes_k V & & \end{array}$$

commutes

(resp.: same for R)

and

(ii) $$V \xrightarrow{\hat{\sigma}} S \otimes_k V \xrightarrow{\hat{\beta} \otimes 1_V} V$$ is the identity

(resp.: same for R).

In the case of a finite-dimensional vector space V, let $\{e_i\}_{1 \le i \le n}$ be a basis of V, and let $\hat{\sigma}(e_i) = \sum a_{ij} \otimes e_j$. Then the elements $a_{ij} \in \Gamma(G, \varrho_G)$ define a morphism from G to $\mathbf{A}^{n^2}$ and conditions (i) and (ii) state that the image is contained in the open subset $GL(n)$ of $\mathbf{A}^{n^2}$ and that the morphism is a homomorphism. In other words, a dual action on a finite-dimensional vector space is simply a translation of the concept of a representation. In case of a ring R, on the other hand, a dual representation is simply a translation of the concept of an action of G on the scheme Spec (R). We shall see that this dual point of view is quite convenient.

Definition 1.3. Let $\hat{\sigma}$ be a dual action of G on V (resp. on R). Then a vector space $W \subset V$ (resp. $W \subset R$) is *invariant* under the action of G if $\hat{\sigma}(W) \subset S \otimes_k W$. Moreover, $x \in V$ (resp. $x \in R$) is *invariant* if $\hat{\sigma}(x) = 1 \otimes x$.

A very important, although elementary, observation is:

Lemma*. If $\hat{\sigma}$ is a dual action of G on an arbitrary vector space V, then V is the union of finite-dimensional invariant subspaces

Proof. Let $S = \Gamma(G, \varrho_G)$, let $\hat{\alpha}: S \to S \otimes S$ be the law of group multiplication, and let $\hat{\beta}: S \to k$ correspond to the inclusion of the identity in G. Then, to prove the lemma, let $V_1 \subset V$ be any finite-dimensional vector space. It suffices to construct a finite-dimensional invariant subspace $V_2 \subset V$ such that $V_1 \subset V_2$.

* This lemma was pointed out to me by Cartier.

Let $S^* = \mathrm{Hom}_k(S, k)$, and let γ be the composition:

$$S^* \otimes V \xrightarrow{1_{S^*} \otimes \hat{\sigma}} S^* \otimes S \otimes V \xrightarrow{\langle\rangle \otimes 1_V} V$$

where $\langle\,\rangle$ stands for "contraction". Let $V_2 = \gamma(S^* \otimes V_1)$.

(a) V_2 *is finite-dimensional.* For, if $\{v_i\}$ is a basis of V_1, and $\hat{\sigma}(v_i) = \sum_j a_{ij} \otimes v_{ij}$, then V_2 is contained in the subspace spanned by the elements $\{v_{ij}\}$.

(b) $V_1 \subset V_2$. For $\hat{\beta}$ is an element of S^*, and γ satisfies the identity $\gamma(\hat{\beta} \otimes v) = v$, where $v \in V$.

(c) $\hat{\sigma}(V_2) \subset S \otimes V_2$. To prove this, it suffices to prove $\hat{\sigma}[\gamma(u^* \otimes v)] \in S \otimes V_2$ for every $u^* \in S^*$, $v \in V_1$. But this follows if, for every $u'^* \in S^*$, we have

$$\langle u'^*, \hat{\sigma}[\gamma(u^* \otimes v)]\rangle \in V_2.$$

But since $\hat{\sigma}$ is a dual action, $(\hat{\alpha} \otimes 1_V) \circ \hat{\sigma} = (1_S \otimes \hat{\sigma}) \circ \hat{\sigma}$. Therefore it follows that:

$$\langle u'^*, \hat{\sigma}[\gamma(u^* \otimes v)]\rangle = \gamma[\hat{\alpha}^*(u'^* \otimes u^*) \otimes v]$$

where $\hat{\alpha}^*$ is dual to $\hat{\alpha}$. The latter is obviously in V_2. *QED.*

Definition 1.4. An algebraic group G is *reductive* if its radical is a torus and *linearly reductive* if every representation of G is completely reducible, i.e. if the action of G on $\mathbf{A}^n$ leaves invariant some $\mathbf{A}^{n'} \subset \mathbf{A}^n$ (where $\mathbf{A}^{n'}$ is a linear subspace through the origin), then it leaves invaraint a complementary $\mathbf{A}^{n-n'}$.

If G is linearly reductive, and $\sigma: G \times \mathbf{A}^n \to \mathbf{A}^n$ is a representation of G, then clearly the affine space $\mathbf{A}^n$can be decomposed into a product

$$\mathbf{B}_1 \times \mathbf{B}_2 \times \cdots \times \mathbf{B}_m$$

of affine spaces such that G leaves each factor invariant and such that $\mathbf{B}_i$ is irreducible under the action of G. Suppose the factors are so numbered that G acts trivially on $\mathbf{B}_1, \ldots, \mathbf{B}_t$ and non-trivially on $\mathbf{B}_{t+1}, \ldots, \mathbf{B}_m$ Then G acts trivially on

$$\mathbf{B}_1 \times \cdots \times \mathbf{B}_t \times (0) \times \cdots \times (0) \subset \mathbf{A}^n$$

and this subspace contains every other subspace on which G acts trivially. The projection of $\mathbf{A}^n$ onto this subspace which annihilates $\mathbf{B}_{t+1} \times \cdots \times \mathbf{B}_m$ is all-important in the sequel. Actually, it is more convenient on the dual level:

Definition 1.5. Let G be a linearly reductive algebraic group, and let $S = \Gamma(G, \varrho_G)$. Let $\hat{\sigma}$ be a dual action of G in a vector space V. Then a ***Reynolds operator*** is a homomorphism $E: V \to V$ such that:

(i) E commutes with $\hat{\sigma}$, i.e. $\hat{\sigma} \circ E = (1_S \otimes E) \circ \hat{\sigma}$.

(ii) $E^2 = E$.

(iii) $Ex = x$ if and only if $\hat{\sigma}(x) = 1 \otimes x$.

If V is finite-dimensional, it follows immediately from the definition of "reductive" that E exists and is unique. The same holds for a general V

by virtue of the above lemma, (compare CARTIER [6], exposé 7). Because of the canonical nature of E, whenever we have dual actions of G on two vector spaces V_1 and V_2, and a linear map $\phi: V_1 \to V_2$ commuting with the dual actions, then $\phi \circ E_1 = E_2 \circ \phi$ where E_1 and E_2 are the Reynolds operators on V_1 and V_2 respectively.

Suppose V is actually a ring R, and σ is an action on the ring. Then the image of E is the subspace of invariant elements of R, which is a *subring* R_0 of R. Although E is not a ring homomorphism, it satisfies the well-known Reynolds identity:

(*) If $x \in R$, $y \in R_0$, then $E(x \cdot y) = (Ex) \cdot y$.

(cf. CARTIER [6], 7—08). To prove this, set $R_1 = \ker(E)$. Then $R \cong R_0 \oplus R_1$, and (*) is equivalent to the 2 statements $R_0 \cdot R_0 \subset R_0$ and $R_0 \cdot R_1 \subset R_1$. To prove the latter, let $g \varepsilon R_0$, and let $V \subset R_1$ be a (finite dimensional) irreducible invariant subspace. Then either $g \cdot V = 0$ or multiplication by g sets up an isomorphism of V with an irreducible invariant subspace $g \cdot V$. Since g is an invariant, the representations of G on V and $g \cdot V$ are isomorphic and nontrivial. Therefore $g \cdot V \subset R_1$.

The linearly reductive algebraic groups have been classified by M. NAGATA [28]. His result is this; a) in characteristic p ($\neq 0$), the only linearly reductive groups G are those whose connected component G_0 is a torus $(G_m)^r$ and such that the order of G/G_0 is prime to p. (b) in characteristic 0, then G is linearly reductive if and only if it is reductive. In that case, it is well-known that G_0 is isogenous to a direct product of a torus T, and a semi-simple algebraic group G_0' which may be taken as the commutator subgroup of G_0 (cf. [28], lemma 10, or [8], exposé 1,6 lemma 2).

§ 2. The affine case

With these preliminaries disposed of, we can now take up the first case of the theory — the action of a reductive algebraic group on an affine scheme X. To keep straight the techniques in the proof, we do not assume that X is an algebraic scheme over k.

Theorem 1.1. Let X be an affine scheme over k, let G be a reductive algebraic group, and let $\sigma: G \times X \to X$ be an action of G on X. Then a uniform categorical quotient (Y, ϕ) of X by G exists, ϕ is universally submersive, and Y is an affine scheme. Moreover, if X is algebraic, then Y is algebraic over k.

If char. $(k) = 0$, (Y, ϕ) is a universal categorical quotient. Moreover X noetherian implies Y noetherian.

Proof.* Let $R = \Gamma(X, O_X)$; then G acts dually on R. Let $R_0 \subset R$ be the ring of invariants, let $Y = \text{Spec}(R_0)$, and let $\phi: X \to Y$ be induced

* Here we give the proof only when char. $(k) = 0$. Modifications to deal with char. $(k) = p$ are discussed in Appendix 1B.

by the inclusion of R_0 in R. Then the first part of the theorem is a consequence of the algebraic facts:

(1) If S_0 is an R_0-algebra, then S_0 is the ring of invariants in $R \underset{R_0}{\otimes} S_0$,

(2) If $(\mathfrak{A}_i)_{i \in I}$ is a set of invariant ideals in R, then

$$(\sum \mathfrak{A}_i) \cap R_0 = \sum (\mathfrak{A}_i \cap R_0).$$

In fact, we can use Remark 6, p. 8 to prove that (Y, ϕ) is a categorical quotient and ϕ is submersive. For (1) implies that for all affine open subsets $U \subset Y$, $\Gamma(U, \varrho_Y)$ is the ring of invariants in $\Gamma(U, \phi_*(\varrho_X))$; hence ϱ_Y is the sub-sheaf of invariant sections of $\phi_*(\varrho_X)$ — condition (ii) of Remark 6, § 0,2. To put (2) in geometric form, let $W_i \subset X$ be the closed subset defined by $\mathfrak{A}_i$. Then (2) asserts:

$$\text{Closure}\left\{\phi\left(\bigcap_i W_i\right)\right\} = \bigcap_i \text{Closure}\left(\phi(W_i)\right).$$

But, applying this to the case $W_1 =$ any closed invariant subset of X, $W_2 = \phi^{-1}(y)$, where y is any closed point of Y, we conclude that $\phi(W_1)$ is closed. This being so, (2) implies also:

$$\phi\left(\bigcap_i W_i\right) = \bigcap_i \phi(W_i)$$

which is condition (iii) of Remark 6, p. 8, Now to conclude that (Y, ϕ) is actually a *universal* categorical quotient, we must consider base extensions $Y' \to Y$. By remark 8, § 0.2, we need only consider the case where Y' is affine, and show that Y' is a categorical quotient of $X \underset{Y}{\times} Y'$ by G. But by (1), $\Gamma(Y', \varrho_Y)$ is still the ring of invariants in $\Gamma(X \underset{Y}{\times} Y', \varrho_{X \underset{Y}{\times} Y'})$, and we are reduced to the case just considered.

To prove the statement (1), let $E: R \to R$ be the Reynolds operator, and let $R_1 = \ker(E)$. Then, by the Reynolds identity, $R \cong [R_0 \oplus R_1]$ *as an R_0-module*. Therefore $\left[R \underset{R_0}{\otimes} S_0\right] \cong S_0 \oplus \left[R_1 \underset{R_0}{\otimes} S_0\right]$, and in particular, S_0 is a subring of $R \underset{R_0}{\otimes} S_0$. Now suppose $\sum a_i \otimes b_i$ is an invariant element of $R \underset{R_0}{\otimes} S_0$, where $a_i \in R$, $b_i \in S_0$. Letting E also denote the Reynolds operator for $R \underset{R_0}{\otimes} S_0$, we deduce:

$$\begin{aligned} \sum(a_i \otimes b_i) &= E\{\sum (a_i \otimes 1) \cdot (1 \otimes b_i)\} \\ &= \sum E(a_i \otimes 1) \cdot (1 \otimes b_i) \quad \text{(via Reynolds identity)} \\ &= \sum E a_i \otimes b_i \\ &= 1 \otimes (\sum E a_i \cdot b_i) \\ &\in S_0. \end{aligned}$$

This proves (1).

To prove statement (2), let $(\mathfrak{A}_i)_{i\epsilon I}$ be any set of invariant ideals in R. Suppose $f \varepsilon (\sum \mathfrak{A}_i) \cap R_0$. Then $f = \sum_{i\epsilon I} f_i$, where $f_i \varepsilon \mathfrak{A}_i$, and all but a finite number of f_i are 0. It follows that:

$$f = Ef = \left[\sum_{i\epsilon I} Ef_i\right] \varepsilon \sum_{i\epsilon I} (\mathfrak{A}_i \cap R_0).$$

We now pass to the second half of the theorem. Suppose X is noetherian. Note that, if $\mathfrak{A} \subset R_0$ is any ideal, statement (1) applied to $S_0 = R_0/\mathfrak{A}$ implies that $(\mathfrak{A} \cdot R) \cap R_0 = \mathfrak{A}$. This implies that the partially ordered set of ideals in R_0 is a sub partially ordered set of the set of ideals in R. Hence the a.c.c. for the ideals in R implies that the a.c.c. for ideals in R_0, and we conclude that Y is noetherian.

Now suppose X is of finite type over k, i.e. R is an algebra of finite type over k. Suppose first of all that R is a *graded* algebra over k, and that the action of G preserves the gradation. Then R_0 is a sub-graded ring of R, and it is automatically finitely generated over k since we have already shown it to be noetherian.

Secondly we can reduce the general case to the graded case: choose a finite dimensional invariant subspace $V \subset R$ which contains a set of generators of R. Let R' be the symmetric algebra on the vector space V. Then the action of G in V extends to a gradation preserving action of G on R'; and R together with the action of G on R can be identified with the quotient $R'/\mathfrak{A}$ of R' by an invariant ideal $\mathfrak{A}$. Then if R'_0 is the ring of invariants in R', R'_0 is finitely generated: therefore it suffices to prove that R_0 is the image of R'_0 in R. This follows from the general algebraic fact:

(3). If $\mathfrak{A}$ is an invariant ideal in R, then $R_0/\mathfrak{A} \cap R_0$ is the ring of invariants in $R/\mathfrak{A}$.

Proof of (3): let $E: R \to R$ be the Reynolds operator on R, and let $\overline{E}: R/\mathfrak{A} \to R/\mathfrak{A}$ be the Reynolds operator on $R/\mathfrak{A}$. Then one has:

$$(Ef) \bmod \mathfrak{A} = \overline{E}(f \bmod \mathfrak{A})$$

for all $f \varepsilon R$. In particular, if $f \varepsilon R$ is such that $(f \bmod \mathfrak{A})$ is an invariant in $R/\mathfrak{A}$, then Ef represents the same element of $R/\mathfrak{A}$ as f. In other words, every invariant of $R/\mathfrak{A}$ is in the image of R_0. *QED.*

Corollary 1.2 (of proof): If W_1 and W_2 are two closed disjoint invariant subsets of X, then they are separated by an invariant $f \varepsilon \Gamma(X, \mathfrak{o}_X)$ which is 0 on W_1 and 1 on W_2.

Proof. Let $\mathfrak{A}_i$ correspond to W_i. Using statement (2), we have $1 \varepsilon (\mathfrak{A}_1 + \mathfrak{A}_2) \cap R_0 = \mathfrak{A}_1 \cap R_0 + \mathfrak{A}_2 \cap R_0$. Therefore $1 = f + g$ where $f \varepsilon \mathfrak{A}_1 \cap R_0$, $g \varepsilon \mathfrak{A}_2 \cap R_0$, and f is the required invariant. *QED.*

This corollary is, for many purposes, the only really important geometric property implied by the reductivity of G.

Amplification 1.3. In the situation of Theorem 1.1, (Y, ϕ) is a geometric quotient of X by G if and only if the action of G on X is closed. Moreover, in char. 0, (Y, ϕ) is actually a universal geometric quotient.

Proof. If (Y, ϕ) is a geometric quotient, then σ is closed by lemma 0.6. Now assume σ is closed. Suppose $\Psi(G \times X)$ were a proper subset of $X \underset{Y}{\times} X$. Then in a suitable algebraically closed field Ω, there is a pair of geometric points x_1, x_2 of X such that $\phi(x_1) = \phi(x_2)$, but $0(x_1), 0(x_2)$ are disjoint subsets of $\overline{X}$. But then $0(x_1), 0(x_2)$ are, by assumption, *closed* invariant disjoint subsets of $\overline{X}$: therefore, there is an invariant $f \varepsilon \Gamma(\overline{X}, \varrho_{\overline{X}}) = \Gamma(X, \varrho_X) \underset{k}{\otimes} \Omega$ which is 0 at x_1, and 1 at x_2. Since the ring of invariants in $\Gamma(\overline{X}, \varrho_X)$ is of the form $R_0 \underset{k}{\otimes} \Omega$, where R_0 is the ring of invariants in $\Gamma(X, \varrho_X)$, there is an invariant $f \varepsilon \Gamma(X, \varrho_X)$ such that $f(x_1) \neq f(x_2)$. Therefore $\phi(x_1) \neq \phi(x_2)$ which is a contradiction.

Now the whole situation is preserved by affine base extension $Y' \to Y$, because of (1) in the proof of Theorem 1.1. According to Remark 3 of § 0.2, this proves that (Y, ϕ) is a universal geometric quotient. *QED.*

§ 3. Linearization of an invertible sheaf

Before passing to the analysis of the action of a reductive algebraic group G on an arbitrary algebraic pre-scheme X, we must relate the actions of G on X with invertible sheaves L on X.

Definition 1.6. Let G, X, L, σ be an algebraic group, an algebraic pre-scheme, an invertible sheaf on X, and an action of G on X respectively. Then a *G-linearization* of L consists of an isomorphism:

$$\phi : \sigma^* L \xrightarrow{\sim} p_2^* L$$

of sheaves on $G \times X$, satisfying the co-cycle condition:

(*) let $\mu : G \times G \to G$ be the group law. Note that p_{23}, $\mu \times 1_X$, and $1_G \times \sigma$ all map $G \times G \times X$ to $G \times X$ The condition is the commutativity of:

$$\begin{array}{ccc}
[\sigma \circ (1_G \times \sigma)]^* L & \xrightarrow{(1_G \times \sigma)^* \phi} & [p_2 \circ (1_G \times \sigma)]^* L \\
\| & & /\!/ \\
\| \quad [\sigma \circ (p_{23})]^* L & \xrightarrow{p_{23}{}^* \phi} & ([p_2 \circ p_{23}]^* L \\
\| & & \| \\
[\sigma \circ (\mu \times 1_X)]^* L & \xrightarrow{(\mu \times 1_X)^* \phi} & [p_2 \circ (\mu \times 1_X)]^* L
\end{array}$$

To understand this concept, suppose the ground field is algebraically closed and let G_k be the group of geometric points of G, in k. Then for all $\alpha \varepsilon G_k$, we can restrict ϕ to $\{\alpha\} \times X \subset G \times X$: then if $T_\alpha : X \to X$

is the automorphism of X given by

$$x \to \sigma(\alpha, x),$$

ϕ restricts to an isomorphism $\phi_\alpha\colon T_\alpha^* L \xrightarrow{\sim} L$ The co-cycle condition then implies:

$$\begin{array}{ccc} T_{\alpha\beta}^* L & \xrightarrow{(\phi_{\alpha\beta})} & L \\ & {\scriptstyle (T_\beta^*\phi_\alpha)} \searrow \quad \nearrow {\scriptstyle (\phi_\beta)} & \\ & T_\beta^* L & \end{array}$$

commutes, for $\alpha, \beta \in G_k$, i.e. $\phi_{\alpha\beta} = \phi_\beta \circ T_\beta^* \phi_\alpha$.

Another way to consider the definition is by means of the line bundle $\boldsymbol{L}$ corresponding to L. Let $\pi\colon \boldsymbol{L} \to X$ be the projection. Then isomorphisms such as ϕ correspond canonically to bundle isomorphisms of the line bundles over $G \times X$:

$$(G\times X) \underset{X}{\times} \boldsymbol{L} \quad \overset{\Phi}{\underset{\sim}{\leftarrow}} \quad (G\times X) \underset{X}{\times} \boldsymbol{L}$$

$$\begin{pmatrix} \text{product via} \\ \sigma: G\times X \to X \end{pmatrix} \begin{pmatrix} \text{product via} \\ p_2: G\times X \to X \end{pmatrix}.$$

(Recall that the transition from sheaves to bundles is a *contra*-variant functor). These correspond canonically to morphisms $\Sigma = p_2 \circ \Phi$ such that:

$$\begin{array}{ccc} G\times \boldsymbol{L} & \xrightarrow{\Sigma} & \boldsymbol{L} \\ {\scriptstyle 1_G \times \pi} \downarrow & & \downarrow {\scriptstyle \pi} \\ G\times X & \xrightarrow{\sigma} & X \end{array}$$

commutes, and such that Σ is a bundle isomorphism of the line bundles $G\times \boldsymbol{L}$ over $G\times X$ and $\boldsymbol{L}$ over X. Then the co-cycle condition translates readily into the commutativity of the following cube:

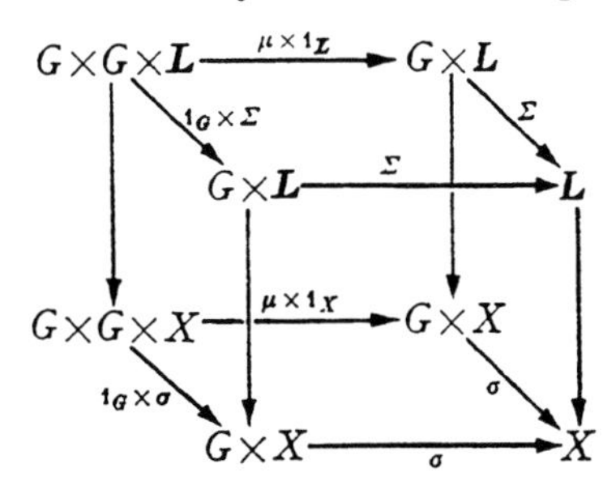

That is to say, the morphism Σ corresponds to a ϕ satisfying (*) if and only if Σ is an *action* of G on $\boldsymbol{L}$. Therefore, a G-linearization of L is a lifting of the action of G on X to a "bundle action" of G on $\boldsymbol{L}$.

For example, suppose $X = \text{Spec}(k)$, $L = \mathfrak{o}_X = k$, $\mathbf{L} = \mathbf{A}^1$. Then a G-linearization of L is nothing other than a character $\chi: G \to \mathbf{G}_m$, since $\mathbf{G}_m$ is the group of automorphisms of $\mathbf{A}^1$ as a line bundle over Spec (k). To be explicit, χ defines the action Σ of G on $\mathbf{A}^1$ such that

$$\Sigma(\alpha, z) = \chi(\alpha) \cdot z$$

for $\alpha \in G_k$, $z \in (\mathbf{A}^1)_k$. On the other hand, it defines $\phi: \mathfrak{o}_G \to \mathfrak{o}_G$ given by

$$\phi(f) = \bar{\chi}^{-1} \cdot f$$

where $\bar{\chi} \in \Gamma(G, \mathfrak{o}_G^*)$ is the function corresponding to the morphism χ. (The inverse results from the contra-variant relation between sheaves and bundles.)

There is still another kind of action induced by a G-linearization of L, and, for the sequel, this is the most important: a dual action of G on $H^0(X, L)$. This action is given by the composition:

$$\begin{array}{ccccc} H^0(X, L) & \xrightarrow{\sigma^*} & H^0(G \times X, \sigma^* L) & \xrightarrow{\phi} & H^0(G \times X, p_2^* L) \\ & & & & \wr\| \\ & & & & H^0(G, \mathfrak{o}_G) \otimes H^0(X, L). \end{array}$$

(The last isomorphism follows from the Kunneth formula.) The conditions for a dual action result from the co-cycle condition on ϕ. Therefore ϕ allows us to speak of *invariant sections* of L.

Notice that the tensor product of 2 G-linearized invertible sheaves, and that the inverse of 1 G-linearized invertible sheaf both carry canonical G-linearizations. Therefore, the set of G-linearized invertible sheaves modulo isomorphism, forms an abelian group We will denote this group by:

$$Pic^G(X).$$

Moreover, if $f: X \to Y$ is a G-linear morphism of pre-schemes on which G acts, there is an induced homomorphism:

$$f^*: Pic^G(Y) \to Pic^G(X).$$

In particular, if G is acting trivially on Y, then each invertible sheaf on Y possesses a trivial G-linearization Therefore, we obtain an induced map from $Pic(Y)$ to $Pic^G(X)$. In case (Y, f) is a geometric quotient of X by G, and the action of G is free, it follows from Proposition 0.9, and the theory of descent of SGA 8, that $Pic^G(X)$ is *isomorphic* to $Pic(Y)$.

An interesting example of this concept is given by the action of $PGL(n+1)$ on projective space $\mathbf{P}_n$. Recall that $PGL(n+1)$ is the open subset of

$$\mathbf{P}_{n^2+2n} \cong \text{Proj}\, k\,[a_{00}, \ldots, a_{0n}; a_{10}, \ldots, a_{1n}; \ldots; a_{n0}, \ldots, a_{nn}]$$

where $\det(a_{ij}) \neq 0$, and that $\mathbf{P}_n = \text{Proj}\, k[X_0, \ldots, X_n]$. Then the morphism:

$$\sigma: PGL(n+1) \times \boldsymbol{P}_n \rightarrow \boldsymbol{P}_n$$

can be defined (cf. EGA 2, § 4.2) by the condition:

$$\sigma^*\left(\underline{o}_{\boldsymbol{P}_n}(1)\right) \cong p_1^*\left[\underline{o}_{\boldsymbol{P}_{n^2+2n}}(1)\right] \otimes p_2^*\left[\underline{o}_{\boldsymbol{P}_n}(1)\right]$$

$$\sigma^*(X_i) = \sum_{j=0}^{n} p_1^*(a_{ij}) \otimes p_2^*(X_j).$$

Unfortunately, $\underline{o}_{\boldsymbol{P}_{n^2+2n}}(1)$ is not trivial on the open set $PGL(n+1)$: in fact, its order in $Pic[PGL(n+1)]$ is $n+1$. Therefore, $\underline{o}_{\boldsymbol{P}_n}(1)$ admits *no* $PGL(n+1)$-linearization!

To obtain $PGL(n+1)$-linearizations, let $\boldsymbol{L}$ be the line bundle on $\boldsymbol{P}_n$ corresponding to $\underline{o}_{\boldsymbol{P}_n}(1)$. Then it is well-known that the nth exterior power of the cotangent bundle to $\boldsymbol{P}_n$ is isomorphic to the $(n+1)^{st}$ tensor power of $\boldsymbol{L}$. Any action on $\boldsymbol{P}_n$ lifts to an action on the cotangent bundle, hence to $\boldsymbol{L}^{n+1}$. Therefore, $\underline{o}_{\boldsymbol{P}_n}(n+1)$ does admit a $PGL(n+1)$-linearization.

On the other hand, let $\overline{\omega}: SL(n+1) \rightarrow PGL(n+1)$ be the canonical isogeny and consider the induced action $\tau = \sigma \circ (\overline{\omega} \times 1_{\boldsymbol{P}_n})$ of $SL(n+1)$ on $\boldsymbol{P}_n$. Then $\underline{o}_{\boldsymbol{P}_n}(1)$ does admit a $SL(n+1)$-linearization. For it is well-known that $SL(n+1)$ acts on the affine cone $\boldsymbol{A}^{n+1}$ over $\boldsymbol{P}_n$, so that the projection

$$\pi: \boldsymbol{A}^{n+1} - (0) \rightarrow \boldsymbol{P}_n$$

is $SL(n+1)$-linear. But $\boldsymbol{L}$ is obtained from $\boldsymbol{A}^{n+1}$ by blowing up (0), and the projection from $\boldsymbol{L}$ to $\boldsymbol{P}_n$ is obtained from π: therefore $SL(n+1)$ acts on $\boldsymbol{L}$, compatibly with its projection to $\boldsymbol{P}_n$.

We now consider the general theory of G-linearizations:

Proposition 1.4. Let a connected algebraic group G act on an algebraic pre-scheme X. Assume that there is no homomorphism of $\bar{G}$ onto $\bar{G}_m$, and that X is geometrically reduced. Then each invertible sheaf L on X has at most one G-linearization.

Proof. If some L had 2 G-linearizations, then the induced sheaf $\bar{L}$ on $\bar{X} = X \underset{k}{\times} K$, for an algebraically closed over-field $K \supset k$, would still have 2 G-linearizations. Therefore we can assume without loss of generality that k is algebraically closed. Also, since G is connected, G-linearizations of an L on the different components of X are independent of each other; therefore, in order to prove the Proposition, we can assume that X is connected.

The Proposition is equivalent to the statement that the canonical map: $Pic^G(X) \rightarrow Pic(X)$ is injective. Since this is a homomorphism it suffices to ask whether $\underline{o}_X$ admits a non-trivial G-linearization. Let

$\phi: \sigma^*(\varrho_X) \to p_2^*(\varrho_X)$ be a G-linearization. Then let ϕ take the unit section of ϱ_X to $f \in H^0(\varrho_{G\times X})$. Since ϕ is an isomorphism, $f \in H^0(\varrho^*_{G\times X})$. Moreover the co-cycle condition on ϕ implies that $f = 1$ on the subscheme $\{e\}\times X$ of $G\times X$. Now by a result of ROSENLICHT (Theorem 2, p. 986, [32]), we know that $H^0(\varrho^*_{G\times X})$ is spanned by the images of $H^0(\varrho^*_G)$ and $H^0(\varrho^*_X)$.† Moreover, by Theorem 3 of the same paper, we know that $H^0(\varrho^*_G) \cong k^*$. Therefore

$$H^0(\varrho^*_{G\times X}) \xleftarrow[p_2^*]{\sim} H^0(\varrho^*_X).$$

It follows that f, being equal to 1 on $\{e\}\times X$, must be identically 1. Therefore, ϕ is the identity. *QED.*

Proposition 1.5. Let a connected linear algebraic group G act on an algebraic variety X, proper over k. Let L be an invertible sheaf on X, and let λ be the k-rational point of the Picard scheme $Pic(X/k)$ defined by L. Then some power L^n of L is G-linearizable if and only if some multiple $n\lambda$ of λ is left fixed by G.

Proof. The only if is clear. Conversely, suppose $n\lambda$ is left fixed by G. Then I claim first that for some m, the two sheaves $\sigma^*(L^{nm})$ and $p_2^*(L^{nm})$ on $G\times X$ are isomorphic. To see this, consider the see-saw exact sequence:

$$0 \to H^1(\varrho^*_G) \to H^1(\varrho^*_{G\times X}) \to H^0(G, R^1p_{1*}(\varrho^*_{G\times X})).$$

Since $H^1(\varrho^*_G)$ is a finite group (Séminaire Chevalley, [9], 5—21), it is enough to show that the image of $\sigma^*(L^n) \otimes p_2^*(L^n)^{-1}$ in $H^0(G, R^1p_{1*}(\varrho^*_{G\times X})$ is zero. But, by the functorial definition of $Pic(X/k)$, (cf. Ch. 0, § 5, (d))

$$H^0(G, R^1p_{1*}(\varrho^*_{G\times X})) \subset \mathrm{Hom}_k(G, Pic(X/k))$$

But, as in the proof of proposition 1.4, $H^0(\varrho^*_{G\times X}) \cong H^0(\varrho^*_G)$, and the latter is just $k^* \times M$, where M is the set of homomorphismus $\chi: G \to G_m$. Choose an isomorphism $\phi: \sigma^*(L^{nm}) \xrightarrow{\sim} p_2^*(L^{nm})$ which is the identity on $\{e\}\times X$. Then $p_{23}^*\phi \circ (1_G\times\sigma)^*\phi$ and $(\mu\times 1_X)^*\phi$ differ by a factor of the form $a p_1^*(\chi_1)\, p_2^*(\chi_2)$, $a \in k^*$, $\chi_i \in M$. Restricting to $\{e\}\times G\times X$, we see that $a = 1$, $\chi_2 \equiv 1$ and restricting to $G\times\{e\}\times X$ shows that $\chi_2 \equiv 1$. *QED.*

† ROSENLICHT only stated his result in the case where X is a variety. The generalization to an arbitrary reduced algebraic pre-scheme is immediate.

Corollary 1.6. Let G, X, and L be as above. Then if $\bar{X}$ is a normal variety, some power L^n of L is always linearizable.

Proof. According to a result of CHEVALLEY [10], in this case all components of $Pic(X/k)$ are proper over k, hence all reduced components are abelian varieties. Therefore the connected linear group G, being birational to projective space as a variety, must act trivially on $Pic(X/k)$. *QED.*

The following connection between G-linearizations and projective embeddings will be needed:

Proposition 1.7. Let X be an algebraic pre-scheme, let G be an algebraic group acting on X, and let L be a G-linearized invertible sheaf on X such that the sections of L have no common zeroes. Then there exists a morphism $I: X \to \boldsymbol{P}_n$, an action of G on $\boldsymbol{P}_n$, and a G-linearization of $\varrho_{\boldsymbol{P}}(1)$ such that I is a G-linear and such that L together with its G-linearization is induced via I from $\varrho_{\boldsymbol{P}}(1)$ and its G-linearization. Moreover, if X is proper over k, one may take I to be the morphism. associated with the complete linear system $H^0(X, L)$; and if L is very ample, we may take I to be an immersion.

Proof.* Let $\phi: \sigma^* L \xrightarrow{\sim} p_2^* L$ be the given linearization. Then ϕ defines a dual action of G on $H^0(X, L)$.

Let $V_1 \subset H^0(X, L)$ be any finite-dimensional subspace such that the sections in V_1 have no common zeroes. If L is very ample, take V_1 so that the morphism from X to $\boldsymbol{P}(\check{V}_1)$ is an immersion; if X is proper over k, take $V_1 = H^0(X, L)$. In any case, by the lemma of § 1, there is a finite-dimensional vector space V such that $V_1 \subset V \subset H^0(X, L)$, and such that V is invariant under the action of G. Then we take I to be the morphism from X to $\boldsymbol{P}(V)$. Then, by definition, $H^0(\boldsymbol{P}(V), \varrho_{\boldsymbol{P}}(1)) = V$, so we have also a dual action of G on $H^0(\boldsymbol{P}(V), \varrho_{\boldsymbol{P}}(1))$. But I claim:

(*) There is a natural equivalence between the set of all dual actions of G on $H^0(\boldsymbol{P}(V), \varrho_{\boldsymbol{P}}(1))$, and the set of all actions of G on $\boldsymbol{P}(V)$ plus G-linearizations of $\varrho_{\boldsymbol{P}}(1)$. (*Proof omitted.*)

This gives us all the required actions. Their compatibility follows in a straightforward way also. *QED.*

We will require in Chapter 2 a strengthening of this result. Although this involves concepts which we only define in the next section, we include it here since its proof is related to the proof just given:

* See KAMBAYASHI [338], SUMIHIRO [319].

Amplification 1.8. Assume L is ample. Then, for large N, the G-linear immersion I of Proposition 1.7 such that $L^N \cong I^*(\mathcal{O}_{P_n}(1))$ can be so chosen that:

$$X^s_{(0)}(L) = I^{-1}\{P^s_{n,(0)}(\mathcal{O}_{P_n}(1))\}.$$

Proof. By definition, $X^s_{(0)}(L)$ is covered by a finite number of affine open sets X_{s_i}, where s_i is an invariant section of L^{n_i}, and where the action of G on X_{s_i} has only 0-dimensional stabilizers. Let

$$\Gamma(X_{s_i}, \mathcal{O}_X) = k\,[f^{(i)}_1, \ldots, f^{(i)}_{m_i}].$$

Replacing each s_i by a suitable power of itself, we can assume:

i) each s_i is a section of L^N, some fixed N,

ii) for all j between 1 and m_i, $f^{(i)}_j \cdot s_i$ is a section of L^N.

Then suppose, in the previous proof, we select the vector space V_1 so that $s_i \in V_1$ and $f^{(i)}_j \cdot s_i \in V_1$ for all i, j. It follows that if $I: X \to P_n$ is the resulting immersion, then $s_i = I^*(X_i)$ for a suitable invariant section X_i of $\mathcal{O}_{P_n}(1)$: hence $X_{s_i} = I^{-1}[P_{n,X_i}]$. Moreover, by (ii) we conclude that I maps X_{s_i} isomorphically onto *a closed* subscheme of $(P_n)_{X_i}$. Therefore, the orbit of every geometric point of X_{s_i} is closed in $(P_n)_{X_i}$. By Amplification 1.11, this implies that every point of $I(X_{s_i})$ is properly stable in P_n. Hence

$$X^s_{(0)}(L) = \bigcup_i X_{s_i} \subset I^{-1}\{P^s_{n,(0)}(\mathcal{O}_{P_n}(1))\}.$$

The other inclusion "$\supset$" is proven in Proposition 1.18. *QED.*

§ 4. The general case

We now proceed to analyze the general case: G is a reductive algebraic group, acting via σ, on an arbitrary algebraic pre-scheme X. The key concepts will be these:

Definition 1.7. Let x be a geometric point of X. Then:

(a) x is *pre-stable* (with respect to σ) if there exists an invariant affine open subset $U \subset X$ such that x is a point of U, and the action of G on U is closed.

Now suppose L is an invertible sheaf on X, and ϕ is a G-linearization of L. Then:

(b) x is *semi-stable* (with respect to σ, L, ϕ) if there exists a section $s \in H^0(X, L^n)$ for some n, such that $s(x) \neq 0$, X_s is affine, and s is invariant, i.e. if $\phi_n: \sigma^*(L^n) \to p_2^*(L^n)$ is induced by ϕ, then $\phi_n(\sigma^*(s)) = p_2^*(s)$.

(c) x is *stable* (with respect to σ, L, ϕ) if there exists a section $s \in H^0(X, L^n)$ for some n, such that $s(x) \neq 0$, X_s is affine, s is invariant, and the action of G on X_s is closed.

Note that the set of geometric points with any of these 3 properties is the set of geometric points of an open subset of X. The corresponding 3 open subsets will be written:

$$X^s(\text{Pre})$$
$$X^{ss}(L)$$
$$X^s(L)^*.$$

(1) We note immediately that, for each invariant open $U \subset X$ such that U is connected and affine and the action of G on U is closed, the stabilizers of all the geometric points of U have the same dimension — cf. Amplification 1.3 and Remark 5 of § 0.2. Therefore we can write

$$X^s(\text{Pre}) = X^s_{(0)}(\text{Pre}) \cup \cdots \cup X^s_{(r)}(\text{Pre})$$
$$X^s(L) = X^s_{(0)}(L) \cup \cdots \cup X^s_{(r)}(L)$$

where the right hand side represents a *disjoint* union of open sets, and where the dimension of the stabilizer of every geometric point of $X^s_{(i)}(\text{Pre})$ and of $X^s_{(i)}(L)$ is i.

Definition 1.8. $X^s_{(0)}(L)$ is the set of *properly stable* points (with respect to σ, L, ϕ).

(2). Note that if X is proper over k, and L is ample, then the condition that X_s be affine is redundant. Moreover, in any case, L will be ample when restricted to $X^{ss}(L)$ (cf. condition (b), Theorem 4.5.2 of EGA, Chapter 2).

(3). Recall that the set of all sections $s \in H^0(X, L^n)$ such that X_s is affine is a linear subspace of $H^0(X, L^n)$: this follows, for example, from condition (b) of Theorem 4.5.1, EGA 2. Therefore so is the set of invariant sections such that X_s is affine. Call this V_n. Moreover $\sum_{n=0}^{\infty} V_n$ is a ring, since $X_{st} = X_s \cap X_t$, and this is affine if both X_s and X_t are affine.† This ring will take the place of the affine ring of invariants in the general case.

(4) The 4th concept "pre-semi-stable", is not so useful, but we note SUMIHIRO's result [319]: if a torus T acts on a normal variety X, then every $x \in X$ has a T-invariant affine open neighborhood.

Proposition 1.9. Let X be an algebraic pre-scheme over k, and let G be a reductive algebraic group acting on X. Then a uniform geometric quotient (Y, ϕ) of $X^s(\text{Pre})$ by G exists. Moreover, ϕ is affine and Y is an algebraic pre-scheme. Conversely, if $U \subset X$ is any invariant open set such that a geometric quotient (Z, ψ) of U by G exists, and such that ψ is affine, then $U \subset X^s(\text{Pre})$. If char. $(k) = 0$, then (Y, ϕ) is a universal geometric quotient.

* To justify omitting ϕ in the notation, see Proposition 1.17.

† This is true even though X is not necessarily a scheme. For, if $s \in H^0(X, L^n)$, $t \in H^0(X, L^m)$, then X_{st} is the open subset of the affine space X_s where t^n/s^m is not zero.

The proof of this is omitted, as it is easy from what has already been proved and is never used in the sequel. Incidentally, in view of Proposition 0.5, the condition that ψ is affine in the converse can be replaced by the stronger hypothesis: "G acts properly on U", which may seem a more reasonable assumption since it is topological in nature.

The problem with this theorem is that even when X is a projective scheme, Y can be a very "pre" pre-scheme. For this reason, we will find much more useful:

Theorem 1.10. Let X be an algebraic pre-scheme over k, and let G be a reductive algebraic group acting on X. Suppose L is a G-linearized invertible sheaf on X. Then a uniform categorical quotient (Y, ϕ) of $X^{ss}(L)$ by G exists. Moreover:

(i) ϕ is affine and universally submersive;

(ii) there is an ample invertible sheaf M on Y such that $\phi^*(M) \cong L^n$ for some n; hence Y is a quasi-projective algebraic scheme;

(iii) there is an open subset $\tilde{Y} \subset Y$ such that $X^s(L) = \phi^{-1}(\tilde{Y})$ and such that $(\tilde{Y}, \phi \mid X^s(L))$ is a uniform geometric quotient of $X^s(L)$ by G.

Proof. Since X is noetherian, there exists an N, and a finite set $s_1, \ldots, s_n$ of invariant sections of L^N such that $U_i = X_{s_i}$ is affine and $X^{ss}(L) = \bigcup_{i=1}^{n} U_i$. By Theorem 1.1, there exists a uniform categorical quotient (V_i, ϕ_i) of U_i by G. Our first step is to patch the affine schemes V_i together into a pre-scheme. But for every pair of integers (i, j) $(1 \leq i, j \leq n)$, consider the quotient s_j/s_i in the open set U_i where $s_i \neq 0$: it is an invariant element of the ring $\Gamma(U_i, \mathfrak{o}_X)$. Now since V_i is a categorical quotient of U_i by G, $\Gamma(V_i, \mathfrak{o}_{V_i})$ is exactly the ring of invariants in $\Gamma(U_i, \mathfrak{o}_X)$. Therefore, s_j/s_i is induced by a function $\sigma_{ij} \varepsilon \Gamma(V_i, \mathfrak{o}_{V_i})$. Let $V_{ij} = V_i - \{y \mid \sigma_{ij}(y) = 0\}$. Then:

$$\phi_i^{-1}(V_{ij}) = U_i \cap U_j = \phi_j^{-1}(V_{ji}).$$

Since V_i (resp. V_j) is a *uniform* categorical quotient of U_i (resp. U_j), it follows that both V_{ij} and V_{ji} are categorical quotients of $U_i \cap U_j$ — therefore, there is a *unique* isomorphism $\psi_{ij}: V_{ij} \xrightarrow{\sim} V_{ji}$ such that:

$$\begin{array}{ccc} & U_i \cap U_j & \\ {}^{\phi_i}\swarrow & & \searrow^{\phi_j} \\ V_{ij} & \xrightarrow{\psi_{ij}} & V_{ji} \end{array}$$

commutes. From the uniqueness, it follows easily that this set of identifications patches $\{V_i\}$ into a pre-scheme Y containing each V_i as an affine open subset. Moreover, it is clear that the ϕ_i patch together into an affine and universally submersive morphism ϕ from $X^{ss}(L)$ to Y, such that $\phi^{-1}(V_i) = U_i$.

The second step is to notice that the collections of functions $\sigma_{ij} \mid V_{ij}$ forms a Cech 1-cocycle for the covering $\{V_i\}$ of Y and in the sheaf $\underline{o}_Y^*$. Therefore these functions define an invertible sheaf M on Y. Moreover, it is clear that $L^N = \phi^*(M)$. *I* claim that M is an ample sheaf on Y in the sense of Theorem 4.5.2 Chapter 2, EGA. To prove this, we use condition a') of that theorem, i.e. the collection of functions σ_{ij} *for fixed* j, and variable i, is a collection of functions, one on each V_i such that on $V_{i_1} \cap V_{i_2}$,

$$\sigma_{i_2,j} = \sigma_{i_1,j} \cdot \sigma_{i_2,i_1}.$$

Therefore, it defines a section t_j of M. Moreover, it is clear that $s_j = \phi^*(t_j)$. Therefore, for every point $x \in X^{ss}(L)$,

$$s_j(x) = 0 \Leftrightarrow t_j(\phi(x)) = 0.$$

Therefore, $V_j = Y_{t_j} = Y - \{y \mid t_j(y) = 0\}$. Since V_i is affine, the condition a') referred to is verified. By the remark following Theorem 4.5.2, this implies that Y is in fact a scheme, and, of course, M is ample on Y.

Finally, by enlarging the set of sections s_i is necessary, we may assume that the action σ of G on U_i is closed if $i \in I \subset \{1, \ldots, n\}$, and that

$$X^s(L) = \bigcup_{i \in I} U_i.$$

Then $X^s(L) = \phi^{-1}\left(\bigcup_{i \in I} V_i\right)$. Then if $\tilde{Y} = \bigcup_{i \in I} V_i$, the rest of the Theorem to the effect that Y (resp. $\tilde{Y}$) is a uniform categorical quotient (resp. uniform geometric quotient) of $X^{ss}(L)$ (resp. $X^s(LL)$) by G, follows from remark 7, § 0.2. *QED.*

Amplification 1.11. In the situation of the above theorem, let x be a geometric point of $X^{ss}(L)$. Then the following are equivalent:

1) x is a point of $X^s(L)$,

2) x is regular for the action of G, and its orbit $0(x)$ is closed in $\overline{X^{ss}(L)} = X^{ss}(L) \times_k \Omega$,

3) x is regular for the action of G and there is an invariant section $s \in H^0(L^n)$ such that $s(x) \neq 0$, X_s is affine, and $0(x)$ is closed in $\overline{X}_s$.

Moreover, if $x_1, \ldots, x_n$ is a finite set of geometric points of $X^{ss}(L)$, then there is an invariant $s \in H^0(X, L^m)$ (for some m) such that X_s is affine and $s(x_i) \neq 0$, all i. If $x_1, \ldots, x_n$ are all *stable*, then we can even assume that the action of G on X_s is closed.

Proof. 1) $\Rightarrow$ 2). First of all, since a geometric quotient of $X^s(L)$ by G exists, it follows that x is regular for the action of G (cf. Remark 4, § 0.2). Secondly, in $\overline{X^{ss}(L)}$, *I* claim $0(x) = \phi^{-1}(\phi(x))$: for $\overline{X^s(L)} = \phi^{-1}(\overline{\tilde{Y}})$, hence $\phi^{-1}(\phi(x)) \subset \overline{X^s(L)}$; and if $y \in X^s(L)$, then $\phi(x) = \phi(y)$ if and only if $y \in 0(x)$ since $\overline{\tilde{Y}}$ is a geometric quotient of $\overline{X^s(L)}$ by G. Therefore, $0(x)$ is closed in $\overline{X^{ss}(L)}$.

2) $\Rightarrow$ 3). obvious.

3) $\Rightarrow$ 1). Let $s \in H^0(L^n)$ be an invariant section as in (3). We shall construct an invariant function $f \in \Gamma(X_s, \mathfrak{o}_X)$ such that $f(x) \neq 0$, and such that the action of G on $(X_s)_f$ is closed. Then, for some m, $s^m \cdot f$ will be an invariant section of L^{nm} and $(X_s)_f = X_{s^m \cdot f}$ will be contained in $X^s(L)$, which proves (1). To construct f, introduce the closed subsets S_r of X_s consisting of points whose stabilizer has dimension at least r. Suppose $x \in \overline{S}_r - \overline{S}_{r+1}$. Put

$$Z_1 = S_{r+1} \cup \{\text{Closure of } S_0 - S_r\}$$

$$\overline{Z}_2 = 0(x).$$

Since x is regular, $\overline{Z}_1$ and $\overline{Z}_2$ are disjoint, closed, invariant subsets of $\overline{X}_s$. Therefore there is an invariant $f \in \Gamma(\overline{X}_s, \mathfrak{o}_{\overline{X}})$ which is 0 on $\overline{Z}_1$ and 1 on $\overline{Z}_2$, Write $\overline{f} = \sum_{i=1}^{N} f_i \otimes \alpha_i$, where $\alpha_1, \ldots, \alpha_N$ are elements of Ω, linearly independent over k, and where $f_1, \ldots, f_N$ are elements of $\Gamma(X_s, \mathfrak{o}_X)$. It follows that each f_i is invariant. Moreover, since $\overline{Z}_1$ is the extension to Ω of the closed subset Z_1 of X_s, it follows that every f_i is 0 on Z_1. Finally, since $f(x) = 1$, it follows that for some i, $f_i(x) \neq 0$. This f_i can be taken as the sought-for f. One need only notice that, since $f = 0$ on Z_1, every orbit of G in $(X_s)_f$ has the same dimension, hence the action of G on $(X_s)_f$ is closed.

To prove the last statement, note that by definition there is a set of invariant sections $s_i \in H^0(X, L^N)$ such that X_{s_i} is affine, and $s_i(x_i) \neq 0$. If the ground field k is infinite, then there are constants $\alpha_i \in k$ so that the section $\Sigma\, \alpha_i s_i$ is not zero at any of the points x_i; if k is finite, there is still some homogeneous polynomial $P(X_1, \ldots, X_n)$ such that the section $P(s_1, \ldots, s_n)$ is not zero at any of the points x_i. In any case, $X_{\Sigma \alpha_i s_i}$ or $X_{P(s_1,\ldots,s_n)}$ is still affine by Remark 3 at the beginning of the section. Finally, if all the x_i are stable, then we can assume that the action σ on X_{s_i} is closed: then the dimension of the stabilizers of points of X_{s_i} is constant on the connected components of each X_{s_i}. Therefore, this is also true on $\cup X_{s_i}$, hence on $X_{\Sigma \alpha_i s_i}$ or $X_{P(s_1,\ldots,s_n)}$. Therefore the action of these sets is closed also. *QED.*

Another interesting point in connexion with Theorem 1.10 is that, if X *is proper over* k and if L *is ample* on X, then the categorical quotient Q of $X^{ss}(L)$ by G is actually projective over k. Therefore, this categorical quotient is a compactification of the topologically more significant geometric quotient of $X^s(L)$ by G. We omit the proof, except to say that it follows from interpreting Q as Proj (R_0), where R_0 is the subring of $\sum_{n=0}^{\infty} H^0(X, L^n)$ of invariant sections. In fact (see [305]), $X^{ss}(L)/G$ can be regarded as the quotient of $X^{ss}(L)$ by the equivalence relation $x \sim y \Leftrightarrow \overline{O(x)} \cap \overline{O(y)} \neq \emptyset$.

Converse 1.12. Let G be a reductive algebraic group acting on an algebraic pre-scheme X. Then if a categorical quotient (Y, ϕ) of X by G exists, and if ϕ is affine and Y is quasi-projective, it follows that for some $L \varepsilon Pic^G(X)$, $X = X^{ss}(L)$. Moreover, if (Y, ϕ) is a geometric quotient of X by G, then $X = X^s(L)$.

Proof. Let M be an ample invertible sheaf on Y. Then, as pointed out in § 3, $L = \phi^*(M)$ carries a canonical G-linearization with the property that all sections $\phi^*(s)$, $s \varepsilon H^0(Y, M^n)$, are invariant sections of L^n. Since Y is covered by affine open sets Y_s, for suitable sections s of M^n, X is covered by the affine open sets $\phi^{-1}(Y_s) = X_{\phi^*(s)}$. Therefore $X = X^{ss}(L)$. If (Y, ϕ) is a geometric quotient, the action of G on X is closed by lemma 0.6; therefore $X = X^s(L)$. *QED.*

In applications, however, this converse is usually too weak. Suppose one is given a pre-scheme X, on which G is acting. The interesting question is to classify those invariant open subsets U in X such that a quotient U/G exists. The above converse relates this to G-linearized sheaves L *on* $\underline{U}$, and the open sets $U^s(L)$. Much more interesting is to relate this with the open sets $X^s(L)$, which may sometimes be computed without a prior analysis of all possible invariant open sets U. In this direction, the most useful result is:

Converse 1.13. Let X be a connected algebraic pre-scheme smooth over k, and let G be a connected reductive algebraic group acting on X. If U is an invariant open subset of X, then the following are equivalent:

i) for some $L \varepsilon Pic^G(X)$, $U \subset X^s(L)$,

ii) there is a geometric quotient (Y, ϕ) of U by G, ϕ is affine and Y is quasi-projective.

If the stabilizer of the generic point of X is 0-dimensional, then i) and ii) are also equivalent to:

iii) the action of G on U is proper, a geometric quotient (Y, ϕ) of U by G exists and Y is quasi-projective.

Proof. iii) implies ii) by Proposition 0.7; ii) implies iii) by Converse 1.12 and Corollary 2.5 (still to be proven). i) implies ii) by Theorem 1.10. It remains to prove that ii) implies i). But first of all, by Converse 1.12, there is a G-linearized invertible sheaf L on U such that $U = U^s(L)$. Then, in any case, since X is smooth, there is some invertible sheaf on X extending L: choose one and write it L also. Let $D_1, \ldots, D_k$ be the components of $X - U$ of codimension 1. For a very large integer N, to be fixed later, let

$$L' = L\left[N \sum_{i=1}^{k} D_i\right].$$

This is also an invertible sheaf, since X is smooth. Note that the G-linearization $\phi: \sigma^* L \to p_2^* L$ on $G \times U$ extends to a G-linearization of L on all of $G \times X$. For ϕ corresponds to a non-zero section ψ of $p_2^* L \otimes (\sigma^* L)^{-1}$

over $G \times U$. Since G is connected, the only irreducible divisors in $G \times X - G \times U$ are $G \times D_1, \ldots, G \times D_k$. Suppose the order of ψ at $G \times D_i$ is k_i. Then ψ extends to a non-zero section of $p_2^* L \otimes (\sigma^* L)^{-1} \left[- \sum_{i=1}^{k} k_i (G \times D_i) \right]$. Therefore ϕ extends to an isomorphism of $\sigma^* L$ and $p_2^* L \left[- \sum_{i=1}^{k} k_i (G \times D_i) \right]$. But, restricting to $\{e\} \times X$, it follows that it extends to an isomorphism of L and $L \left[- \sum_{i=1}^{k} k_i D_i \right]$. Since ϕ satisfies the co-cycle condition, its restriction to $\{e\} \times U$ must be the indentity homomorphism from L to L. Therefore all the integers k_i equal 0. It follows that the original ϕ extends to an isomorphism of $\sigma^* L$ and $p_2^* L$; and, as ϕ satisfies the co-cycle condition generically, ϕ must satisfy it over all of $G \times X$. For similar reasons, ϕ extends to a G-linearization of L' also.

We claim now that if N is sufficiently large, $X^s(L') \supset U$. Let x be a geometric point of U. Then there is an invariant section $s \in H^0(U, L^n)$, for some n, such that $s(x) \neq 0$, U_s is affine, and the action of G on U_s is closed. But if N is large enough, s extends to a section $t \in H^0(X, L'^n)$. And, increasing N further, t must be 0 on all the divisors D_i. Since s is invariant, it follows that t is invariant. Finally, I claim that $X_t = U_s$. Certainly, we have

$$U_s \subset X_t \subset X - \bigcup_{i=1}^{k} D_i.$$

But, in general, if V is an affine open sbuset of a scheme Z smooth/k then the components of $Z - V$ all have codimension 1. Therefore, all the components of $X_t - U_s$ are subsets of $X - U$ of codimension 1. Since all the D_i are outside X_t, there are no such subsets, i.e. $X_t = U_s$. Therefore, x is a stable point of X with respect to L'. *QED.*

In the last three Propositions of § 4, we shall show that stability is independent of some alterations of X, G, L, and k.

Proposition 1.14. Let X be an algebraic pre-scheme over k, let G be a reductive group acting on X, and let L be a G-linearized invertible sheaf on X. Let $K \supset k$ be any over-field. Then if $\bar{X} = X \times_k K$, and if $\bar{L}$ is the sheaf induced by L on X, we have:

$$\bar{X}^s(\bar{L}) = \overline{X^s(L)}$$

$$\bar{X}^{ss}(\bar{L}) = \overline{X^{ss}(L)}.$$

Proof. By amplification 1.11, $X^s(L)$, as a subset of $X^{ss}(L)$, is characterized by a geometric condition, i.e. part (2). Therefore, if the result is proven for $X^{ss}(L)$, it follows for $X^s(L)$.

To prove it for $X^{ss}(L)$, it is clearly sufficient to treat the two cases:

(i) K algebraically closed, and K/k separable, (ii) K/k purely inseparable. We treat first case (i). Let $V_n \subset H^0(X, L^n)$ be the subspace of invariant sections s such that X_s is affine; let $\bar{V}_n$ be the analogous subspace of $H^0(\bar{X}, \bar{L}^n)$. Since K/k is flat, it follows that

$$H^0(\bar{X}, \bar{L}^n) \cong H^0(X, L^n) \underset{k}{\otimes} K.$$

Then the subspace $\bar{V}_n$ has a least field of definition L, where $K \supset L \supset k$. But if τ is any automorphism of K over k, then τ commutes with the action of G, and the G-linearization of L, hence certainly maps invariant sections of $H^0(\bar{X}, \bar{L}^n)$ into invariant sections. Also, if $\bar{X}_s$ is affine, then $(\bar{X})_{s^\tau} = (\bar{X}_s)^\tau$ is affine. Therefore, τ maps $\bar{V}_n$ into itself. According to Cartier [7], Chapter 1, this implies that τ leaves L pointwize fixed. By our assumptions on K/k, this implies that $L = k$. Therefore, for some subspace $V'_n \subset H^0(X, L^n)$, $\bar{V}_n = V'_n \underset{k}{\otimes} K$. But if $s \varepsilon H^0(X, L^n)$, then s is invariant and X_s is affine if and only if s is invariant as a section of $H^0(\bar{X}, \bar{L}^n)$ and $(\bar{X})_s = (X_s) \underset{k}{\times} K$ is affine, (cf. SGA 8, Cor. 5.6). Therefore $V'_n = V_n$. Finally, $X - X^{ss}(L)$ (resp. $\bar{X} - \bar{X}^{ss}(\bar{L})$) is the set of common zeroes of all sections of all the spaces V_n (resp. $\bar{V}_n$). Since V_n generates $\bar{V}_n$, the result follows.

Now suppose K/k is purely inseparable. The result is then immediate because if $s \varepsilon H^0(\bar{X}, \bar{L}^n)$ then for some ν, s^{p^ν} is a section of $L^{n \cdot p^\nu}$ over X, and because:

$$(\bar{X}_s) = (X_{s^{p^\nu}}) \underset{k}{\times} K. \quad QED.$$

Proposition 1.15. Let the reductive algebraic group G act via σ on an algebraic pre-scheme X, and let L be a G-linearized invertible sheaf on X. Then, if G_0 is the connected component of $e \varepsilon G$, the open set of stable points (resp. semi-stable points) is the same for the action of G and of G_0.

Proof. By Proposition 1.14, it suffices to look at the case where the ground field k is algebraically closed. Moreover, by amplification 1.11, it suffices to prove this for semi-stability. Let U (resp. U_0) be the open set of semi-stable points with respect to G (resp. G_0). Clearly $U \subset U_0$. Secondly, note that U_0 is invariant under the action of G as well as of G_0. In fact, if $\alpha \varepsilon G_k$, and if $s \varepsilon H^0(X, L^n)$ is G_0-invariant, we may define s^α by means of the G-linearization of L, and s^α will still be G_0-invariant since G_0 is a normal subgroup of G. Then, since $\sigma(\alpha, X_s) = X_{(s^\alpha)}$, it follows that U_0 is invariant under α, hence under G.

Now let x be any k-rational point of U_0. Let $\alpha_1, \ldots, \alpha_N$ be representatives for the cosets of G/G_0. Applying the second part of amplification 1.11 to the finite set of points $\sigma(\alpha_i, x)$,we find that there is a G_0-invariant section s of L^n (for some n), which is not zero at any of the points $\sigma(\alpha_i, x)$, and such that X_s is affine. Then put $s' = \prod_{i=1}^{N} s^{\alpha_i}$. s' is a G-invariant section of L^{nN}, $s'(x) \neq 0$, and

$$X_{s'} = \bigcap_{i=1}^{N} \sigma(\alpha_i, X_s)$$

is affine. Therefore $x \varepsilon U$. *QED.*

Proposition 1.16. Let the reductive algebraic group G act via σ on an algebraic pre-scheme X, and let L be a G-linearized invertible sheaf on X. Let $f: X_{\text{red}} \rightarrow X$ be the canonical immersion. Then

$$X^{ss}(L) = X^{ss}_{\text{red}}(f^* L),$$
$$X^{s}(L) = X^{s}_{\text{red}}(f^* L).$$

Proof.* The "$\subset$" is obvious. To prove the inclusion "$\supset$", start with an invariant section $s \varepsilon H^0(X_{\text{red}}, f^* L^n)$ such that $[X_{\text{red}}]_s$ is affine. It will suffice to prove that s^k lifts to an invariant section t in $H^0(X, L^{nk})$, for some k, since it follows automatically that X_t is affine. But by lemma 4.5.13.1 of EGA 2, there is a positive integer k such that s^k lifts to some section t of L^{nk} over X. Now we are given a dual action of G on $H^0(X, L^{nk})$: therefore there is a Reynolds operator E on this vector space, and Et is an invariant section that still restricts to s^k on X_{red}. *QED.*

Corollary 1.17. In the situation of the above Proposition, if there are no homomorphisms of the connected component of $\overline{G}$ onto $\overline{G}_m$, then $X^{ss}(L)$ and $X^s(L)$ are independent of the G-linearization of L.

Proof. By Propositions 1.4, 1.14, 1.15, and 1.16.

§ 5. Functorial properties

In this section, we are concerned with the following situation: $f: X \rightarrow Y$ is a G-linear morphism between two algebraic pre-schemes on which G is acting. Moreover, $M \varepsilon Pic^G(Y)$. Then what is the relation between:

$$X^s_{(0)}(f^*M) \quad \text{and} \quad f^{-1}\{Y^s_{(0)}(M)\}.$$

Many of our results are also valid for semi-stability, and stability but for the sake of simplicity we will only consider the open sets of properly stable points.

Proposition 1.18. In the above situation, if f is quasi-affine, then

$$X^s_{(0)}(f^*M) \supset f^{-1}\{Y^s_{(0)}(M)\}.$$

Proof.* By Proposition 1.14, it suffices to prove this under the added assumption that the ground field k is algebraically closed. Now let x

* Modifications to extend this proof, and the proofs of 1.18, 1.19, to char. p are given in Appendix 1 B.

be a closed point of $f^{-1}\{Y^s_{(0)}(M)\}$: we must prove that $x \in X^s_{(0)}(f^*M)$. Since $f(x)$ is properly stable, there is an invariant section $t \in H^0(Y, M^n)$, for some n, such that Y_t is affine, $f(x) \in Y_t$, and all the stabilizers of the k-rational points of Y_t are 0-dimensional. Let $s = f^*t$ be the induced element of $H^0(X, (f^*M)^n)$. s is still invariant, $x \in X_s$, and all stabilizers of k-rational points of X_s are 0-dimensional — but X_s is not necessarily affine. In fact, $X_s = f^{-1}(Y_t)$ is merely quasi-affine in general. But let:

$$R = \Gamma(X_s, \varrho_X)$$

$$\tilde{X} = \mathrm{Spec}\,(R).$$

Then we have the commutative diagram:

$$\begin{array}{ccc} x \varepsilon X_s & \xrightarrow{I} & \tilde{X} \\ & {\scriptstyle f}\searrow \quad \swarrow{\scriptstyle \tilde{f}} & \\ & Y_t & \end{array}$$

where I is an open immersion (Prop. 5.1.2, part (b) in EGA 2). Moreover, since G acts on X_s, there is a dual action of G on R, hence an action of G on $\tilde{X}$. It is clear that all the above morphisms are G-linear. Notice first that as $\tilde{f}$ is G-linear, it follows that all stabilizers of all geometric points of $\tilde{X}$ are 0-dimensional: therefore the action of G on $\tilde{X}$ is closed. Put:

$$Z_1 = \tilde{X} - I(X_s),$$

$$Z_2 = 0(x).$$

Then Z_1 and Z_2 are disjoint closed invariant subsets of $\tilde{X}$. Therefore, by Corollary 1.2, there is an invariant $f \in R$ which is 0 on Z_1, and 1 at x. It follows that $(\tilde{X})_f \subset X_s$, and that $(\tilde{X})_f$ is affine. But by Theorem 1.3.1, EGA 1, there is an integer k such that the section $s^k \cdot f$ of $(f^*M)^{nk}$ over X_s extends to a section s' of $(f^*M)^{nk}$ over X. Finally, let E be the Reynolds operator on $H^0(X, (f^*M)^{nk})$. Then Es' is invariant, and Es' still equals the invariant section $s^k \cdot f$ over the open set X_s Then $(s \cdot Es')$ is 0 both (a) outside X_s, and (b) at points of X_s where $f = 0$. Therefore $X_{(s \cdot Es')}$ is an affine neighborhood of x contained in X_s: since $s \cdot Es'$ is invariant and every stabilizer in X_s is 0-dimensional, x is properly stable. *QED*.

The next question is: when does $X^s_{(0)}(f^*M)$ actually equal $f^{-1}\{Y^s_{(0)}(M)\}$? A reasonable suggestion would be that they are equal at least if f is finite. Unfortunately, this is false. One need only consider the following case: $x \in X_k$ and $f = \psi_x : G \to X$. Then if $0(x)$ is closed, and $S(x)$ is 0-dimensional, it follows from lemma 0.3 that f is finite. However, $\psi^*_x(L) \cong \varrho_G$, and it is clear that

$$G^s_{(0)}(\varrho_G) = G.$$

But we will see from almost every example in Chapter 4 that there are plenty of closed orbits, with trivial stabilizers, that are not stable. The following is true, however:

Theorem 1.19. In the above situation, assume f is finite, X is proper over k, and M is ample on Y (but Y is not assumed proper over k!). Then

$$X^s_{(0)}(f^*M) = f^{-1}\{Y^s_{(0)}(M)\}.$$

Proof. By Proposition 1.7, there is an embedding $I: Y \to \boldsymbol{P}_n$, an action of G on $\boldsymbol{P}_n$ and a G-linearization of $\varrho_{\boldsymbol{P}}(1)$ such that I is G-linear, and $M^r = I^*(\varrho_{\boldsymbol{P}}(1))$ (as a G-linearized sheaf). Using Proposition 1.18, it is clear that if we prove the Theorem for $I \circ f$, then it follows for f.

Now suppose $Y = \boldsymbol{P}_n$, $M = \varrho_{\boldsymbol{P}}(1)$. Let R be the homogeneous coordinate ring of $\boldsymbol{P}_n$, let S be the homogeneous coordinate ring of X, and let $f: X \to \boldsymbol{P}_n$ be defined by a graded R-algebra structure on S. Since f is finite, S is a finite R-module. The actions of G on X and $\boldsymbol{P}_n$, and the G-linearizations of $f^*(\varrho(1))$ and $\varrho_{\boldsymbol{P}}(1)$ define dual actions of G on R and S, compatible with the R-module structure on S. Let E and F be the Reynolds operators on R and S respectively. Now suppose x is a properly stable geometric point of X. Then there is an invariant element

$$s \in H^0(X, f^*(\varrho_{\boldsymbol{P}}(n)) = S_n$$

such that $s(x) \neq 0$, the orbit of x is closed in X_s, and the stabilizer of x is 0-dimensional. Since S is a finite R-module, there is an equation of integral dependence:

$$s^m + \bar{a}_1 \cdot s^{m-1} + \cdots + \bar{a}_m = 0$$

where a_i is a homogeneous element of R, and $\bar{a}_i$ denotes its image in S. Applying F, we obtain:

$$\begin{aligned} 0 &= F\{s^m + \bar{a}_1 \cdot s^{m-1} + \cdots + a_m\} \\ &= s^m + F\bar{a}_1 \cdot s^{m-1} + \cdots + F\bar{a}_m \\ &= s^m + \overline{Ea_1} \cdot s^{m-1} + \cdots + \overline{Ea_m}. \end{aligned}$$

Since $s(x) \neq 0$, it follows that for some i, $Ea_i(f(x)) \neq 0$. Therefore, $f(x) \in (\boldsymbol{P}_n)_{Ea_i}$. But this implies that

$$x \in X_{f^*(Ea_i)} = f^{-1}[(\boldsymbol{P}_n)_{Ea_i}].$$

By amplification 1.11, the orbit of x in $X_{f^*(Ea_i)}$ is closed; since f is proper, the orbit of $f(x)$ in $(\boldsymbol{P}_n)_{Ea_i}$ is closed.Since f is actually finite, the dimension of the orbits of x and $f(x)$ is the same; therefore the stabilizer of $f(x)$ is 0-dimensional. By amplification 1.11, $f(x)$ is properly stable. This proves $X^s_{(0)}(f^*\varrho_{\boldsymbol{P}}(1)) \subset f^{-1}\{(\boldsymbol{P}_n)^s_{(0)}(\varrho_{\boldsymbol{P}}(1))\}$. Using Proposition 1.18, the result follows. *QED.*

Corollary 1.20. Let a reductive algebraic group G act on a scheme X, proper over k. If there are no homomorphisms of the connected component of $\bar{G}$ onto $\bar{G}_m$, and if $L \varepsilon Pic^G(X)$ is ample, then $X^s_{(0)}(L)$ depends only on the polarization containing L.

Proof. This means two things:

(i) $X^s_{(0)}(L_1^p) = X^s_{(0)}(L_1^q)$, for positive integers p and q, and ample L_1 in $Pic^G(X)$.

(ii) If L_1 and L_2 are ample sheaves in $Pic^G(X)$, which are algebraically equivalent (in $Pic(X)$), then $X^s_{(0)}(L_1) = X^s_{(0)}(L_2)$.

The first is obvious. To prove the second, we can make preliminary reductions to the case: k algebraically closed, X reduced, and G connected by Propositions 1.14, 1.15 and 1.16. In fact, we may assume that X is normal: for let $\pi: X' \to X$ be the finite morphism from the normalization of X to X. Then applying Theorem 1.19 to π, and noting the algebraically equivalent sheaves on X are algebraically equivalent on X', we are reduced to proving the result for X'.

Now let L_i define the k-rational point $\lambda_i \varepsilon Pic(X/k)$, and let P be the reduced scheme of the connected component of $Pic(X/k)$ containing λ_1 and λ_2. Let $\mathcal{L}$ be an invertible sheaf on $X \times P$ which restricts on each closed fibre $X \times \{p\}$ to the sheaf L_p corresponding to p. Replacing $\mathcal{L}$ by $\mathcal{L} \otimes p_2^*(M)$, for a sufficiently ample sheaf M on P, we may assume that $\mathcal{L}$ is ample on $X \times P$. By a result of Chevalley [10], P is a principal homogeneous space over an abelian variety; therefore the action of the linear group G on P induced from its action on X is trivial. Then G acts on $X \times P$ by the product of its action on X and the trivial action on P. By Corollary 1.6, some power $\mathcal{L}^k$ of $\mathcal{L}$ admits a G-linearization. For every k-rational point $p \varepsilon P$, we can apply Theorem 1.19 to the G-linear morphism

$$f_p: X \xrightarrow{\sim} X \times \{p\} \subset X \times P.$$

We conclude that

$$[(X \times P)^s_{(0)}(\mathcal{L})] \cap [X \times \{p\}] = X^s_{(0)}(L_p).$$

In particular, for any point $x \varepsilon X_k$, it follows that the set of $p \varepsilon P_k$ such that x is properly stable for L_p is the set of k-rational points of an open subset $U \subset P$. Now suppose that for x, U is non-empty. Then I claim that $U = P$: for if p is a k-rational point of $P - U$, then there is a finite collection of points $p_1, \ldots, p_N$ in U such that

$$p_1 + \cdots + p_N = N \cdot p$$

in the group scheme $Pic(X/k)$. But then:

$$\begin{aligned} X^s_{(0)}(L_p) &= X^s_{(0)}(L_p^N) \\ &= X^s_{(0)}(L_{p_1} \otimes \cdots \otimes L_{p_N}) \\ &\supset \bigcap_{i=1}^{N} X^s_{(0)}(L_{p_i}). \end{aligned}$$

Since $x \in X^s_{(0)}(L_q)$ for every $q \in U_k$, it follows that $x \in X^s_{(0)}(L_p)$, i.e. $p \in U_k$. Therefore $U = P$.

This implies that $X^s_{(0)}(L_p)$ is independent of p; in particular, $X^s_{(0)}(L_1) = X^s_{(0)}(L_{\lambda_1}) = X^s_{(0)}(L_{\lambda_2}) = X^s_{(0)}(L_2)$. *QED.*

For the sake of applications in the next chapter, we must make one mention of semi-stability. We note that Proposition 1.18 is trivially valid, if f is *affine*, for X^{ss} and Y^{ss}; and that Theorem 1.19 is likewise valid, if Y *is proper* over k too, for X^{ss} and Y^{ss}. The proof of the latter can be read word for word from the above proof, omitting the final steps. Actually, Theorem 1.19 is valid in all cases for X^{ss} and Y^{ss}, using a small additional argument: but we will not use this fact.

Chapter 2

Analysis of stability

It might seem that the various concepts of stability, introduced in the previous chapter, are merely accidental and unworkable notions. We hope to show here that this is not so. In the first place, there is a strong numerical criterion for stability. Almost all of our later examples will use this criterion. And in the second place, this criterion leads to a description of structures in the group which are associated naturally to points where stability breaks down: these structures are "convex sets in the flag complex", and this association is the explanation of a well-known intuition which runs as follows:

> if $PGL(n)$ is acting on a set of cycles or subschemes in $\mathbf{P}_n$, and we attempt to form projective invariants for these objects, i.e. to construct the quotient of the set of objects by $PGL(n)$, then we cannot do this until we first discard those cycles or subschemes which have a singularity at a *flag* or order of contact with a *flag* which is too "bad".

Examples of this will be seen in Chapters 3 and 4.

Since we are dealing with the questions of stability for a fixed geometric point, there is no loss in assuming from the start that the ground field k is *algebraically closed.* Since all pre-schemes X will be of finite type over k, the set X_k will be the set of closed points of X.

§ 1. A numerical criterion

Definition 2.1. A 1-*parameter subgroup* of G is a homomorphism λ: $\mathbf{G}_m \to G$. We abbreviate this to: λ is a 1-*PS* of G.

Now suppose we are given an action σ of an algebraic group G on an algebraic scheme X, *proper* over k. Let x be a closed point of X, and let λ be a 1-*PS* of G.

Consider the morphism $\psi_x \circ \lambda$ from G_m to X. Identifying G_m with Spec $k[\alpha, \alpha^{-1}]$, we may embed G_m in the affine line $A^1 = \text{Spec}\, k[\alpha]$. Then $\psi_x \circ \lambda$ extends uniquely to a morphism $f: A^1 \to X$. This follows from the valuative criterion for the properness of X over k (cf. EGA 2, § 7), since the local ring of A^1 at the origin (0) is a valuation ring. The closed point $f(0)$ in X will be called the specialization of $\sigma(\lambda(\alpha), x)$ when $\alpha \to 0$. Clearly $f(0)$ is fixed under the action of G_m on X induced by λ. Now if $L \in Pic^G(X)$, we can consider the induced G_m-linearization of L restricted to the fixed point $f(0)$. As remarked in § 3 of Chapter 1, this is given by a character of G_m: say $\chi(\alpha) = \alpha^r$, for $\alpha \in (G_m)_k$. With all this preparation, we can make the key definition:

Definition 2.2. If G acts on the algebraic scheme X, proper over k, if x is a closed point of X, λ is a 1-*PS* of G, and $L \in Pic^G(X)$, then

$$\mu^L(x, \lambda) = -r.$$

The functorial properties of μ are:

(i) $\mu^L(\sigma(\alpha, x), \lambda) = \mu^L(x, \alpha^{-1} \cdot \lambda \cdot \alpha)$, if $\alpha \in G_k$,

ii) for fixed x and λ, $\mu^L(x, \lambda)$ defines a homomorphism from $Pic^G(X)$ to $\mathbf{Z}$ as L varies.

iii) If $f: X \to Y$ is a G-linear morphism of schemes on which G acts, $L \in Pic^G(Y)$ and $x \in X_k$, then

$$\mu^{f^*L}(x, \lambda) = \mu^L(fx, \lambda).$$

iv) If $\sigma(\lambda(\alpha), x) \to y$ as $\alpha \to 0$, then $\mu^L(x, \lambda) = \mu^L(y, \lambda)$. (Proofs Immediate.)

The theorem which we shall ultimately prove is:

Theorem 2.1. Let a reductive group G act on a scheme X, proper over k. Let $L \in Pic^G(X)$, and assume L is ample. Then if $x \in X_k$:

$$x \in X^{ss}(L) \Leftrightarrow \mu^L(x, \lambda) \geq 0 \text{ for all 1-PS's } \lambda,$$

$$x \in X^s_{(0)}(L) \Leftrightarrow \mu^L(x, \lambda) > 0 \text{ for all 1-PS's } \lambda.$$

The essential idea behind this proof stems from Hilbert [14], where the case $G = SL(n)$, $X = P^n$ is analyzed.

There are two approaches to the proof of this theorem. Either we choose a G-linear immersion $X \subset P_{n-1}$, and reduce to the case $X = P_{n-1}$ by Theorem 1.19; or we can put Y equal to the normalization of the closure of $0(x)$, and via the canonical $\pi: Y \to X$, reduce to the case where $0(x)$ is dense in X by Theorem 1.14. We shall follow the former which is more down to earth. Therefore we may assume $X = P_{n-1}$. Now let

$$V = H^0(P_{n-1}, \underline{o}(1))$$

$$A^n = V(V): \text{ the affine cone over } P_{n-1}.$$

There is a natural projection from $A^n - (0)$ to P_{n-1}. We shall say that a closed point $x^* \in A^n$ *lies over* x, if $x^* \neq (0)$, and if x^* projects to x in

$\mathbf{P}_{n-1}$. As remarked in the proof of Proposition 1.7, the action of G on $\mathbf{P}_{n-1}$ and the G-linearization of $\varrho_{\mathbf{P}_{n-1}}(1)$ together define also:

i) a dual action of G on V,

ii) a linear action σ^* of G on $\mathbf{A}^n$, compatible with σ [with respect to the projection $\mathbf{A}^n - (0) \to \mathbf{P}_{n-1}$].

The first step is:

Proposition 2.2. x is semi-stable if and only if (0) is not in the closure of the orbit $0(x^*)$ for one (and hence all) closed points x^* lying over x; x is properly stable if and only if ψ_{x^*} is proper for one (and hence all) closed points x^* lying over x.

Proof. By definition, x is semi-stable if and only if there is an invariant $s \varepsilon H^0(\mathbf{P}_{n-1}, \varrho_{\mathbf{P}}(k))$ such that $s(x) \neq 0$. This is the same as asking whether there is an invariant homogeneous polynomial function F on $\mathbf{A}^n$ such that $F(x^*) \neq 0$. But if such an F exists, then F equals a non-zero constant on $0(x^*)$, hence (0) cannot be in the closure of $0(x^*)$. Conversely, suppose (0) is not in the closure of $0(x^*)$. Then if:

$$Z_1 = \text{closure of } 0(x^*),$$

$$Z_2 = \{(0)\},$$

Z_1 and Z_2 are disjoint invariant closed subsets of $\mathbf{A}^n$. By corollary 1.2, as G is reductive, there must be some invariant function F' on $\mathbf{A}^n$ such that $F' = 0$ on Z_2, and $F' = 1$ on Z_1. Suppose we write:

$$F' = F_{k_1} + F_{k_2} + \cdots + F_{k_l}$$

where each F_{k_i} is homogeneous of degree k_i. Then each k_i is positive, i.e. each F_{k_i} is 0 on Z_2. Moreover *some* F_{k_i} is not zero at x^*. Let F be that F_{k_i}.

By definition, x is properly stable if and only if its stabilizer is finite, and for one (and hence all) $s \varepsilon H^0(\mathbf{P}_{n-1}, \varrho_{\mathbf{P}}(k))$ as above, $0(x)$ is closed in $[\mathbf{P}_{n-1}]_s$ (cf. amplification 1.11 and lemma 0.3). That is to say,

$$\psi'_x : G \to [\mathbf{P}_{n-1}]_s$$

is proper. But let s correspond, as above, to the invariant homogeneous function F on $\mathbf{A}^n$. Then, for some non-zero $\alpha \varepsilon k$, $0(x^*)$ is contained in the closed subscheme Z_α of $\mathbf{A}^n$ defined by $F = \alpha$. Therefore

$$\psi_{x^*} : G \to \mathbf{A}^n$$

is proper if and only if:

$$\psi'_{x^*} : G \to Z_\alpha$$

is proper. But let the projection from $\mathbf{A}^n - (0)$ to $\mathbf{P}_{n-1}$ define the morphism $\pi : Z_\alpha \to [\mathbf{P}_{n-1}]$. Then, in fact: $\pi(Z_\alpha) \subset [\mathbf{P}_{n-1}]_s$, and if π' denotes π with image taken as $[\mathbf{P}_{n-1}]_s$, then π' is proper [i.e. let

$R = \Gamma(\boldsymbol{A}^n, \underline{o}_{\boldsymbol{A}})$, and let k be the degree of F. Then $Z_\alpha = \text{Spec}\,(R/(F-\alpha))$ and $[\boldsymbol{P}_{n-1}]_s \simeq \text{Spec}\,(R(k)/(F-\alpha))$. But $R/(F-\alpha)$ is a finite module over $R(k)/(F-\alpha)$.] Finally, $\psi'_x = \pi' \circ \psi'_{x^*}$, hence ψ'_x is proper if and only if ψ'_{x^*} is proper. *QED*.

The next step is to interpret the function μ in the case where the ambient space X is $\boldsymbol{P}_{n-1}$. We recall that a linear action of a torus on affine space can be diagonalized (cf. Seminaire Chevalley, [8], exposé 4): i.e. for a suitable coordinate system in $\boldsymbol{A}^n$, the closed points $(\alpha) \in (\boldsymbol{G}_m)^r_k$ act via diagonal matrices $[\chi_i(\alpha) \cdot \delta_{ij}]$, for characters $\chi_1, \ldots, \chi_n$ of $(\boldsymbol{G}_m)^r$. Moreover, recall that if $(\alpha) = (\alpha_1, \ldots, \alpha_r)$, then every character χ of $(\boldsymbol{G}_m)^r$ is of the form

$$\chi(\alpha) = \prod_{i=1}^{r} \alpha_i^{m_i}$$

for suitable integers m_i. In particular, suppose we are given any action of $\boldsymbol{G}_m$ on $\boldsymbol{P}_{n-1}$, plus a $\boldsymbol{G}_m$-linearization of $\underline{o}(1)$. Then this gives a linear action of $\boldsymbol{G}_m$ on the cone $\boldsymbol{A}^n$, and, for suitable coordinates, the action of $\alpha \in (\boldsymbol{G}_m)_k$ is given by the matrix $(\alpha^{r_i} \cdot \delta_{ij})$, for fixed integers $r_1, \ldots, r_n$.

Proposition 2.3. Let x be a closed point of $\boldsymbol{P}_{n-1}$, let λ be a 1-*PS* of G, and let $\sigma(\lambda(\alpha), x)$ specialize to y when $\alpha \to 0$. Let x^* be a closed homogeneous point over x. Fix coordinates in $\boldsymbol{A}^n$ so that the action of $\boldsymbol{G}_m$ induced via λ is diagonalized as above. Then

$$\mu^{o(1)}(x, \lambda) = \max\{-r_i \mid i \text{ such that } x_i^* \neq 0\}$$

where $x^* = (x_1^*, \ldots, x_n^*)$. Moreover, [$\sigma^*(\lambda(\alpha), x^*)$ has no specialization, resp. some specialization, resp. specialization (0), in $\boldsymbol{A}^n$, when $\alpha \to 0$] $\Leftrightarrow$ [$\mu(x;\lambda) > 0$, resp. $\mu(x,\lambda) = 0$, resp. $\mu(x,\lambda) < 0$].

Proof. Let $\mu_1 = \mu(x;\lambda)$ and $\mu_2 = \max\{-r_i \mid x_i^* \neq 0\}$. Note that $\sigma^*(\lambda(\alpha), x^*)$ has coordinates

$$(\alpha^{r_1} \cdot x_1^*, \ldots, \alpha^{r_n} \cdot x_n^*).$$

Therefore, $\alpha^{\mu_2} \cdot x^*$ has a finite non-zero specialization y^* in $\boldsymbol{A}^n$ when $\alpha \to 0$. Therefore, the last assertion is obvious, once we have proven that $\mu_1 = \mu_2$. On the other hand, since $\alpha^{\mu_2} \cdot x^*$ has a non-zero specialization in $\boldsymbol{A}^n$, this specialization lies over some point of $\boldsymbol{P}_{n-1}$, and this must be the specialization of $\sigma(\lambda(\alpha), x)$. Therefore y^* lies over y. Now if $y^* = (y_1^*, \ldots, y_n^*)$, one sees that $y_i^* = 0$ if either $x_i^* = 0$ or $r_i > -\mu_2$. Therefore,

$$\sigma^*\big(\lambda(\alpha), y^*\big) = \alpha^{-\mu_2} \cdot y^*.$$

In other words, the trivial action of $\boldsymbol{G}_m$ on y, plus the $\boldsymbol{G}_m$-linearization of $\underline{o}(1)$ restricted to y, correspond to the linear representation on the cone $\{\beta \cdot y^* \mid \beta \in k\}$ over y given by

$$\alpha : y^* \to \alpha^{-\mu_2} \cdot y^*.$$

Here the affine line through y^* is canonically the line bundle over y corresponding to the invertible sheaf $\mathcal{O}(1) \otimes \varkappa(y)$. Therefore, as we saw in § 3 of Chapter 1, the character of G_m corresponding to the G-linearization of $\mathcal{O}(1)$ over the fixed point y is:

$$\chi(\alpha) = \alpha^{-\mu_2}.$$

Therefore, $\mu_2 = \mu_1$ according to Definition 2.2. *QED.*

We fix the following notation:

$$R = k[[t]] \quad (= \text{formal power series ring in } t)$$

$$K = k((t)) \quad (= \text{quotient field of } R).$$

We shall be interested in the groups of R and K-valued points of the reductive group G. Let:

$$G(R) = \mathrm{Hom}_k(\mathrm{Spec}\, R, G) = \text{group of } R\text{-valued points},$$

$$G(K) = \mathrm{Hom}_k(\mathrm{Spec}\, K, G) = \text{group of } K\text{-valued points}.$$

Note that $G(R)$ is a subgroup of $G(K)$, i.e. via the natural morphism $\mathrm{Spec}\, K \to \mathrm{Spec}\, R$; moreover, there is a natural map $\overline{\omega}$ from $G(R)$ to G_k induced by the morphism $\mathrm{Spec}\, k \to \mathrm{Spec}\, R$. In classical language, $G(R)$ is the subgroup of points $\phi \in G(K)$ which have a specialization in G, when $t \in K$ specializes to 0, and $\overline{\omega}(\phi)$ is this specialization.

If λ is a 1-*PS* of G, then a canonical K-valued point of G is defined by λ; it can be defined as the composition:

$$\begin{array}{ccccc} \mathrm{Spec}\, K & \xrightarrow{A} & G_m & \xrightarrow{\lambda} & G \\ & & \wr\Vert & & \\ & & \mathrm{Spec}\, k[\alpha, \alpha^{-1}], & & \end{array}$$

where A, in turn, is defined by the k-homomorphism

$$\begin{cases} \tilde{A} : k[\alpha, \alpha^{-1}] \to K \\ \tilde{A}(\alpha) = t. \end{cases}$$

Note that, when t specializes to 0, α specializes to 0. This point of $G(K)$ will be denoted $\langle\lambda\rangle$. We are now in a position to state:

Theorem (Iwahori). Let G be a semi-simple algebraic group over k of adjoint type. Every double coset of $G(K)$ with respect to the subgroup $G(R)$ is represented by a point of the type $\langle\lambda\rangle$; for some 1-*PS* λ.

In fact, Iwahori has proved this [16] for more general rings R, for the "Tohoku" or adjoint groups of Chevalley. On the other hand, suppose G is any reductive group and the characteristic is 0, and $\pi: G \to G'$ is the homomorphism of G to its adjoint group. Since the characteristic is 0, π is smooth; therefore the induced map $G(R) \to G'(R)$ is surjective, since R is a hensel local ring. Now starting with a K-valued point of G, we see that multiplying on the left and right by R-valued points, we can assume that its image in G' becomes a K-valued point of a subgroup $\lambda(G_m) \subset G'$. Then the point in G becomes a K-valued point of the subgroup $\pi^{-1}(\lambda(Gm))$: this is an extension of torus by a finite group. By the result for such a subgroup, the theorem is proven for G.†

This result will be used in the last step of the proof:

(i) ψ_{x^*} is not proper if and only if for some non-trivial 1-*PS* λ of G, $\sigma^*(\lambda(\alpha), x^*)$ has a specialization when $\alpha \to 0$.

(ii) (0) is in the closure of $0(x^*)$ if and only if for some 1-*PS* λ of G, $\sigma^*(\lambda(\alpha), x^*)$ specializes to (0) when $\alpha \to 0$.

Proof. The two "if" statements are obvious. Next consider the "only if" in (i): assume ψ_{x^*} is not proper. Then by the valuative criterion for properness (EGA 2, § 7), there is a K-valued point ϕ: Spec $(K) \to G$ such that ϕ is not induced by an R-valued point of G, but such that $\psi_{x^*} \circ \phi$ is an R-valued point of A^n. By Iwahori's theorem, then, we know that ϕ has the form:

$$\phi = \psi_1 \cdot \langle\lambda\rangle \cdot \psi_2$$

for some 1-*PS* λ of G, and for $\psi_1, \psi_2 \in G(R)$. Moreover, λ is non-trivial since ϕ is not itself an R-valued point of G. Let $b_i \in G_k$ be the k-valued point $\overline{\omega}(\psi_i)$ obtained from ψ_i when t specializes to 0. By choosing suitable coordinates in A^n, we may assume that the action of $b_2^{-1} \cdot \lambda \cdot b_2$ on A^n is diagonalized. Suppose $b_2^{-1} \cdot \lambda(\alpha) \cdot b_2$ acts via the matrix $(\alpha^{r_i} \cdot \delta_{ij})$, for $\alpha \in (G_m)_k$. Now

$$\begin{aligned} \psi_{x^*} \circ \phi &= \sigma^*(\phi, x^*) \\ &= \sigma^*\left[(\psi_1 \cdot b_2) \cdot (b_2^{-1} \cdot \langle\lambda\rangle \cdot b_2) \cdot (b_2^{-1} \cdot \psi_2), x^*\right] \end{aligned}$$

(Here k-valued points of G and of A^n are identified with the K-valued points obtained by the base extension $K \supset k$).

Therefore, since σ^* is a group action:

$$(*) \qquad \sigma^*\left((\psi_1 \cdot b_2)^{-1}; \psi_{x^*} \circ \phi\right) = \sigma^*\left[(b_2^{-1} \cdot \langle\lambda\rangle \cdot b_2); \sigma^*(b_2^{-1} \cdot \psi_2; x^*)\right].$$

But $(\psi_1 \cdot b_2)^{-1}$ is an R-valued point of G, and $\psi_{x^*} \circ \phi$ is an R-valued point of A^n; therefore the term on the left is an R-valued point of A^n.

Now, if f is a Spec (A)-valued point of A^n, for some ring A, let us denote by $X_i(f)$ the element of A which is its ith coordinate. Then, recalling the definition of $\langle\lambda\rangle$, and that $b_2^{-1} \cdot \lambda \cdot b_2$ is diagonalized,

---

† For the case of char. $(k) = p$, see Appendix 2A. Note that when $G = GL(n)$, the result simply says that every nxn matrix A over $k((t))$ can be transformed to $B = (\delta_{ij} t^{n_i})$ by elementary row and column operations over $k[[t]]$.

equation (*) reads, in coordinates:

$$(*)' \qquad X_i\{\sigma^*((\psi_1 b_2)^{-1}; \psi_{x^*} \circ \phi)\} = t^{r_i} \cdot X_i\{\sigma^*(b_2^{-1} \cdot \psi_2; x^*)\}$$

and, in particular:

$$\text{(A)} \qquad X_i\{\sigma^*(b_2^{-1} \cdot \psi_2; x^*)\} \ \varepsilon \ t^{-r_i} \cdot R.$$

But $b_2^{-1}\psi_2$ is an R-valued point of G, whose specialization, when $t \to 0$, is the identity $e \,\varepsilon\, G$; therefore $\sigma^*(b_2^{-1}\psi_2, x^*)$ is an R-valued point of A^n whose specialization, when $t \to 0$, is x^*; therefore

$$\text{(B)} \qquad X_i\{\sigma^*(b_2^{-1}\psi_2, x^*)\} = X_i(x^*) + t(Z_i)$$

for some $Z_i \,\varepsilon\, R$.

Combining (A) and (B), it follows that $r_i \geq 0$, whenever $X_i(x^*) \neq 0$. Therefore, $\mu(x, b_2^{-1} \cdot \lambda \cdot b_2) \leq 0$ and (i) is proven, by Proposition 2.3.

The "only if" in (ii) is proven in the same way: assume (0) ε closure $0(x^*)$. As a first step, we obtain a K-valued point ϕ of G such that (0) is the specialization of $\psi_{x^*} \circ \phi$ when $t \to 0$. We decompose ϕ exactly as before, and obtain (*)'. But now the term on the left hand side of (*)' is actually in $t \cdot R$ since $\sigma^*((\psi_1 b_2)^{-1}, \psi_{x^*} \circ \phi)$ actually specializes to (0) when $t \to 0$. Therefore, instead of (A), we obtain:

$$(\tilde{\text{A}}) \qquad X_i\{\sigma^*(b_2^{-1}\psi_2, x^*)\} \ \varepsilon \ t^{-r_i+1} \cdot R.$$

(B) is the same as before, hence combining ($\tilde{\text{A}}$) and (B) it follows that $r_i > 0$ whenever $X_i(x^*) \neq 0$. Therefore, $\mu(x, b_2^{-1} \cdot \lambda \cdot b_2) < 0$ and (ii) is proven. *QED.*

A second application of Iwahori's theorem is:

Proposition 2.4. Let a reductive algebraic group G over k act via σ on an algebraic scheme X over k. Then the action of G is proper if and only if for every non-trivial 1-*PS* $\lambda: G_m \to G$, the induced action of G_m on X is proper.

Proof. The only if is clear To prove the if, suppose the action of G is not proper, i.e. $\Psi: G \times X \to X \times X$ is not proper. Then, by the valuative criterion for properness, there are K-valued points ϕ and ξ of G and X respectively such that $\Psi(\phi \times \xi)$ is an R-valued point of $X \times X$, but $\phi \times \xi$ is not an R-valued point of $G \times X$ In other words, ξ and $\sigma(\phi, \xi)$ are R-valued points of X, but ϕ is not an R-valued point of G. By Iwahori's theorem, we can write

$$\phi = \psi_1 \cdot \langle\lambda\rangle \cdot \psi_2$$

for R-valued points ψ_1 and ψ_2 of G, and some 1-*PS* λ. Moreover λ is non-trivial since ϕ is not an R-valued point of G, But then

$$\xi' = \sigma(\psi_2, \xi)$$

and

$$\sigma(\langle\lambda\rangle, \xi') = \sigma(\psi_1^{-1}, \sigma(\phi, \xi))$$

are both R-valued points of X. In other words, the K-valued, but not R-valued point $\langle\lambda\rangle\times\xi'$ of $G\times X$ is mapped by Ψ to an R-valued point of $X\times X$. Since by definition, $\langle\lambda\rangle$ is induced from a K-valued point of $\boldsymbol{G_m}$ via $\lambda:\boldsymbol{G_m}\to G$, this means that the composite morphism $\Psi\circ(\lambda\times 1_X):\boldsymbol{G_m}\times X\to X\times X$ is not proper, i.e. the induced action of $\boldsymbol{G_m}$ is not proper. *QED.*

Corollary 2.5.* Let a reductive algebraic group G act on an algebraic pre-scheme X. Let $L \varepsilon \mathit{Pic}^G(X)$. Then G acts properly on $X^s_{(0)}(L)$.

Proof. Without loss of generality, we may replace X by $X^s_{(0)}(L)$: then L is ample on X, and X is a scheme. By the above proposition, it suffices to prove that for every non-trivial 1-*PS* λ of G, the induced action of $\boldsymbol{G_m}$ is proper. Fix some λ. Let U be the open set in X of points properly stable for the induced action of $\boldsymbol{G_m}$: then I claim $U=X$. For if x is a geometric point of X, then x has an affine neighborhood X_s, where s is an invariant section of L^N, and such that all stabilizers of points of X_s are 0-dimensional. But then s is invariant under $\boldsymbol{G_m}$, and the action of $\boldsymbol{G_m}$ on X_s is still closed. Therefore, by definition, x is a point of U.

Therefore, a quasi-projective geometric quotient $X/\boldsymbol{G_m}$ exists. But this implies that the action of $\boldsymbol{G_m}$ on X is separated. Therefore, by lemma 0.5, the action of $\boldsymbol{G_m}$ on X is proper. *QED.*

§ 2. The flag complex

In the last section, we have related the concept of stability with respect to actions of G to the function $\mu^L(x;\lambda)$ involving a 1-parameter subgroup λ of G. In this section, we shall consider the dependence of μ on λ. The analysis leads naturally to an ungainly but remarkable metric space investigated extensively by J. Tits, which we will call the *flag complex* of G. The first step is:

Definition 2.3/Proposition 2.6. Let G be a reductive algebraic group and let λ be a 1-*PS* of G. Then there is a unique algebraic subgroup $P(\lambda)\subset G$ such that:

$$\gamma \,\varepsilon\, P(\lambda)_k \Leftrightarrow \lambda(\alpha)\cdot\gamma\cdot\lambda(\alpha^{-1}) \text{ has a specialization in } G \text{ when } \alpha \,\varepsilon\, (\boldsymbol{G_m})_k \text{ specializes to } 0.$$

Moreover, $P(\lambda)$ is a parabolic subgroup of G, λ is a 1-*PS* of the radical of $P(\lambda)$, and the specialization γ' of $\lambda(\alpha)\cdot\gamma\cdot\lambda(\alpha^{-1})$, as $\alpha\to 0$, centralizes λ, for all $\gamma\,\varepsilon\,P(\lambda)_k$.

Proof. Most of these facts can be seen most easily by means of a faithful representation $\phi: G\to GL(n)$. Recall that the image $\phi(G)$ is *closed* in $GL(n)$. Therefore, $P(\lambda)=\phi^{-1}(P(\phi\circ\lambda))$, at least as a set of closed points. Now by composing ϕ with a suitable inner automorphism

* This is also proven by Luna [187], p. 90.

of $GL(n)$, we may assume that $\phi \circ \lambda$ is diagonalized, i.e.

$$\phi \circ \lambda(\alpha) = \{\alpha^{r_i} \cdot \delta_{ij}\}$$

for $\alpha \in (G_m)_k$ and for suitable integers r_i. We may even assume that $r_1 \geq r_2 \geq \cdots \geq r_n$. But now, if $\gamma = \{a_{ij}\} \in GL(n)_k$,

$$[\phi \circ \lambda(\alpha)] \cdot \gamma \cdot [\phi \circ \lambda(\alpha^{-1})] = \{\alpha^{r_i - r_j} \cdot a_{ij}\}.$$

This has a finite specialization as $\alpha \to 0$ if and only if $a_{ij} = 0$ when $r_i < r_j$, i.e. γ is of the form:

$$\begin{pmatrix} * & * & * & * \\ 0 & * & * & * \\ 0 & 0 & * & * \\ 0 & 0 & 0 & * \end{pmatrix}$$

This is well-known to be a parabolic subgroup $P \subset GL(n)$. Therefore, at least $P(\lambda)$ exists. Secondly, the radical of P is precisely the set of $\{a_{ij}\}$ of the above form for which the diagonal blocks are multiples of the identity, i.e. $a_{ij} = a_{ii} \cdot \delta_{ij} = a_{jj} \cdot \delta_{ij}$ if $r_i = r_j$. Then λ is a 1-*PS* of this subgroup, hence it is also a 1-*PS* of the radical of $P(\lambda)$. Thirdly, the specialization of $[\phi \circ \lambda(\alpha)] \cdot \gamma \cdot [\phi \circ \lambda(\alpha^{-1})]$ as $\alpha \to 0$ is precisely $\gamma' = \{a'_{ij}\}$ where

$$\begin{aligned} a'_{ij} &= a_{ij}, \quad \text{if} \quad r_i = r_j \\ &= 0, \quad \text{if} \quad r_i \neq r_j. \end{aligned}$$

This certainly centralizes λ.

It remains to verify that $P(\lambda)$ is parabolic in G. Let T be a maximal torus of G containing $\lambda(G_m)$. We use the notation of CHEVALLEY [8], exposé 9, § 5:

Definition 2.4. If T is a torus, then:

$$\Gamma^{\mathbf{Q}}(T) = \operatorname{Hom}(G_m, T) \otimes \mathbf{Q}$$

$$\Gamma^{\mathbf{R}}(T) = \operatorname{Hom}(G_m, T) \otimes \mathbf{R}.$$

Now λ defines a point $\bar{\lambda} \in \Gamma^{\mathbf{R}}(T)$. Suppose $-\bar{\lambda}$ is in the closure of the Weyl chamber $W_1 \subset \Gamma^{\mathbf{R}}(T)$. There is a canonical correspondence between the Weyl chambers $W \subset \Gamma^{\mathbf{R}}(T)$ and the Borel subgroups B of G containing T: namely, each W corresponds to the B spanned by T and those subgroups $G_a \subset G$ defining roots of T which are positive on W — cf. Séminaire Chevalley [8], exposé 11, Theorem 1. Let the above W_1

correspond to B_1. Then it is easy to check that $B_1 \subset P(\lambda)$, which proves that $P(\lambda)$ is parabolic: namely, T itself is certainly in $P(\lambda)$ since it centralizes λ. Suppose $G_a \subset B_1$ corresponds to the root $\chi: T \to G_m$, i.e.

$$(\alpha^{-1} \cdot \gamma \cdot \alpha) = \chi(\alpha) \cdot \gamma$$

for $\alpha \,\varepsilon\, T_k$, $\gamma \,\varepsilon\, (G_a)_k$. Let χ define the linear functional $\bar{\chi}: \Gamma^{\boldsymbol{R}}(T) \to \boldsymbol{R}$. Then $\bar{\chi}(w) \geq 0$ if $w \,\varepsilon\, W_1$, by definition of the correspondence between W_1 and B_1. Therefore $-\bar{\chi}(\lambda) \geq 0$. But

$$\begin{aligned} \lambda(\alpha) \cdot \gamma \cdot \lambda(\alpha^{-1}) &= \chi(\lambda(\alpha^{-1})) \cdot \gamma \\ &= \alpha^{-\bar{\chi}(\bar{\lambda})} \cdot \gamma \end{aligned}$$

for $\gamma \,\varepsilon\, (G_a)_k$. Therefore $\lambda(\alpha) \cdot \gamma \cdot \lambda(\alpha^{-1})$ has a finite specialization when $\alpha \to 0$, i.e. $G_a \subset P(\lambda)$. Therefore $B_1 \subset P(\lambda)$. *QED.*

Proposition 2.7. Let a reductive group G act via σ on a scheme X, proper over k. Then for all $x \,\varepsilon\, X_k$, $L \,\varepsilon\, Pic^G(X)$, and 1-*PS*'s λ of G,

$$\mu^L(x, \lambda) = \mu^L(x, \gamma^{-1} \cdot \lambda \cdot \gamma)$$

if $\gamma \,\varepsilon\, P(\lambda)_k$.

Proof. Let y be the specialization of $\sigma(\lambda(\alpha), x)$ as $\alpha \to 0$. Suppose we write $y = \lim_{\alpha\to 0} \sigma(\lambda(\alpha), x)$. In this notation, we calculate:

$$\begin{aligned} \lim_{\alpha\to 0} \sigma(\gamma^{-1} \cdot \lambda(\alpha) \cdot \gamma; x) &= \lim_{\alpha\to 0} \sigma((\gamma^{-1} \cdot \lambda(\alpha) \cdot \gamma \cdot \lambda(\alpha^{-1}); \sigma[\lambda(\alpha), x]) \\ &= \sigma\Big(\gamma^{-1} \cdot \lim_{\alpha\to 0} [\lambda(\alpha) \cdot \gamma \cdot \lambda(\alpha^{-1})]; y\Big). \end{aligned}$$

Let $\lim_{\alpha\to 0} \lambda(\alpha) \cdot \gamma \cdot \lambda(\alpha^{-1}) = \gamma'$. According to Proposition 2.6, γ' centralizes λ. Now $\mu(x, \gamma^{-1} \cdot \lambda \cdot \gamma)$ is calculated by looking at $\lim_{\alpha\to 0} \sigma(\gamma^{-1} \cdot \lambda(\alpha) \times \gamma; x)$. Therefore:

$$\mu(x, \gamma^{-1} \cdot \lambda \cdot \gamma) = \mu(\sigma[\gamma^{-1} \cdot \gamma'; y], \gamma^{-1} \cdot \lambda \cdot \gamma).$$

By property (i) of μ (§ 1 above), we have:

$$\begin{aligned} \mu(\sigma(\gamma^{-1} \cdot \gamma', y), \gamma^{-1} \cdot \lambda \cdot \gamma) &= \mu(\sigma(\gamma', y), \lambda) \\ &= \mu(y, \gamma'^{-1} \cdot \lambda \cdot \gamma') \\ &= \mu(y, \lambda). \end{aligned}$$

But $\mu(y, \lambda) = \mu(x, \lambda)$ by property (iv) of μ. Putting these together, $\mu(x, \gamma^{-1} \cdot \lambda \cdot \gamma) = \mu(x, \lambda)$. *QED.*

In view of the last Proposition, we have reason to define the complex referred to above as follows:

Definition 2.5. Let G be a reductive algebraic group. The rational flag complex $\Delta(G)$ is the set of non-trivial 1-*PS*'s λ of G modulo the equivalence relation:

(*) $\lambda_1 \sim \lambda_2$ if there are positive integers n_1 and n_2 and a point $\gamma \varepsilon P(\lambda_1)_k$ such that

$$\lambda_2(\alpha^{n_2}) = \gamma^{-1} \cdot \lambda_1(\alpha^{n_1}) \cdot \gamma$$

for all $\alpha \varepsilon (G_m)_k$.

The point of $\Delta(G)$ defined by λ will be denoted $\Delta(\lambda)$.

Note first of all that if $\lambda_1 \sim \lambda_2$, then $P(\lambda_1) = P(\lambda_2)$. Therefore we can talk of $P(\delta)$ for a point $\delta \varepsilon \Delta(G)$. Now, by means of a simple normalization of μ, we can obtain from μ a function on $\Delta(G)$. In fact, note that if:

$$\lambda_1(\alpha^{n_1}) = \lambda_2(\alpha^{n_2})$$

for all $\alpha \varepsilon (G_m)_k$, and for positive integers n_1 and n_2 then the specializations of $\sigma(\lambda_1(\alpha), x)$ and of $\sigma(\lambda_2(\alpha), x)$ as $\alpha \to 0$ are the same. Therefore we can readily show:

$$\mu^L(x, \lambda_1) = \frac{n_2}{n_1} \mu^L(x, \lambda_2).$$

Now suppose T is a maximal torus in G, and let $N(T)$ be the normalizer of T. Then $N(T)/T$ is a finite group — the Weyl group — and it acts on $\Gamma^{\mathbf{R}}(T)$ via inner automorphisms. There certainly exists at least one positive definite symmetric bilinear form $\langle x, y\rangle$ on $\Gamma^{\mathbf{R}}(T)$ which is invariant under this group and is rational on the rational subspace $\Gamma^{\mathbf{Q}}(T)$. *Fix such a form.* We can then define a norm for a 1-*PS* λ of G:

for each λ, there is a $\gamma \varepsilon G_k$ such that $\gamma \cdot \lambda \cdot \gamma^{-1}$ is a 1-*PS* of T. Let $\gamma \cdot \lambda \cdot \gamma^{-1}$ define $\overline{\gamma \cdot \lambda \cdot \gamma^{-1}} \varepsilon \Gamma^{\mathbf{R}}(T)$. Then put

$$||\lambda|| = \sqrt{\langle \overline{\gamma \cdot \lambda \cdot \gamma^{-1}}, \overline{\gamma \cdot \lambda \cdot \gamma^{-1}} \rangle}.$$

We must check that this is independent of the choice of γ. By the invariance of $\langle x, y\rangle$ under the Weyl group, this amounts to:

Lemma 2.8. Suppose $\gamma \varepsilon G_k$, and suppose λ and $\gamma \cdot \lambda \cdot \gamma^{-1}$ are both 1-*PS*'s of T. Then there is a $\gamma' \varepsilon N(T)_k$ such that $\gamma \cdot \lambda \cdot \gamma^{-1} = \gamma' \cdot \lambda \cdot \gamma'^{-1}$.

Proof*. Let Z be the connected component of the centralizer of $\gamma \cdot \lambda \cdot \gamma^{-1}$. Since λ is a 1-*PS* of T, $\gamma \cdot \lambda \cdot \gamma^{-1}$ is a 1-*PS* of $\gamma \cdot T \cdot \gamma^{-1}$. Therefore both T and $\gamma \cdot T \cdot \gamma^{-1}$ are maximal tori in Z. By the conjugacy theorem for maximal tori, there is a $\beta \varepsilon Z_k$ such that

$$\beta \cdot [\gamma \cdot T \cdot \gamma^{-1}] \cdot \beta^{-1} = [T].$$

Put $\gamma' = (\beta \cdot \gamma)$. *QED.*

Definition 2.6. Let G be a reductive algebraic group. Fix a norm $||\lambda||$ on the 1-*PS*'s of G as above. Then if G acts on the scheme X, proper

* I have taken this proof from a lecture of A. Borel.

over k, if $L \varepsilon Pic^G(X)$, if $x \varepsilon X_k$, and if $\delta = \Delta(\lambda)$, put:

$$\nu^L(x; \delta) = \mu^L(x, \lambda)/||\lambda||.$$

It follows immediately from the above that the left hand side is a function only of $\Delta(\lambda)$, so that this definition is meaningful.

The rest of this section will be devoted to examining the structure of $\Delta(G)$. Intuitively, $\Delta(G)$ may be considered as the set of rational points at ∞ on G. It depends only on the Dynkin diagram of G, and can be constructed purely formally. $\Delta(G)$ is extremely rich in structure. For example, although we will not use this, via the norms $||\lambda||$ on the set of 1-*PS*'s of G, we can define the canonical *metrics* on $\Delta(G)$. In fact, we shall see below that if δ and ε are any 2 points of $\Delta(G)$, then there are two 1-*PS*'s λ and μ of G such that $\delta = \Delta(\lambda)$, $\varepsilon = \Delta(\mu)$ and λ and μ commute with each other. Then $\lambda \cdot \mu$ is also a 1-*PS* of G, and we can define:

$$\varrho(\delta, \varepsilon) = \text{arc}\cos\frac{1}{2}\left\{\frac{||\lambda \cdot \mu||^2}{||\lambda|| \cdot ||\mu||} - \frac{||\lambda||}{||\mu||} - \frac{||\mu||}{||\lambda||}\right\}.$$

To bring out the structure of $\Delta(G)$, we require:

Lemma 2.9. If $\delta \varepsilon \Delta(G)$, and $T \subset G$ is a maximal torus, then

$$T \subset P(\delta) \Leftrightarrow \exists \text{ a 1-}PS\ \lambda \text{ of } T \text{ such that}$$
$$\delta = \Delta(\lambda).$$

Therefore, if δ_1, δ_2 are 2 points of $\Delta(G)$, there is some maximal torus T such that both δ_1 and δ_2 are represented by 1-*PS*'s of T.

Proof. If λ is a 1-*PS* of T, then $T \subset P(\lambda)$ by the definition of $P(\lambda)$. This establishes the implication $\Leftarrow$. Conversely, if $T \subset P(\delta)$ and T is a maximal torus, then any 1-*PS* of $P(\delta)$ is conjugate by some $\sigma \varepsilon P(\delta)_k$ to a 1-*PS* of T. Since any λ representing δ is a 1-*PS* of $P(\delta)$, this establishes the implication $\Rightarrow$. The last claim now follows from the fact that the intersection of any 2 parabolic subgroups contains some maximal torus (cf. [34]). *QED.*

Now suppose G and H are 2 reductive groups, and suppose that $\phi: H \to G$ is a homomorphism with finite kernel. Composition with ϕ maps non-trivial 1-*PS*'s of H to non-trivial 1-*PS*'s of G. It is easy to check that this induces a map $\phi_*: \Delta(H) \to \Delta(G)$.

Proposition 2.10. ϕ_* is injective.

Proof. Let ε_1 and ε_2 be any 2 points of $\Delta(H)$. By lemma 2.6, ε_1 and ε_2 are represented by 1-parameter subgroups λ_1 and λ_2 of some torus $T \subset H$. If $\varepsilon_1 \neq \varepsilon_2$, then there do not exist positive integers n and m such that $\lambda_1(\alpha^n) = \lambda_2(\alpha^m)$, all $\alpha \varepsilon \mathbf{G}_m$. But then there is a character χ of T such that $\chi \circ \lambda_1 = 0$, (the trivial homomorphism from $\mathbf{G}_m$ to $\mathbf{G}_m$), and $\chi \circ \lambda_2 \neq 0$. Since the kernel of ϕ is finite, there is an integer N such that χ^N is induced by a character χ' of $\phi(T)$. Now suppose that

$\phi_* \varepsilon_1 = \phi_* \varepsilon_2$. Then $P(\phi \circ \lambda_1) = P(\phi_* \varepsilon_1) = P(\phi_* \varepsilon_2) = P(\phi \circ \lambda_2)$. Call this group P, and let T'' be the intersection of $\phi(T)$ with the radical of P. On the one hand, notice that $\phi \circ \lambda_1$ and $\phi \circ \lambda_2$ are 1-PS's in $\phi(T)$ **and** in the radical of P, hence in T''. On the other hand, the set of all characters of P defines homomorphism

$$\psi : P \to T'''$$

(T''' a torus) which is an isogeny when restricted to any maximal torus in the radical of P (compare the remarks in Chapter 1, § 1). In particular T''' has a character χ''' such that $\chi''' \circ \psi$ on T'' is a power $(\chi')^M$ of χ'. Now by assumption,

$\Delta(\phi \circ \lambda_1) = \Delta(\phi \circ \lambda_2)$, hence there is a $\gamma \in P_k$ and positive integers n_1, n_2 such that

$$\phi \circ \lambda_1 (\alpha^{n_1}) = \gamma \cdot \phi \circ \lambda_2 (\alpha^{n_2}) \cdot \gamma^{-1}$$

all $\alpha \in (\mathbf{G}_m)_k$. Therefore

$$\psi \circ \phi \circ \lambda_1 (\alpha^{n_1}) = \psi \circ \phi \circ \lambda_2 (\alpha^{n_2}) .$$

But

$$\chi''' \circ \psi \circ \phi \circ \lambda_1 = (\chi')^M \circ \phi \circ \lambda_1 = \chi^{MN} \circ \lambda_1 = 0$$

while

$$\chi''' \circ \psi \circ \phi \circ \lambda_2 = (\chi')^M \circ \phi \circ \lambda_2 = (\chi)^{MN} \circ \lambda_2 \neq 0.$$

This is a contradiction, so therefore $\phi_* \varepsilon_1 \neq \phi_* \varepsilon_2$. *QED.*

Corollary 2.11. If ϕ is an isogeny, ϕ_* is an isomorphism.

Proof. It remains to show that ϕ_* is surjective. But if λ is any 1-parameter subgroup of G, then for some positive integer n, and for some 1-parameter subgroup μ of H, $\phi \circ \mu(\alpha) = \lambda(\alpha^n)$, $\alpha \in \mathbf{G}_m$. *QED.*

Suppose we apply Proposition 2.10 to the inclusion of a maximal torus T in G. This induces an inclusion of $\Delta(T)$ in $\Delta(G)$. But $\Delta(T)$ is very simple — it is essentially the set of rational points on a sphere. In fact, it follows immediately from the definition that $\Delta(T)$ is the set of rays in $\Gamma^{\mathbf{R}}(T)$ which contain a point of $\Gamma^{\mathbf{Q}}(T)$. Therefore, in view of lemma 2.9, $\Delta(G)$ can be viewed as the result of pasting together spheres, one for each $T \subset G$, with sufficient identifications so that any two points are both on at least one sphere. The subsets $\Delta(T)$ will be called *skeletons* of $\Delta(G)$. We can push this structure further in 2 ways: first the spheres $\Delta(T)$ can be broken up by considering the function $P(\delta)$:

Definition 2.7. For all parabolic subgroups P, let $\Delta_P(G)$ be the set of $\delta \in \Delta(G)$ such that $P(\delta) \supset P$.

To describe $\Delta_P(G)$ explicitly, let T be a maximal torus of P. Then $\Delta_P(G) \subset \Delta(T)$ by lemma 2.9. Suppose P is spanned by T and by additive subgroups $\mathbf{G}_a \subset P$ such that

$$\alpha^{-1} \cdot \gamma \cdot \alpha = \chi(\alpha) \cdot \gamma$$

for $\alpha \in T_k$, $\gamma \in (G_a)_k$, and a suitable character χ of T. Then one checks that if $\delta \in \Delta(T)$ is induced by $\bar{\lambda} \in \Gamma^{\mathbf{R}}(T)$, and if χ induces the linear functional $\bar{\chi}$ on $\Gamma^{\mathbf{R}}(T)$, then

$$P(\delta) \supset P \iff \forall \text{ such } \chi, \bar{\chi}(\lambda) \leq 0 .$$

Therefore, $\Delta_P(G)$ is simply the set of rational points on an intersection of hemi-spheres, (each determined by a rational linear functional, in fact). If G is actually semi-simple, it can be shown that $\Delta_P(G)$ is even the set of rational points on a spherical *simplex*, and hence the collection of $\Delta_P(G)$ constitutes a "triangulation", and $\Delta(G)$ is the complex formed from the simplices $\Delta_P(G)$. This is the motivation of the terminology "flag complex".*

A second direction in which we can make explicit this structure on $\Delta(G)$ is to show that the skeletons are pasted together in a sufficiently nice way so that various constructions which can be carried out in each skeleton can also be carried out in $\Delta(G)$. The principal one we have in mind is drawing a great circle through two points. One approach to this is to complete the complex $\Delta(G)$ with respect to the metric ϱ mentioned above: then Tits has shown that the geodesic segments on this completion are precisely the segments of great circles of length $\leq \pi$, on the true spheres obtained by completing the skeletons $\Delta(T)$. Since we are avoiding this metric, we define these lines, and other auxiliary concepts, directly:

Definition 2.8. A pair of points $\delta_1, \delta_2 \in \Delta(T)$ is ***antipodal*** if there is a 1-*PS* λ of G such that $\delta_1 = \Delta(\lambda)$ and $\delta_2 = \Delta(\lambda^{-1})$.

If G is a torus, then every point of $\Delta(G)$ has a unique antipodal point, i.e. its antipodal on the sphere is the usual sense. More generally, if δ_1 is in the skeleton $\Delta(T)$ of $\Delta(G)$, for some maximal torus T, then the antipodal δ_2 of δ_1 as a point of $\Delta(T)$, is one of its antipodal points in $\Delta(G)$. In fact, this is the only point antipodal to δ_1 in $\Delta(T)$:

Lemma 2.12. If δ_1, δ_2 are two points in the skeleton $\Delta(T)$ of $\Delta(G)$, for some maximal torus T, then δ_1 and δ_2 are antipodal as points of $\Delta(G)$ if and only if they are antipodal as points of the sphere $\Delta(T)$.

Proof. Suppose $\delta_1 = \Delta(\lambda)$ and $\delta_2 = \Delta(\lambda^{-1})$ for some 1-*PS* λ of G. Then λ is a 1-*PS* of $P(\lambda) \cap P(\lambda^{-1}) = P(\delta_1) \cap P(\delta_2)$; and T is a maximal torus of $P(\delta_1) \cap P(\delta_2)$ by lemma 2.9. Therefore there is some $\gamma \in [P(\delta_1) \cap P(\delta_2)]_k$ such that $\gamma^{-1} \cdot \lambda \cdot \gamma$ is a 1-*PS* of T. Then $\delta_1 = \Delta(\gamma^{-1} \cdot \lambda \cdot \gamma)$ and $\delta_2 = \Delta(\gamma^{-1} \cdot \lambda^{-1} \cdot \gamma)$, hence δ_1 and δ_2 are antipodal in the group T. *QED.*

---

* In fact, if G is semi-simple, Tits defines his complex as follows: start with a single point Δ_P for each **maximal** parabolic subgroup $P \subset G$. Then, if $P_1, \ldots, P_n$ are maximal parabolic subgroups such that $P_1 \cap \cdots \cap P_n = P$ is parabolic, join the points $\Delta_{P_1}, \ldots, \Delta_{P_n}$ with an $(n-1)$-dimensional simplex Δ_P.

Now suppose δ_1 and δ_2 are not antipodal. Suppose δ_1 and δ_2 are both in $\Delta(T)$. Then we make:

Definition 2.9/Proposition 2.13. Suppose $\delta_i = \Delta(\lambda_i)$ for a 1-*PS* λ_i of T. Then the set of points $\Delta(\lambda_1^{n_1} \cdot \lambda_2^{n_2})$, where n_1, n_2 are non-negative integers, is called the *line joining* δ_1 *and* δ_2. It is independent of the choice of T.

Before proving the last statement, we make precise the notation $\lambda_1^{n_1} \cdot \lambda_2^{n_2}$:

$$\lambda_1^{n_1} \cdot \lambda_2^{n_2}(\alpha) = \lambda_1(\alpha)^{n_1} \cdot \lambda_2(\alpha)^{n_2}$$

for all $\alpha \in (G_m)_k$. Note that, since δ_1 and δ_2 are not antipodal, none of the 1-*PS*'s $\lambda_2^{n_1} \cdot \lambda_2^{n_2}$ is trivial. Therefore $\Delta(\lambda_1^{n_1} \cdot \lambda_2^{n_2})$ is meaningful. Moreover, this "line" is also the set of rational points on the geodesic segments between δ_1 and δ_2, as points on the sphere obtained by completing $\Delta(T)$ (in a standard metric).

Proof of Prop. 2.13. First note the obvious remark:

(*) If δ_1, δ_2 are in $\Delta(T_1) \cap \Delta(T_2)$, then there is a $\gamma \in [P(\delta_1) \cap P(\delta_2)]_k$ such that $T_1 = \gamma^{-1} \cdot T_2 \cdot \gamma$.

Proof. T_1 and T_2 are both maximal tori in $P(\delta_1) \cap P(\delta_2)$ by lemma 2.9. Therefore they are conjugate in $P(\delta_1) \cap P(\delta_2)$. *QED.*

Now suppose δ_1, δ_2 are, in fact, in $\Delta(T_1) \cap \Delta(T_2)$: we must show that the lines joining them in $\Delta(T_1)$ and in $\Delta(T_2)$ are the same. Let $T_1 = \gamma^{-1} \cdot T_2 \cdot \gamma$ where $\gamma \in [P(\delta_1) \cap P(\delta_2)]_k$ and suppose λ_1, λ_2 are 1-*PS*'s of T_2 representing δ_1, δ_2, respectively. Then $\gamma^{-1} \cdot \lambda_1 \cdot \gamma$, $\gamma^{-1} \cdot \lambda_2 \cdot \gamma$ are 1-*PS*'s of T_1 representing δ_1, δ_2 respectively. Therefore, we must show that the sets of points

$$\{\Delta(\lambda_1^{n_1} \cdot \lambda_2^{n_2})\} \quad \text{and} \quad \{\Delta((\gamma^{-1} \cdot \lambda_1 \cdot \gamma)^{n_1} \cdot (\gamma^{-1} \cdot \lambda_2 \cdot \gamma)^{n_2})\}$$

are the same. But this is clear provided that $\Delta(\lambda_1^{n_1} \cdot \lambda_2^{n_2}) = \Delta(\gamma^{-1} \lambda_1^{n_1} \cdot \lambda_2^{n_2} \gamma)$; hence it follows if $P(\lambda_1) \cap P(\lambda_2) \subset P(\lambda_1^{n_1} \cdot \lambda_2^{n_2})$ — so that $\gamma \in P(\lambda_1^{n_1} \cdot \lambda_2^{n_2})$.

Therefore, we are reduced to proving that if λ_1, λ_2 are 1-*PS*'s of G which centralize each other, then

$$(**) \qquad P(\lambda_1 \cdot \lambda_2) \supset P(\lambda_1) \cap P(\lambda_2).$$

To prove this, it is convenient to represent G in $GL(n)$ by some closed immersion $\phi: G \to GL(n)$. Then, exactly as in the proof of Proposition 2.6, we have reduced the proof of (**) to the case $G = GL(n)$. In this case, we can simultaneously diagonalize two commuting 1-*PS*'s: say $\lambda_1(\alpha)$ is the matrix $\{\alpha^{r_i} \cdot \delta_{ij}\}$, and $\lambda_2(\alpha)$ is the matrix $\{\alpha^{s_i} \cdot \delta_{ij}\}$, for $\alpha \in (G_m)_k$. Then, as in Proposition 2.6,

$$P(\lambda_1) = \{(a_{ij}) \mid a_{ij} = 0 \quad \text{if} \quad r_i < r_j\}$$
$$P(\lambda_2) = \{(a_{ij}) \mid a_{ij} = 0 \quad \text{if} \quad s_i < s_j\}$$
$$P(\lambda_1 \cdot \lambda_2) = \{(a_{ij}) \mid a_{ij} = 0 \quad \text{if} \quad r_i + s_i < r_j + s_j\}.$$

Since $r_i + s_i < r_j + s_j$ implies that $r_i < r_j$ *or* $s_i < s_j$, it is immediate that $P(\lambda_1 \cdot \lambda_2) \supset P(\lambda_1) \cap P(\lambda_2)$. *QED.*

Definition 2.10. A subset $C \subset \Delta(G)$ is *semi-convex* if it contains the line joining any pair of points $\delta_1, \delta_2 \varepsilon C$ provided δ_1 and δ_2 are not antipodal; C is *convex* if, in addition, it contains no pair of antipodal points.

According to (**) above, it follows for example that the subsets $\Delta(T)$ and $\Delta_P(G)$ are all semi-convex, for maximal tori T, and parabolic P. Actually, if G is semi-simple, it can be shown that $\Delta_P(G)$ is always convex.

§ 3. Applications

In this section we return to the functions $\mu^L(x, \lambda)$ of 1-*PS*'s λ and $\nu^L(x, \delta)$ of $\delta \varepsilon \Delta(G)$ in order to apply the general theory to group actions.

First of all, consider $\mu^L(x, \lambda)$ for a fixed action of G on some proper X, a fixed ample $L \varepsilon Pic^G(C)$, a fixed $x \varepsilon X_k$, but where λ varies among the 1-*PS*'s of a maximal torus $T \subset G$. We use Proposition 2.3 to describe μ: namely, there exists a morphism

$$\phi : X \to \mathbf{P}_{n-1}$$

for some n, plus an action of G on $\mathbf{P}_{n-1}$ and a G-linearization of $\underline{o}_{\mathbf{P}}(1)$ such that ϕ is G-linear and, for some N,

$$L^N = \phi^*(\underline{o}(1)).$$

Therefore: $\mu^L(x, \lambda) = \frac{1}{N}\mu^{\underline{o}(1)}(\phi x, \lambda)$.

As in § 1, the action of G on $\mathbf{P}_{n-1}$ is induced by a linear representation of G in the affine cone $\mathbf{A}^n$ over $\mathbf{P}_{n-1}$. Suppose we choose coordinates in $\mathbf{A}^n$ so that the representation of T is diagonalized: i.e. $\alpha \varepsilon T_k$ acts via the matrix $\{\chi_i(\alpha) \cdot \delta_{ij}\}$ for suitable characters $\chi_1, \ldots, \chi_n$ of T. Let χ_i induce the linear functional $\bar{\chi}_i$ on $\Gamma^{\mathbf{R}}(T)$, and as usual let a 1-*PS* λ of T induce the point $\bar{\lambda} \varepsilon \Gamma^{\mathbf{R}}(T)$. Then if $(\phi x_1^*, \ldots, \phi x_n^*)$ are coordinates of a homogeneous point lying over ϕx, Proposition 2.3 asserts:

$$\mu^{\underline{o}(1)}(\phi x, \lambda) = \max\{-\bar{\chi}_i(\bar{\lambda}) \mid i \text{ such that } \phi x_i^* \neq 0\}.$$

We conclude:

Proposition 2.14. Let a reductive G act on X, proper over k. Let $T \subset G$ be a maximal torus, and let $L \varepsilon Pic^G(X)$ be ample. Then there is a finite set of linear functionals $l_1, \ldots, l_n$ on $\Gamma^{\mathbf{R}}(T)$ which are rational on $\Gamma^{\mathbf{Q}}(T)$ with the following property:

for every $x \varepsilon X_k$, there is a subset $I \subset \{1, \ldots, n\}$ such that

$\mu^L(x, \lambda) = \max\{l_i(\bar{\lambda}) \mid i \varepsilon I\}$ for all 1-*PS*'s λ of T.

Corollary 2.15. If λ_1, λ_2 are 1-*PS*'s of T and $x \in X_k$, then

$$\mu^L(x, \lambda_1 \cdot \lambda_2) \leq \mu^L(x, \lambda_1) + \mu^L(x, \lambda_2).$$

Proof. Immediate, since $\overline{\lambda_1 \cdot \lambda_2} = \bar{\lambda}_1 + \bar{\lambda}_2$.

Corollary 2.16. Let a reductive G act on S, proper over k. Let $L \in Pic^G(X)$ be ample, and let $x \in X_k$. Then $\{\delta \mid \nu^L(x, \delta) \leq 0\}$ is semi-convex, and $\{\delta \mid \nu^L(x, \delta) < 0\}$ is convex.

Proof. Immediate.

It follows from this Corollary and from Theorem 2.1, that if $x \in X_k$ is not *semi-stable*, then we can associate to x the convex set C of $\delta \in \Delta(G)$ for which $\nu^L(x, \delta) < 0$. Roughly speaking, we can say that a parabolic subgroup P is "responsible" for this breakdown of semi-stability if $C \cap \Delta_P(G) \neq \phi$. Since C is convex, not too many parabolic subgroups P are responsible; according to a conjecture of Tits, there is even a natural way to find one P which is *most* responsible, i.e. $P(\delta)$ for $\delta =$ the "center" of C.

In fact, Tits conjectures that any convex subset C of a flag complex has a natural *center* δ. For the convex set $C = \{\delta \mid \nu^L(x, \delta) < 0\}$, Kempf and Rousseau discovered quite simply that $\nu^L(x, \delta)$ is strictly convex on the line joining any 2 points $\delta_1, \delta_2 \in C$, and takes on a unique minimum at a single point $\delta_0 \in C$ (see Appendix 2B). This $P(\delta_0)$ may be considered the "worst" parabolic subgroup. As a Corollary, they deduce:

if x is rational over k, and not semi-stable for the action of G, then there is a 1-*PS* λ, rational over k, such that $\mu(x, \lambda) < 0$.

(This is a natural generalization of Godement's conjecture on compact fundamental sets.)

Another consequence of Proposition 2.14 is:

Proposition 2.17. Let a reductive G act via σ on X, proper over k. Let $L \in Pic^G(x)$ be ample. Then there is a constant K such that

$$|\nu^L(x, \delta)| \leq K$$

for all $x \in X_k$ and $\delta \in \Delta(G)$. Moreover, for every subset $S \subset X_k^1$, there is a $\delta_0 \in \Delta(G)$ and $x_0 \in S$ such that

$$\nu^L(x_0, \delta_0) \leq \nu^L(x, \delta)$$

for every other $\delta \in \Delta(G)$.

Proof. Let $T \subset G$ be a maximal torus. Let $\langle x, y \rangle$ be the form on $\Gamma^R(T)$ which defines the norm $\|\lambda\|$. Let $l_1, \ldots, l_n$ be the linear functionals on $\Gamma^R(T)$ given by Proposition 2.14. For any $x \in X_k$, let $I(x)$ be the subset $\{1, \ldots, n\}$ given by Proposition 2.14. To estimate $\nu^L(x, \delta)$, say $\delta = \Delta(\lambda)$. There is always some $\gamma \in G_k$ such that $\gamma \cdot \lambda \cdot \gamma^{-1}$ is a

1-*PS* of T. Then:

$$\begin{aligned} \nu^L(x, \delta) &= \frac{\mu^L(x,\lambda)}{\|\lambda\|} \\ &= \frac{\mu^L(\sigma(\gamma, x), \gamma\cdot\lambda\cdot\gamma^{-1})}{\|\gamma\cdot\lambda\cdot\gamma^{-1}\|} \\ &= \max\left\{\frac{l_i(\overline{\gamma\cdot\lambda\cdot\gamma^{-1}})}{\sqrt{\langle\overline{\gamma\cdot\lambda\cdot\gamma^{-1}}, \overline{\gamma\cdot\lambda\cdot\gamma^{-1}}\rangle}} \quad i \in I(\sigma(\gamma, x))\right\}. \end{aligned}$$

But, for each i, $|l_i(z)|/\sqrt{\langle z, z\rangle}$ is bounded when z varies over all non-zero points of $\Gamma^{\mathbf{R}}(T)$. Therefore $\nu^L(x, \delta)$ is bounded when x and δ vary arbitrarily. Now suppose that for $x \in S \subset X_k$, we seek to minimize $\nu^L(x, \delta)$. According to the above,

$$\inf_{\substack{\delta\in\Delta(G)\\ x\in S}} [\nu^L(x, \delta)] = \inf_{\substack{\gamma\in G_k\\ x\in S}}\left[\inf_{z\in\Gamma^{\mathbf{Q}}(T)}\left[\max_{i\in I(\sigma(\gamma,x))} \frac{l_i(z)}{\sqrt{\langle z, z\rangle}}\right]\right].$$

But the first "inf" is essentially a minimum over the finite set of subsets $I \subset \{1, \ldots, n\}$ which occur as subsets $I(\sigma(\gamma, x))$. Therefore, this "inf" will always be attained for some x and γ. As for the inner "inf", if we replace the variable by $z \in \Gamma^{\mathbf{R}}(T)$, then the "inf" will certainly be attained for some z — for the expression to be minimized is invariant under the transformation $z \to \alpha z$, $\alpha > 0$, and the set of rays in $\Gamma^{\mathbf{R}}(T)$ is *compact*. We must check that this "inf" is actually attained by a rational point z. But each linear functional l_i is rational, and the inner product $\langle x, y\rangle$ on $\Gamma^{\mathbf{R}}(T)$ is rational: then it is an exercise in elementary calculus to check that this is so. *QED*.

An application of this is:

Proposition 2.18. Let G be a reductive group acting on algebraic schemes X and Y. Let $f: X \to Y$ be a G-linear morphism, let $L \in Pic^G(Y)$ and $M \in Pic^G(X)$. Then if M is relatively ample for f and L is ample on Y, there is an n_0 such that:

$$n \geq n_0 \Rightarrow X^s_{(0)}(M \otimes f^*L^n) \supset f^{-1}\{Y^s_{(0)}(L)\}.$$

Proof. We shall first reduce this Proposition to the case where X and Y are proper over k. First of all, there is an open immersion $I_Y: Y \subset Y_1$, where Y_1 is proper over k, and there is an action of G on Y_1 extending the given one of G on Y, and there is an $L_1 \in Pic^G(Y_1)$, ample on Y_1 which extends L^N for some N, such that:

$$Y^s_{(0)}(L) = I_Y^{-1}\{Y^s_{1,(0)}(L_1)\}$$

(cf. Proposition 1.8). Secondly, for some N_0, $M \otimes f^*L^{N_0}$ is ample on X by EGA 2, 4.6.13, (ii). Then there is also an open immersion $I_X: X \subset X_1$, where X_1 is proper over k, and there is an action of G on X_1 extending the given one on X, and there is an $M_1 \in Pic^G(X_1)$, ample on X_1, which extends $M^N \otimes f^* L^{N_0 N}$ for some N. (This N may be taken as the same N as above.) Then we have the diagram:

$$\begin{array}{ccc} X & \xrightarrow{(I_X, I_Y \circ f)} & X_1 \times Y_1 \\ {\scriptstyle f}\downarrow & & \downarrow{\scriptstyle p_2} \\ Y & \underset{I_Y}{\hookrightarrow} & Y_1 \end{array}$$

Now if the result is proven when X and Y are proper, we can apply it to p_2: $X_1 \times Y_1 \to Y_1$, and the sheaves $M_1 \otimes \varrho_{Y_1}$ on $X_1 \times Y_1$ and L_1 on Y_1. Then

$$\begin{aligned} f^{-1}\{Y^s_{(0)}(L)\} &= f^{-1}\{I_Y^{-1}[Y^s_{1,(0)}(L_1)]\} \\ &= (I_X, I_Y \circ f)^{-1}\{p_2^{-1}[Y^s_{1,(0)}(L_1)]\} \\ &\subset (I_X, I_Y \circ f)^{-1}\{(X_1 \times Y_1)^s_{(0)}(M_1 \otimes L_1^n)\}, \quad \text{if} \quad n \geq n_0 \\ &\subset X^s_{(0)}[(I_X, I_Y \circ f)^* (M_1 \otimes L_1^n)] \text{ by Prop. 1.18,} \\ &= X^s_{(0)}[M^N \otimes f^* L^{NN_0 + Nn}] \\ &= X^s_{(0)}[M \otimes f^* L^{N_0 + n}]. \end{aligned}$$

Now, if X and Y are proper over k, we note again that $M \otimes f^* L^n$ is ample on X for $n \geq m$. Therefore, we can apply Theorem 2.1 to the computation of $X^s_{(0)}(M \otimes f^* L^n)$. But if $x \in X_k$, $\delta \in \Delta(G)$, then:

$$\begin{aligned} \nu^{M \otimes f^* L^n}(x, \delta) &= \nu^{M \otimes f^* L^m}(x, \delta) + \nu^{f^* L^{n-m}}(x, \delta) \\ &= \nu^{M \otimes f^* L^m}(x, \delta) + (n - m)\nu^L(f(x), \delta). \end{aligned}$$

First of all, apply the first part of Proposition 2.17 to $\nu^{M \otimes f^* L^m}$. We conclude that, for some K:

$$\nu^{M \otimes f^* L^m}(x, \delta) \geq -K.$$

Secondly, apply the second part of Proposition 2.17 to $\nu^L(y, \delta)$, for $y \in S = Y^s_{(0)}(L)_k$. By Theorem 2.1, all these numbers are strictly positive. Therefore, by the Proposition:

$$\nu^L(y, \delta) \geq \varepsilon > 0$$

for all $y \in Y^s_{(0)}(L)_k$, and $\delta \in \Delta(G)$. Therefore, if $n \geq \left[m + \frac{K}{\varepsilon} + 1\right]$, we conclude

$$\nu^{M \otimes f^* L^n}(x, \delta) > 0$$

whenever $x \in f^{-1}\{Y^s_{(0)}(L)\}_k$. Therefore, by Theorem 2.1, all such x are in $X^s_{(0)}(M \otimes f^* L^n)$. *QED.*

Chapter 3

An elementary example

The purpose of this chapter is to give, independently of the foregoing theory, an exhaustive analysis of a single special class of actions. This has two objectives: first of all, the concepts of pre-stable and stable are worked out in a representative non-trivial case. Secondly, by attacking this case directly, we can circumvent the difficulties involved in extending the previous work to semi-simple groups in characteristic p (cf. Preface). And, in fact, we obtain the result over the ring of integers, hence, *a fortiori*, over any field. This will enable us later to construct the so-called arithmetic schemes of moduli of abelian varieties.

We have no use for a ground field in this chapter: all the schemes that we will consider will be of finite type over Spec $(\mathbf{Z})$. $\mathbf{A}^n$, $\mathbf{P}_n$, and $PGL(n)$ stand for the usual schemes *over* $\mathbf{Z}$. The action we shall study is the canonical action:

$$\sigma_m : PGL(n+1) \times (\mathbf{P}_n)^{m+1} \to (\mathbf{P}_n)^{m+1}$$

i.e. of projective transformations on sequences of points in projective space.

§ 1. Pre-stability

The first step is to obtain a large number of invariant affine open subsets in $(\mathbf{P}_n)^{m+1}$. To this end, let $p_0, \ldots, p_m$ denote the projections of $(\mathbf{P}_n)^{m+1}$ onto its $m+1$ factors. Let $L_i = p_i^*(\varrho_{\mathbf{P}}(1))$. Now unfortunately, $\varrho_{\mathbf{P}}(1)$ admits no $PGL(n)$-linearization with respect to the action of $PGL(n+1)$ on $\mathbf{P}_n$. However ,if

$$\overline{\omega} : SL(n+1) \to PGL(n+1)$$

is the canonical isogeny, then with respect to the induced action of $SL(n+1)$, $\varrho_{\mathbf{P}}(1)$ admits an $SL(n+1)$-linearization. This follows because $SL(n+1)$ acts on the affine cone over $\mathbf{P}_n$, compatibly with the given action of $PGL(n)$ on $\mathbf{P}_n$. This linearization induces an $SL(n+1)$-linearization of L_i for each i (for the induced action of $SL(n+1)$), and therefore such a linearization for each product

$$\bigotimes_{i=0}^{m} L_i^{r_i} .$$

On the other hand, the $(n+1)^{st}$ power of the $SL(n+1)$-linearization of $\varrho_{\mathbf{P}}(1)$, which is a linearization of $\varrho_{\mathbf{P}}(n+1)$, is induced, via $\overline{\omega}$, from a $PGL(n+1)$-linearization (cf. Chapter 1, § 2). Therefore, the $SL(n+1)$-linearization of

$$\bigotimes_{i=0}^{m} L_i^{r_i}$$

is a $PGL(n+1)$-linearization if $n+1$ divides each r_i.

In any case, by means of these linearizations we can define invariant sections of all the sheaves $\bigotimes_{i=0}^{m} L_i^{r_i}$. To construct such invariant sections, let $X_0, \ldots, X_n$ be the canonical sections of $\mathcal{O}_{\boldsymbol{P}}(1)$ on $\boldsymbol{P}_n$. Let

$$X_i^{(j)} = p_j^* (X_i)$$

be the induced sections of L_j.

Definition 3.1. For all sequences $\alpha_0, \ldots, \alpha_n$ of integers such that $0 \leq \alpha_i \leq m$, let

$$D_{\alpha_0, \ldots, \alpha_n} = \det_{0 \leq i,j \leq n} [X_i^{(\alpha_j)}]$$

be the section of $L_{\alpha_0} \otimes \cdots \otimes L_{\alpha_n}$ obtained by addition and tensor product as in the determinant. It is evident that $D_{\alpha_0, \ldots, \alpha_n}$ is an *invariant* section of $L_{\alpha_0} \otimes \cdots \otimes L_{\alpha_n}$. The non-vanishing of suitable D's defines the open sets we are looking for.

Definition 3.2. An *R-partition* of $\{0, 1, \ldots, n\}$ is an ordered set of subsets $S_1, S_2, \ldots, S_\nu$ of $\{0, 1, \ldots, n\}$ such that

(i) $S_i \cap (S_{i-1} \cup \cdots \cup S_2 \cup S_1)$ consists of exactly one integer, for $2 \leq i \leq \nu$,

(ii) $\bigcup_{i=1}^{\nu} S_i = \{0, 1, \ldots, n\}$.

Definition 3.3. Given an R-partition $R = \{S_1, \ldots, S_\nu\}$, let $U_R \subset (\boldsymbol{P}_n)^{n+\nu+1}$ be the open subset defined by

(i) $D_{0,1,\ldots,n} \neq 0$,

(ii) for all k between 1 and ν, and for all $i \in S_k$.

$$D_{0,1,\ldots,\hat{i},\ldots,n,n+k} \neq 0.$$

Not only is U_R affine, but the whole structure of the action of $PGL(n)$ on U_R can be described explicitly:

Proposition 3.1. Let $R = \{S_1, \ldots, S_\nu\}$ be an R-partition of $\{0,1,\ldots,n\}$. Let $PGL(n+1)$ act on $PGL(n+1) \times \boldsymbol{A}^{n\nu-n}$ by the product of left translation on itself and the trivial action on the affine space. Then there is a $PGL(n+1)$-linear isomorphism:

$$(*) \qquad U_R \cong PGL(n+1) \times \boldsymbol{A}^{n\nu-n}.$$

Hence U_R is a globally trivial principal fibre bundle with respect to the action of $PGL(n+1)$, with base space $\boldsymbol{A}^{n\nu-n}$.

Proof. We shall construct morphisms from the left hand side of (*) to the right, and vice versa, leaving it to the reader to check that they are mutually inverse. The morphisms will be obviously $PGL(n+1)$-linear. First of all, we identify $\boldsymbol{A}^{n\nu-n}$ with the scheme:

$$\boldsymbol{A}_R = \operatorname{Spec} \frac{\boldsymbol{Z}[x_0^{(n+1)}, \ldots, x_n^{(n+1)}; x_0^{(n+2)}, \ldots, x_n^{(n+2)}; \ldots; x_0^{(n+\nu)}, \ldots, x_n^{(n+\nu)}]}{\text{Ideal generated by } \{x_i^{(n+k)} - 1 \mid i \in S_k\}}$$

since both are affine spaces of dimension $n\nu - n$. Then a morphism

$$S_R : \boldsymbol{A}_R \to U_R$$

is defined by the collection of composite morphisms

$$S_i : \boldsymbol{A}_R \xrightarrow{S_R} U_R \subset (\boldsymbol{P}_n)^{n+\nu+1} \xrightarrow{p_i} \boldsymbol{P}_n$$

and S_i is defined by the conditions (cf. EGA 2, 4.2.3):

$$\begin{aligned} S_i^*\big(\underline{o}(1)\big) &\cong \underline{o}_A \\ S_i^*(X_j) &= \delta_{ij}, \text{ if } 0 \le i \le n \\ S_i^*(X_j) &= x_j^{(i)}, \text{ if } n+1 \le i \le n+\nu+1. \end{aligned}$$

We then define ψ: $PGL(n+1) \times \boldsymbol{A}_R \to U_R$ to be $\sigma_{n+\nu} \circ (1_{PGL(n+1)} \times S_R)$. To define the inverse of ψ, we require:

Definition 3.4. Given an R-partition $R = \{S_1, \ldots, S_\nu\}$, define $\mu(k)$ to be the integer in

$$S_k \cap \{S_{k-1} \cup \cdots \cup S_1\}$$

for $2 \le k \le \nu$. Define $\mu(1)$ to be the smallest integer in S_1. Define $\varkappa(i)$ to be the least integer k for which $i \in S_k$, for $0 \le i \le n$. Finally, define a partial ordering of $\{0, 1, \ldots, n\}$ by:

$$i_1 >_R i_2$$

if $\varkappa(i_1) > \varkappa(i_2)$.

Note that, for any i, there is a canonical sequence of integers:

$$i >_R \mu\big(\varkappa(i)\big) >_R \mu\big(\varkappa(\mu(\varkappa(i)))\big) >_R \cdots$$

which always terminates at $\mu(1)$ since $\{0, 1, \ldots, n\}$ is finite. Notice that i and $\mu(\varkappa(i))$ are both in $S_{\varkappa(i)}$, $\mu(\varkappa(i))$ and $\mu(\varkappa(\mu(\varkappa(i))))$ are both in $S_{\varkappa(\mu(\varkappa(i)))}$, etc. ... This allows us to make the inductive definition:

Definition 3.5. $\lambda_{\mu(1)} = 1$

$$\lambda_j = \lambda_{\mu(\varkappa(j))} \frac{D_{0,1,\ldots,\widehat{\mu(\varkappa(j))},\ldots,n,n+\varkappa(j)}}{D_{0,1,\ldots,\hat{j},\ldots,n,n+\varkappa(j)}}$$

for every $j \in \{0, 1, \ldots, n\}$.

Note that λ_j is a section of $L_j \otimes L_{\mu(1)}^{-1}$ over the open set U_R. We now define

$$\phi : U_R \to PGL(n+1) \times \boldsymbol{A}_R$$

by its two components $p_1 \circ \phi$ and $p_2 \circ \phi$. The projective group $PGL(n+1)$, is, by definition, the affine open subset of

$$\boldsymbol{P}_{n^2+2n} = \operatorname{Proj} \boldsymbol{Z}\,[a_{00}, \ldots, a_{0n}; a_{10}, \ldots, a_{1n}; \ldots; a_{n0}, \ldots, a_{nn}]$$

where $\det(a_{ij}) \ne 0$. Therefore $p_1 \circ \phi$ is determined by

$$\begin{aligned} (p_1 \circ \phi)^*\big(\underline{o}_{\boldsymbol{P}}(1)\big) &= L_{\mu(1)} \\ (p_1 \circ \phi)^*(a_{ij}) &= X_i^{(j)} \otimes (-1)^j \lambda_j^{-1}. \end{aligned}$$

Finally the morphism $p_2 \circ \phi$ into affine space is determined by:

$$(p_2 \circ \phi)^* (x_i^{(n+k)}) = \frac{D_{0,1,\ldots,\hat{i},\ldots,n,n+k}}{D_{0,1,\ldots,\widehat{\mu(k)},\ldots,n,n+k}} \cdot \frac{\lambda_i}{\lambda_{\mu(k)}}$$

(It should be noted that the expression on the right is actually a section of ϱ_{A_R}) Modulo the mechanical verifications that remain, this proves the Proposition. *QED.*

Of the morphisms constructed in the course of the above proof, it is useful to fix for later use the notation:

$$\phi_R = p_2 \circ \phi : U_R \to A^{n\nu-n}$$

(identifying A_R as $A^{n\nu-n}$), so that $(A^{n\nu-n}, \phi_R)$ is a geometric quotient of U_R by $PGL(n)$. Moreover, write σ_R for S_R and note that

$$\sigma_R : A^{n\nu-n} \to U_R$$

is a global section for ϕ_R. The following is obvious:

Corollary 3.2. Suppose $U \subset U_R$ is any invariant open subset. Then $U \cong PGL(n+1) \times \sigma_R^{-1}(U)$, the isomorphism being compatible with the obvious operations of $PGL(n+1)$.

To illustrate these results, which may be somewhat obscured by the cumbersome machinery of R-partitions, look at the case $\nu = 1$, $S_1 = \{0, 1, \ldots, n\}$. The proposition is then the classical result that $n+2$ points in "general" position in $\boldsymbol{P}_n$ can be normalized to the canonical points $(1, 0, \ldots, 0)$, $(0, 1, \ldots, 0), \ldots,$ $(0, 0, \ldots, 1)$, and $(1, 1, \ldots, 1)$ by a unique projective transformation. To give a second example, suppose $n = 2$, $\nu = 2$,

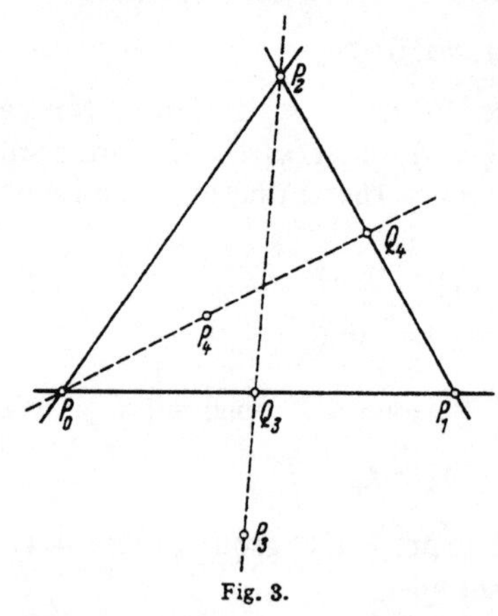

Fig. 3.

$$S_1 = \{0, 1\},\ S_2 = \{1, 2\}.$$

The proposition asserts that, given 5 points P_0, P_1, P_2, P_3 and P_4 in the plane, suppose P_0, P_1 and P_2 are not collinear, that P_3 is not on the line $\overline{P_0P_2}$ or the line $\overline{P_1P_2}$ and that P_4 is not on the line $\overline{P_0P_1}$ or $\overline{P_0P_2}$: then we can uniquely normalize P_0, P_1, P_2 to $(1, 0, 0)$, $(0, 1, 0)$ and $(0, 0, 1)$ respectively, and normalize P_3 to $(1, 1, a)$, P_4 to $(b, 1, 1)$ by a suitable projective transformation (cf. Fig. 3, where $Q_3 = (1, 1, 0)$, $Q_4 = (0, 1, 1)$).

Our next aim is to show that we have enough open sets:

Definition 3.6. For fixed m, let U_{reg} be the union of all the open subsets U_R and their images under permutations of the factors of $(\boldsymbol{P}_n)^{m+1}$, (for R-partitions such that $\nu = m - n$).

Proposition 3.3. Let $x = (x^{(0)}, x^{(1)}, \ldots, x^{(m)})$ be a geometric point of $(\boldsymbol{P}_n)^{m+1}$. Then the following are equivalent:

(1) The stabilizer $S(x)$ is 0 dimensional,

(2) there do not exist disjoint proper linear subspaces L' and L'' in $\bar{\boldsymbol{P}}_n$ such that every $x^{(i)}$ is in either L' or L'',

(3) x is a geometric point of U_{reg}.

Proof. Let k be the algebraically closed field over which x is defined. For simplicity, we shall write $\boldsymbol{P}_n$ for $\boldsymbol{P}_n \times k$, U_R for $U_R \times k$, etc. in the course of this proof. First, the implication 1) $\Rightarrow$ 2) is clear: for, in suitable homogeneous coordinates $\{x_i\}$, one may assume that:

$$L' \subset \{X_0 = X_1 = \cdots = X_r = 0\}$$
$$L'' \subset \{X_{r+1} = \cdots = X_n = 0\}.$$

Then the subgroup of transformations:

$$\begin{array}{c} \uparrow \\ r+1 \\ \downarrow \\ \hline \uparrow \\ n-r \\ \downarrow \end{array} \left[\begin{array}{ccc|ccc} \alpha & & 0 & & & \\ & \ddots & & & 0 & \\ 0 & & \alpha & & & \\ \hline & & & \beta & & 0 \\ & 0 & & & \ddots & \\ & & & 0 & & \beta \end{array}\right]$$

leaves x fixed. Secondly, 3) $\Rightarrow$ 1) is an immediate consequence of Proposition 3.1.

Thirdly, we will prove that 2) $\Rightarrow$ 3). By virtue of 2), all the points $x^{(i)}$ cannot lie in one hyperplane, hence we can choose $n+1$ of the $x^{(i)}$ which are not in one hyperplane, say $x^{(0)}, x^{(1)}, \ldots, x^{(n)}$. Without loss of generality, we may assume that $x^{(i)}$ has homogeneous coordinates $x_j^{(i)} = \delta_{ij}$. Now for each $n+k$ between $n+1$ and m, let S'_k be the set of integers i such that

$$D_{0,1,\ldots,\hat{i},\ldots,n,n+k} \neq 0,$$

i.e. $x^{(n+k)}$ is not in the hyperplane spanned by $x^{(0)}, \ldots, \widehat{x^{(i)}}, \ldots, x^{(n)}$. Then I claim that there is no partition of the set $\{0, 1, \ldots, n\}$ into 2 disjoint subsets T' and T'' such that every S'_k is contained in T' or T''. For if there were, and if one let L' (resp. L'') be the linear subspace defined by $X_i = 0$ for all $i \in T'$ (resp. $i \in T''$), then every point $x^{(k)}$ would be in $L' \cup L''$. It follows immediately that a suitable set of subsets $S_i \subset S'_i$ is an R-partition R and that $x \in U_R$. *QED.*

From this Proposition, it is clear that U_{reg} is precisely the set of pre-stable points of $(P_n)^{m+1}$, in the sense of Chapter 1, § 4.

Our next step is to construct a geometric quotient of U_{reg} by $PGL(n)$: let $U_1, \ldots, U_N$ be the open subsets U_R of U_{reg} and the subsets obtained from these by permuting the coordinates. Let (Z_i, ϕ_i) be the geometric quotient of U_i by $PGL(n+1)$. For all pairs i, j, $U_i \cap U_j$ is an invariant open subset in U_i and in U_j. Therefore, by Corollary 3.2, if $\sigma_i : Z_i \to U_i$ is the global section σ_R of ϕ_i, we know:

$$PGL(n+1) \times \sigma_i^{-1}[U_i \cap U_j] \cong U_i \cap U_j \cong PGL(n+1) \times \sigma_j^{-1}[U_i \cap U_j]$$

In other words, both $\sigma_i^{-1}[U_i \cap U_j]$ and $\sigma_j^{-1}[U_i \cap U_j]$ are geometric quotients of $U_i \cap U_j$ by $PGL(n+1)$; therefore, they are canonically isomorphic. We use this isomorphism to glue together Z_i and Z_j. For any three of the quotients Z_i, Z_j and Z_k, these identifications are obviously compatible. Therefore, we have defined a pre-scheme Z and a morphism Φ: $U_{reg} \to Z$. Clearly, U_{reg} is a locally trivial principle fibre bundle over Z; *a fortiori*, (Z, Φ) is a geometric quotient of U_{reg} by $PGL(n+1)$. However, Z is very far from being a scheme, let alone being quasi-projective.

Although Z is not quasi-projective, it carries various invertible sheaves. To investigate these, we make use of the theory of descent: by SGA 8, § 1, the set of invertible sheaves on Z is isomorphic to the set of invertible sheaves on U_{reg} plus descent data for Φ. But Φ-descent data is precisely the same as a $PGL(n+1)$- linerarization, since:

$$U_{reg} \underset{Z}{\times} U_{reg} \cong PGL(n+1) \times U_{reg}.$$

But L_i^{n+1} admits a $PGL(n+1)$-linearization. Therefore, there is an nvertible sheaf M_i on Z such that

$$L_i^{n+1} \cong \Phi^*(M_i).$$

Moreover, the section:

$$(D_{\alpha_0,\ldots,\alpha_n})^{n+1} \varepsilon\, \Gamma(U_{reg}, L_{\alpha_0}^{n+1} \otimes \cdots \otimes L_{\alpha_n}^{n+1})$$

is invariant in the $SL(n+1)$-linearization of this sheaf, hence in the $PGL(n+1)$-linearization of this sheaf. Therefore, according to SGA 8, § 1, there is a section $E_{\alpha_0,\ldots,\alpha_n}$ of $M_{\alpha_0} \otimes \cdots \otimes M_{\alpha_n}$ such that:

$$(D_{\alpha_0,\ldots,\alpha_n})^{n+1} = \Phi^*(E_{\alpha_0,\ldots,\alpha_n}).$$

§ 2. Stability

The purpose of this section is to show that although $M_0 \otimes \cdots \otimes M_m$ is not ample on Z — which is not even a scheme — it is ample on a fairly large open subset $Z_0 \subset Z$. We first define the subset $\Phi^{-1}(Z_0)$ of U_{reg}:

Definition 3.7/Proposition 3.4. U_{stable} is the open subset of $(\boldsymbol{P}_n)^{m+1}$ whose geometric points $x = (x^{(0)}, \ldots, x^{(m)})$ are those points such that for every proper linear subspace $L \subset \bar{\boldsymbol{P}}_n$:

$$\frac{\text{number of points } x^{(i)} \text{ in } L}{m+1} < \frac{\dim L + 1}{n+1}.$$

Proof. Consider the set of all pairs $\{I, n_0\}$ consisting of a subset $I \subset \{0, 1, \ldots, m\}$ and an integer n_0 between 0 and $n-1$ such that:

$$\frac{\text{cardinality of } I}{m+1} \geq \frac{n_0+1}{n+1}.$$

Then the set of geometric points *not* satisfying the above condition is the (finite) union over all $\{I, n_0\}$ of the sets consisting of those $(x^{(0)}, \ldots, x^{(m)})$ such that:

$$\{x^{(i)} \mid i \in I\} \text{ spans a linear space of dimension } \leq n_0.$$

But these geometric points are precisely those for which all $(n_0+2) \times (n_0+2)$ minors of the matrix of coordinates $\{x_j^{(i)} \mid 0 \leq j \leq n, i \in I\}$ vanish, i.e. those geometric points x at which a certain collection of sections of invertible sheaves vanish. Therefore these are the geometric points of a closed subset of $(\boldsymbol{P}_n)^{m+1}$; hence the set of geometric points violating the stability condition is also the set of points of a closed subset. *QED.*

Proposition 3.5. $U_{\text{stable}} \subset U_{\text{reg}}$.

Proof. Let $x = (x^{(0)}, \ldots, x^{(m)})$ be a stable but not regular geometric point. By condition (b) of Proposition 3.3, there are disjoint linear subspaces L', L'' in $\bar{\boldsymbol{P}}_n$ containing every point $x^{(i)}$. Then:

$$\begin{aligned} 1 &= \frac{\text{number of } x^{(i)} \text{ in } L'}{m+1} + \frac{\text{number of } x^{(i)} \text{ in } L''}{m+1} \\ &< \frac{\dim L' + 1}{n+1} + \frac{\dim L'' + 1}{n+1} \quad \text{(by stability)} \\ &= \frac{\dim \{\text{Join of } L', L''\} + 1}{n+1} \\ &\leq 1. \end{aligned}$$

This contradiction proves the result. *QED.*

The key property of stable geometric points turns out to be the following:

Proposition 3.6. Let x be a geometric point of $(\boldsymbol{P}_n)^{m+1}$. Then x is stable if and only if there are integers $N \geq N_0 > 0$ and there are monomials $P_0, \ldots, P_m$ in the expressions $D_{\alpha_0, \ldots, \alpha_n}$ such that, for $0 \leq i \leq m$:

(i) P_i is a section of $L_0^N \otimes \cdots \otimes L_i^{N-N_0} \otimes \cdots \otimes L_m^N$,

(ii) $P_i(x) \neq 0$.

Proof. Let $E = (\mathbf{R})^{m+1}$ be Euclidean space of dimension $m + 1$. To each $D_{\alpha_0,\ldots,\alpha_n}$ associate a point $d_{\alpha_0,\ldots,\alpha_n} \varepsilon E$ as follows:

$$d_{\alpha_0,\ldots,\alpha_n} = (\varepsilon_0, \varepsilon_1, \ldots, \varepsilon_m)$$
$$\varepsilon_i = 0 \quad \text{if} \quad i \notin \{\alpha_0, \ldots, \alpha_n\}$$
$$\varepsilon_i = 1 \quad \text{if} \quad i \,\varepsilon\, \{\alpha_0, \ldots, \alpha_n\}.$$

Then d determines the invertible sheaf of which D is a section. More generally, to any monomial P in the D's, associate the corresponding sum of the d's. Now define C_x to be the convex cone spanned by those points $d_{\alpha_0,\ldots,\alpha_n}$ such that $D_{\alpha_0,\ldots,\alpha_n}(x) \neq 0$, i.e. such that the components $x^{(\alpha_0)}, \ldots, x^{(\alpha_n)}$ do not lie in a hyperplane. Then it is easy to see that:

> $(1, \ldots, 1)$ ε interior of $C_x \Leftrightarrow$ there are integers $N \geq N_0 > 0$ such that positive integral sums of the d's generating C_x are the points:
>
> $(N - N_0, N, \ldots, N), (N, N - N_0, \ldots, N), \ldots, (N, N, \ldots, N - N_0)$
>
> $\Leftrightarrow P_i$, as in the Proposition, exist.

But on the other hand, $\{(1, 1, \ldots, 1) \notin$ interior of $C_x\}$ is equivalent to the existence of a linear functional l on E such that $l((1, 1, \ldots, 1)) = 0$, and $l(d_{\alpha_0,\ldots,\alpha_n}) \geq 0$ for every d generating C_x. Suppose

$$l((Z_0, \ldots, Z_m)) = \sum_{i=0}^{m} \lambda_i Z_i.$$

Assume, for a moment, that $\lambda_0 \leq \lambda_1 \leq \cdots \leq \lambda_m$. The key point is:

Lemma. If for some set of λ_i such that $\lambda_0 \leq \lambda_1 \leq \cdots \leq \lambda_m$ and $\sum_{i=0}^{m} \lambda_i = 0$, then $l(d_{\alpha_0,\ldots,\alpha_n}) \geq 0$ for every $d_{\alpha_0,\ldots,\alpha_n}$ generating C_x, then, in fact, this happens for some $\{\lambda_i\}$ of the form:

$$\lambda_0 = \lambda_1 = \cdots = \lambda_\beta = -(m - \beta)$$
$$\lambda_{\beta+1} = \cdots = \lambda_m = +(\beta + 1).$$

Proof. The point is that if $\lambda_0 \leq \cdots \leq \lambda_m$, then we can pick, *a priori*, that determinant $D_{\alpha_0,\ldots,\alpha_n}$ which is not zero at x, and for which $l(d_{\alpha_0,\ldots,\alpha_n}) = \sum_{i=0}^{n} \lambda_{\alpha_i}$ assumes its minimum value. Namely, suppose

(i) $x^{(0)} = x^{(1)} = \cdots = x^{(\beta_1 - 1)} \neq x^{(\beta_1)}$,

(ii) $x^{(0)}, x^{(1)}, \ldots, x^{(\beta_2 - 1)}$ all lie on exactly one line, but $x^{(\beta_2)}$ is not on this line.

(iii) $x^{(0)}, x^{(1)}, \ldots, x^{(\beta_3 - 1)}$ all lie on exactly one plane, but $x^{(\beta_3)}$ is not in this plane, etc., ...

(n) $x^{(0)}, \ldots, x^{(\beta_n - 1)}$ all lie in exactly one hyperplane, but $x^{(\beta_n)}$ is not in this hyperplane.

Then $x^{(0)}, x^{(\beta_1)}, \ldots, x^{(\beta_n)}$ do not lie in a hyperplane, hence $D_{0,\beta_1,\beta_2,\ldots,\beta_n}(x) \neq 0$; and whenever $D_{\alpha_0,\alpha_1,\ldots,\alpha_n}(x) \neq 0$, then

$$\alpha_0 \geq 0,\ \alpha_1 \geq \beta_1, \ldots, \alpha_n \geq \beta_n.$$

Therefore, the condition on l is that:

$$(*) \qquad l(d_{0,\beta_1,\ldots,\beta_n}) \geq 0.$$

But it is easy to check that the set of all $\{\lambda\}$'s such that $\lambda_0 \leq \cdots \leq \lambda_m$ and $\sum_{i=0}^{m} \lambda_i = 0$ is a convex set spanned by the particular sets of $\{\lambda\}$'s given in the lemma. Therefore (*) holds for some l if and only if it holds for one of the extreme sets of $\{\lambda_i\}$. *QED.*

Removing the assumption $\lambda_0 \leq \cdots \leq \lambda_m$, and computing l for these particular λ's, we conclude that the existence of the P's in the proposition is equivalent to:

Whenever $\{0, 1, \ldots, m\} = A \cup B$, where A and B are disjoint, then for some $D_{\alpha_0,\ldots,\alpha_n}$, we have

i) $D_{\alpha_0,\ldots,\alpha_n}(x) \neq 0$,

ii) $$\begin{pmatrix}\text{Cardinality}\\ \text{of } A\end{pmatrix} \cdot \begin{pmatrix}\text{number of}\\ \alpha_i \text{ in } B\end{pmatrix} < \begin{pmatrix}\text{Cardinality}\\ \text{of } B\end{pmatrix} \cdot \begin{pmatrix}\text{number of}\\ \alpha_i \text{ in } A\end{pmatrix}.$$

But suppose that the points $x^{(i)}, i \in A$, span a linear space L of dimension n_0. Then either all the points $x^{(i)}$ do not span $\mathbf{P}_n$ at all, in which case x is not stable, *and* the P's do not exist; or it is possible to select $x^{(\alpha_0)}, \ldots, x^{(\alpha_n)}$ spanning $\mathbf{P}_n$ including $n_0 + 1$ points $x^{(i)}$ such that $i \in A$. In other words, if $D_{\alpha_0,\ldots,\alpha_n}(x) \neq 0$, then the maximum number of α_i in A is exactly $n_0 + 1$. Therefore, the inequality (ii) holds for some $D_{\alpha_0,\ldots,\alpha_n}$ such that $D_{\alpha_0,\ldots,\alpha_n}(x) \neq 0$ if and only if

$$\begin{pmatrix}\text{Cardinality}\\ \text{of } A\end{pmatrix} \cdot (n - n_0) < \left[m + 1 - \begin{pmatrix}\text{Cardinality}\\ \text{of } A\end{pmatrix}\right] \cdot (n_0 + 1)$$

$$\text{i.e.} \frac{(\text{Cardinality of } A)}{m+1} < \frac{n_0 + 1}{n + 1}.$$

Now the P's exist if and only if this holds for all A; and this is clearly the same as x being stable. *QED.*

Corollary 3.7. Let x be a stable geometric point of $(\mathbf{P}_n)^{m+1}$. There exists an R-partition R and an invariant affine open neighborhood U_0 of x such that U_0 is defined by $P \neq 0$, where P is a monomial in the D's which is a section of $\{L_0 \otimes \cdots \otimes L_m\}^M$ for some M, and such that $U_0 \subset U_{\text{stable}} \cap U_R$.

Proof. Let $N \geq N_0 > 0$, and $P_0, \ldots, P_m$ be given by the Proposition. Moreover, let Q be the product of all the D's such that $D(x) \neq 0$. Then a suitable monomial $P = Q^{N_0} \otimes P_0^{r_0} \otimes \cdots \otimes P_m^{r_m}$ is a section of the ample invertible sheaf $(L_0 \otimes \cdots \otimes L_m)^M$; hence $P \neq 0$ defines an affine open neighborhood of x. Moreover, if $P(y) \neq 0$ for some geometric point y of $(\mathbf{P}_n)^{m+1}$, then $P_0(y) \neq 0, \ldots, P_m(y) \neq 0$, hence y is stable by Proposition 3.6. Finally, y is contained in any open set U_R that contains x. *QED.*

We can now apply our work to obtain the final result. Let (Z, Φ) be the geometric quotient of U_{reg} by $PGL(n)$. Let $Z_0 \subset Z$ be the open subset such that $\Phi^{-1}(Z_0) = U_{\text{stable}}$. Let Φ_0 be the restriction of Φ to a morphism from U_{stable} to Z_0.

Theorem 3.8. (Z_0, Φ_0) is a geometric quotient of U_{stable} by $PGL(n+1)$, and, in fact, U_{stable} is a locally trivial principal fibre bundle over Z_0 with group $PGL(n+1)$. Moreover, $M_0 \otimes \cdots \otimes M_m$ is ample on Z_0, hence Z_0 is a quasi-projective scheme.

Proof. The first part is clear. To prove the second part, it suffices to show that if y is any geometric point of Z_0, then y has an affine open neighborhood of the form $(Z_0)_s$ for some section s of $(M_0 \otimes \cdots \otimes M_m)^N$, (cf. EGA 2, Remark following Theorem 4.5.2). But let x be a geometric point of U_{stable} over y. The open neighborhood U_0 of Corollary 3.7 is invariant, hence of the form $\Phi_0^{-1}(V)$ for some open neighborhood V of y. Since $U_0 \subset U_R$ for some R, $\Phi_0|U_0$ admits a global section σ_0. Since U_0 is affine, and $V \simeq \sigma_0(V)$ — a closed subscheme of U_0 — therefore V is affine. Moreover, U_0 is defined by $P \neq 0$, where P is an invariant section of $(L_0 \otimes \cdots \otimes L_m)^N$ (for suitable N). Then P^{n+1} is induced by a section s of $(M_0 \otimes \cdots \otimes M_m)^N$ (cf. § 1), and V must be defined by $s \neq 0$. *QED.*

In chapter 7, we will use this result (cf. Proposition 7.7). For the sake of consistency, we will denote the open set U_{stable} at that point by $(\mathbf{P}_n)^{m+1}_{\text{stable}}$.

Chapter 4

Further examples

In this chapter, we shall utilize the results of Chapters 1 and 2 to obtain various quotients. In contrast to Chapter 3, therefore, we are essentially restricted to characteristic 0. We take the ground field to be $\mathbf{Q}$, the field of rational numbers, so that (i) $PGL(n+1)$ and $SL(n+1)$ are reductive, and (ii) the quotients we construct are still as "universal" as possible. As in Chapter 1, we fix an algebraically closed over-field $\Omega \supset \mathbf{Q}$, and write $\overline{X} = X \times_{\mathbf{Q}} \Omega$ whenever X is an algebraic scheme over $\mathbf{Q}$. Moreover, the standard notations $\mathbf{P}_n$, *Grass*, $PGL(n+1)$ etc. will refer to the corresponding schemes over $\mathbf{Q}$.

§ 1. Binary quantics

The simplest of all cases of the theory is the action of $PGL(2)$ on $\mathbf{P}_n$ obtained by identifying $\mathbf{P}_n$ with the nth symmetric power of the line P_1 on which $PGL(2)$ acts canonically. This example is identical with the hoary theory of *Binary Quantics*, which was diligently pursued in the 2nd half of the nineteenth century. Recall that $o_P(2)$ admits one and only one $PGL(2)$-linearization (cf. § 1.3). Then the main object of that theory was to give an explicit description of the scheme

$$(P_n)^{ss}\left(\underline{o}(2)\right)/PGL(2).$$

This is an amazingly difficult job, and complete success was achieved only for $n \leq 6$, the cases $n = 5$ and 6 being one of the crowning glories of the theory.*

Using the present theory, let us at least determine the open sets:

$$(P_n)^{ss}\left(\underline{o}(2)\right) \quad \textit{and} \quad (P_n)^{s}\left(\underline{o}(2)\right) = (P_n)^{s}_{(0)}\left(\underline{o}(2)\right).$$

It is convenient to use the $PGL(2)$-linear morphism:

$$\overline{\omega} : (P_1)^n \to P_n$$

from the n-fold product to the n-fold symmetric product. Let $p_1, \ldots, p_n$ be the projections of $(P_1)^n$ onto its factors. Let $x = (x^{(1)}, \ldots, x^{(n)})$ be a geometric point of $(P_1)^n$. Then for all 1-*PS*'s λ of $\overline{PGL(2)}$:

$$(*) \qquad \begin{aligned} \mu^{\underline{o}(2)}\left[\overline{\omega}x, \lambda\right] &= \mu^{\overline{\omega}^*(\underline{o}(2))}(x, \lambda] \\ &= \mu^{\otimes p_i^*(\underline{o}(2))}[x, \lambda] \\ &= \sum_{i=1}^{n} \mu^{p_i^*(\underline{o}(2))}[x, \lambda] \\ &= \sum_{i=1}^{n} \mu^{\underline{o}(2)}[x^{(i)}, \lambda]. \end{aligned}$$

On the other hand, the action of $\lambda(G_m)$ on $\overline{P}_1$ is the following: there are 2 geometric points $y, z \in \overline{P}_1$, so that if $t \in \Gamma(\overline{P}_1 - z, o_{P_1})$ is an affine coordinate function, 0 at y and ∞ at z, then

$$\begin{gathered} \lambda(\alpha) : (\overline{P}_1 - z) \to (\overline{P}_1 - z) \\ \lambda(\alpha)^*(t) = \alpha^r \cdot t \end{gathered}$$

for some positive integer r. Note that, if w is any closed point of $\overline{P}_1$, then $\sigma(\lambda(\alpha), w)$ specializes to y (resp. z) as $\alpha \to 0$, if $w \neq z$ (resp. $w = z$). Now the line bundle corresponding to $0(2)$, is the cotangent bundle to P_1, and the $PGL(2)$-linearization of the cotangent bundle can be deduced from the action of $PGL(2)$ by differentiating. In fact, dt is a non-zero section of this bundle over $\overline{P}_1 - z$, and, under the action of $\lambda(\alpha^{-1})$, it is

* For $n = 8$, by an extraordinary four de force, Shioda [313] described the scheme $P_8^{ss}/PGL(2)$.

transformed* into

$$d\left(\lambda(\alpha)^* t\right) = \alpha^r \cdot dt.$$

Therefore, $\lambda(\alpha)$ acts on the fibre of the cotangent bundle over y by the character

$$\chi(\alpha) = \alpha^{-r}.$$

Similarly, it acts on the fibre over z by χ^{-1}. Therefore,

$$\begin{aligned} \mu^{\mathcal{Q}(2)}(w, \lambda) &= +r, \quad w \neq z \\ &= -r, \quad w = z. \end{aligned}$$

If we combine this with (*), we obtain:

$$\mu^{\mathcal{Q}(2)}(\overline{\omega} x, \lambda) = 2r\left\{\frac{n}{2} - \begin{pmatrix}\text{number of } i \text{ such} \\ \text{that } x^{(i)} = z\end{pmatrix}\right\}$$

By Theorem 2.1, we conclude:

Proposition 4.1. A geometric point of $\boldsymbol{P}_n$ which is represented by the 0-cycle $\mathfrak{A}$ is stable (resp. semi-stable) if no geometric point z of $\boldsymbol{P}_1$ occurs in $\mathfrak{A}$ with multiplicity $n/2$ or greater (resp. with multiplicity greater than $n/2$).

We can interpret the results of § 2.2 and 2.3 in this case also. $\Delta\left(\overline{PGL(2)}\right)$ has exactly one point δ for each parabolic subgroup of $\overline{PGL(2)}$, and there is one parabolic subgroup for each closed point $z \in \overline{\boldsymbol{P}}_1$. Any pair of points of $\Delta\left(\overline{PGL(2)}\right)$ forms a skeleton, so any paire of points is antipodal. Any subset of $\Delta\left(\overline{PGL(2)}\right)$ is semi-convex, but the only convex sets consist of exactly one point. The function

$$||\lambda|| = 2r$$

(in the above notations) is a norm on 1-PS's of $\overline{PGL(2)}$, and, if $\delta(z)$ is the point of $\Delta\left(\overline{PGL(2)}\right)$ corresponding to $z \in \overline{\boldsymbol{P}}_1$, then the function ν is simply:

$$\nu^{\mathcal{Q}(2)}\left(\overline{\omega} x, \delta(z)\right) = \frac{n}{2} - \begin{pmatrix}\text{number of } i \text{ such} \\ \text{that } x^{(i)} = z\end{pmatrix}$$

Since 2 different points cannot both occur in the 0-cycle $\sum_{i=1}^{n} x^{(i)}$ with multiplicity greater than $n/2$, the set of $\delta(z)$ for which $\nu < 0$ is indeed convex!

A final point: we saw in Chapter 3 that the open set of pre-stable points in $(\boldsymbol{P}_1)^n$ was much bigger than the open set of stable points, so that one can obtain quotients of larger open sets by allowing pre-schemes, and not sticking to quasi-projective schemes. However, for $\boldsymbol{P}_n$ this is not true. The set of pre-stable points equals the sets of points stable with

* n.b. if $T: \overline{\boldsymbol{P}}_1 \to \overline{\boldsymbol{P}}_1$ is the transformation induced by α, then the usual map T_α^* of the cotangent bundle is contravariant to T_α: it is covariant with $T_{\alpha^{-1}}$.

respect to $\mathcal{O}(2)$ In other words, $(\boldsymbol{P}_n)^s(\mathcal{O}(2))$ is the *unique maximal invariant open set* U such that a geometric quotient (Q, ϕ) of U by $PGL(2)$ exists, and such that ϕ is affine. This follows from Converse 1.13, since $\boldsymbol{P}_n$ is non-singular, and its Picard group is generated by $\mathcal{O}(1)$. One can go further, and show that, if U is an invariant open set in $\boldsymbol{P}_n$ such that an *arbitrary* geometric quotient (Q, ϕ) of U by $PGL(2)$ exists, then $U \subset (\boldsymbol{P}_n)^{ss}(\mathcal{O}(2))$. To see this, by passing to an affine open covering of Q, it suffices to prove this when Q *is affine*. But

$$\Gamma(Q, \mathcal{O}_Q) \subset \Gamma(U, \mathcal{O}_{\boldsymbol{P}_n}),$$

hence $\Gamma(U, \mathcal{O}_{\boldsymbol{P}_n})$ contains non-constant functions. This means that $\boldsymbol{P}_n - U$ has *some* component of codimension 1, i.e. there is an invariant hypersurface D such that $U \subset (\boldsymbol{P}_n - D)$. If D is given by the invariant form $F \in H^0(\boldsymbol{P}_n, \mathcal{O}(N))$, it follows that:

$$U \subset (\boldsymbol{P}_n)_F \subset (\boldsymbol{P}_n)^{ss}(\mathcal{O}(2)).$$

§ 2. Hypersurfaces

In classical invariant theory, mathematicians also studied the so-called theory of ternary quantics, quatenary quantics, etc. More generally:

Definition 4.1. The *classical operations* of $PGL\ (n+1)$ are the actions of $PGL(n+1)$ on the linear systems $|mH|$, i.e. on the projective space of one-dimensional subspaces of $H^0(\boldsymbol{P}_n, \mathcal{O}_{\boldsymbol{P}_n}(m))$.

Since $|mH|$ is a projective space, stability has only one sense for $|mH|$. It turns out to be hard, in this case, to describe even the open set of stable points, let alone the actual quotient of this by $PGL(n+1)$.* On the other hand, it is readily computable for n and m reasonably small. We can prove:

Proposition 4.2. If $m \geq 3$, then a non-singular hypersurface in $\overline{\boldsymbol{P}}_n$ represents a stable geometric point of $|mH|$.

Proof. This is a consequence of the existence of the discriminant: suppose F is an element of $H^0(\boldsymbol{P}_n, \mathcal{O}(m))$, i.e. a homogeneous form in $n+1$ variables of degree m. Then there is a definite homogeneous polynomial Δ in its coefficients such that Δ is zero at F if and only if the hypersurface $F = 0$ is singular. Then Δ can be interpreted as a form in the homogeneous coordinates of $|mH|$. Since the divisor $\Delta = 0$ is invariant under $PGL(n+1)$, Δ must be invariant. Moreover, the stabilizer in $PGL\ (n+1)$ of any non-singular hypersurface in $\overline{\boldsymbol{P}}_n$ of degree at least 3 is finite. This follows from the non-existence of global

* In fact, invariant theorists of the nineteenth century described it only for $n = 2$, $m = 3$ ($m < 3$ being trivial).

vector fields on such hypersurfaces (cf. lemma 14.2, [17]). Therefore, the action of $PGL(n+1)$ on $|mH|_\Delta$ is closed, i.e.

$$|mH|_\Delta \subset |mH|^s. \quad QED.$$

Turning to more particular cases, we summarize the results in a table.

Table

(n, m)	not properly stable	not semi-stable
$(1, m)$	$\exists$ a pt. of multiplicity $\geq \frac{m}{2}$	$\exists$ a pt. of multiplicity $> \frac{m}{2}$
$(n, 1)$	always	always
$(n, 2)$	always	$\exists$ a singular pt.
$(2, 3)$	$\exists$ a singular pt.	$\exists$ a triple pt. or cusp
$(2, 4)$	$\exists$ a triple pt., or tacnode	$\exists$ a triple pt. or curve is cubic + inflexional tangent
$(2, 5)$	$\exists$ a 4th order pt., or triple pt. with only one 4th order tangent line, or triple pt. with one simple branch and one tacnode	$\exists$ a 4th order pt., or triple pt. with only one 4th order tangent line, or curve is quartic, plus 4-fold tangent line through a double pt.
$(3, 3)$	$\exists$ unode, binode, or triple pt.	$\exists$ unode, binode of higher type, or triple pt.

They should be interpreted with the help of the following glossary:

k-fold tangent to curve C at x: a line l such that x occurs with multiplicity at least k in $(C \cdot l)$. For $k = 3$, this is an *inflexional tangent.*

cusp: a double pt. of a curve with only one 3-fold tangent line, i.e. whose tangent cone consists of 1 point.

tacnode: a cusp whose 3-fold tangent line is actually 4-fold, i.e. a cusp which is not resolved by a single quadratic transformation.

binode: a double point of a surface whose tangent cone consists of 2 distinct lines.

unode: a double point of a surface whose tangent cone consists of 1 line.

binode of higher type: a binode which is not resolved by a single quadratic transformation.

It is abundantly clear from these examples that a breakdown of stability is always associated to a "bad" flag in $\boldsymbol{P}_n$, as predicted in Chapter 2. To see this concretely, by computing ν, and to indicate the technique for establishing these results, we shall go into detail for one case: $(n, m) = (2, 4)$.

For the sake of simplicity, let

$$\omega : SL(3) \to PGL(3)$$

be the usual isogeny, and consider the induced action of $SL(3)$ on $|4H|$ instead of the action of $PGL(3)$. Then $\underline{o}(1)$ admits a $SL(3)$-linearization; in fact, if we make $H^0(\boldsymbol{P}_2, \underline{o}(4))$ the set of rational points in the *scheme* $\boldsymbol{A}^{15}$, then $\boldsymbol{A}^{15}$ is the affine cone over $|4H|$. And $SL(3)$ is represented in $\boldsymbol{A}^{15}$ by the 4th symmetric power of the dual of its canonical representation in $\boldsymbol{A}^3$. We shall compute the function μ by means of Proposition 2.3.

Fix a maximal torus $T \subset SL(3)$. For suitable coordinates x_0, x_1, x_2, the action of T on its canonical representation space $\boldsymbol{A}^3$ can be diagonalized, i.e. T becomes the group of matrices:

$$\begin{bmatrix} \alpha_0 & 0 & 0 \\ 0 & \alpha_1 & 0 \\ 0 & 0 & \alpha_2 \end{bmatrix}$$

of determinant 1. For any field K, K-valued points of the affine space of the dual representation are given by linear forms $a_0 x_0 + a_1 x_1 + a_2 x_2$, with $a_i \in K$; and K-valued points of the affine space $\boldsymbol{A}^{15}$ of the 4th symmetric power of the dual representation are given by homogeneous forms of degree 4, with coefficients in K. We write these:

$$\begin{aligned} F = {} & a_{40}x_1^4 + a_{31}x_1^3x_2 + a_{22}x_1^2x_2^2 + a_{13}x_1x_2^3 + a_{40}x_2^4 \\ & + a_{30}x_0x_1^3 + a_{21}x_0x_1^2x_2 + a_{12}x_0x_1x_2^2 + a_{03}x_0x_2^3 \\ & + a_{20}x_0^2x_1^2 + a_{11}x_0^2x_1x_2 + a_{02}x_0^2x_2^2 \\ & + a_{10}x_0^3x_1 + a_{01}x_0^3x_2 \\ & + a_{00}x_0^4. \end{aligned}$$

If the K-rational point of T given by $[\alpha_i \cdot \delta_{ij}]_{0 \le i,j \le 2}$ acts on this form, the new form F' has coefficients a'_{ij}, where

$$a'_{ij} = (\alpha_0^{i+j-4} \cdot \alpha_1^{-i} \cdot \alpha_2^{-j}) \cdot a_{ij}.$$

Now suppose that λ is the 1-*PS* of T given by

$$\lambda(\alpha) = \begin{bmatrix} \alpha^{r_0} & 0 & 0 \\ 0 & \alpha^{r_1} & 0 \\ 0 & 0 & \alpha^{r_2} \end{bmatrix}, \quad r_0 + r_1 + r_2 = 0.$$

Then by Proposition 2.3, if C is a quartic curve in $\overline{\boldsymbol{P}}_2$, and if F is a homogeneous form for C, coefficients in Ω, then

(*) $$\mu^{\underline{o}(1)}(c, \lambda) = \max\{(4-(i+j))\,r_0 + ir_1 + jr_2 \mid \text{all pairs } i, j \text{ such that } a_{ij} \ne 0\}.$$

First of all, we can use this formula to check our table in this case: if C' is not stable, then for some λ, this μ is less than or equal 0. But every λ is conjugate in $SL(3)$ to a 1-PS of T, and even one given by the above equations where $r_0 \geq r_1 \geq r_2$. Therefore, C' is not stable if and only if the above μ is $\leqq 0$ for some projective transform C of C', and some integers r_i satisfying these relations. But by (*), $\mu \leq 0$ implies that $a_{00} = a_{10} = a_{20} = a_{01} = a_{11} = 0$. Moreover, if $\mu \leq 0$, and $a_{30} \neq 0$ and $a_{02} \neq 0$, then

$$0 \geq \{r_0 + 3r_1\} + \{2r_0 + 2r_2\} = r_0 + r_1$$

which contradicts our assumptions on the r_i. Now if $a_{02} = 0$, it follows that C has a triple point at (1, 0, 0); if $a_{30} = 0$, it follows that C has a tacnode at (1, 0, 0). Conversely, suppose C has a triple point at (1, 0, 0), hence $a_{02} = 0$. If $r_0 = 2$, $r_1 = r_2 = -1$, then

$$\mu = \max\{8 - 3i - 3j \mid \text{all pairs } i, j \text{ such that } a_{ij} \neq 0\} \leq -1.$$

If C has a tacnode at (1, 0, 0), then $a_{30} = 0$. Put $r_0 = 1$, $r_1 = 0$, $r_2 = -1$. Then

$$\mu = \max\{4 - i - 2j \mid \text{all pairs } i, j \text{ such that } a_{ij} \neq 0\} \leq 0.$$

This establishes the meaning of stability for quartic curves.

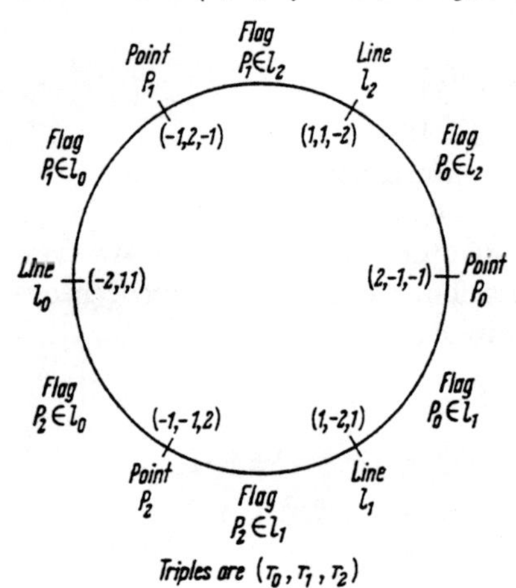

Fig. 4. Skeleton in $\Delta[SL(3)]$

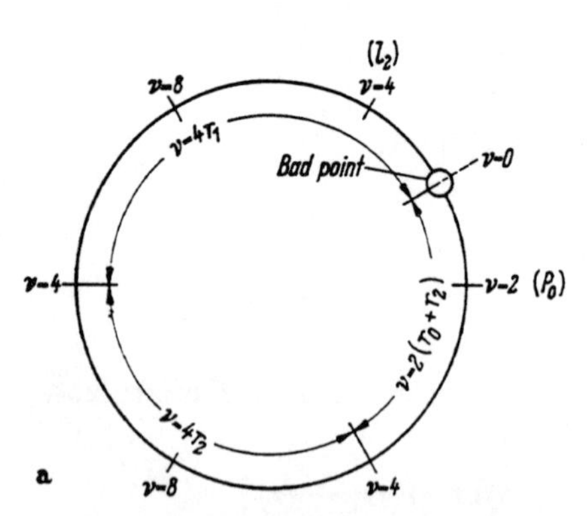

Fig. 5a. ν for generic quartic with tacnode at P_0, tangent l_2

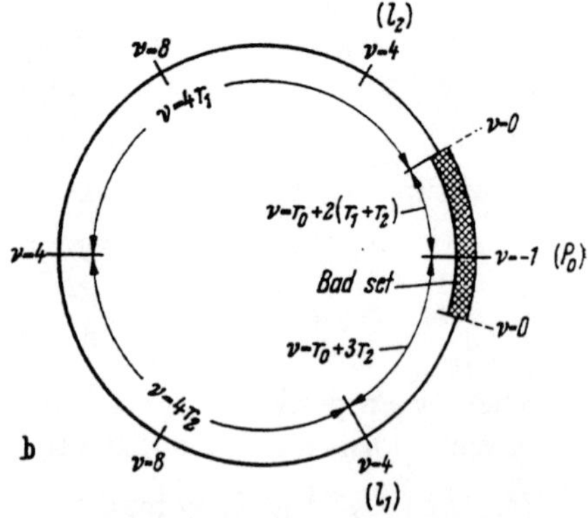

Fig. 5b. ν for generic quartic with P_0 triplept., l_2 a 4-fold tangent

Now consider the skeleton $\Delta(T)$ in $\Delta(\overline{SL(3)})$. It can be represented as the set of points on the circle

$$r_0 + r_1 + r_2 = 0$$
$$r_0^2 + r_1^2 + r_2^2 = 6$$

where the ratios of the r_i's are rational. If l_i is the line in $\boldsymbol{P}_2$ given by $x_i = 0$, and if P_i is the point in $\boldsymbol{P}_2$ with coordinates $x_j = \delta_{ij}$, then proceeding around this circle, the parabolic subgroups correspond to flags as indicated in Fig. 4. Furthermore, for a 1-*PS* λ of T as above, put:

$$\|\lambda\| = \sqrt{(r_0^2 + r_1^2 + r_2^2)/6}.$$

The resulting function $\nu^{\underline{\varrho}^{(1)}}(C, *)$ is indicated in Fig. 5, for some simple curves C.

§ 3. Counter-examples

In this section, we shall consider counter-examples. We would like to show how the reductive groups $SL(n)$ can operate properly on *quasi-projective* schemes X, simple over $\boldsymbol{Q}$, but where there is *no* geometric quotient of X by $SL(n)$; i.e., by Converse 1.12 and Proposition 0.7 there is no L in $Pic^G(X)$ such that $X = X^s(L)$. It should be emphasized that in any such case, there is an analytic space which is a quotient of the analytic space defined by X, by the analytic group $SL(n)$: cf. Palais [31]. The moral of these examples is that the concept of stable points is purely algebraic, and not topological.

The first example is due to Nagata [27]. Here $SL(3)$ is acting on an open set U in $|5H|$, as in § 2. The geometric points of U are precisely those plane quintic curves C such that:

i) C has no 4th order points,

ii) C has no triple points, with tangent cone consisting of only one line, and that line having 5-fold contact with C,

iii) C has no triple points, consisting of a simple branch, and a tacnode,

iv) no component of C is a line.

Nagata has proven that $SL(3)$ acts properly on U. But from the results of § 2, it follows that no geometric quotient of U by $SL(3)$ exists.

The second example is based on an example of Hironaka [15]. Let V_0 be $\boldsymbol{P}_3$, and let γ_1, γ_2 be two conics in V_0, intersecting normally in exactly two points P_1 and P_2. Following Hironaka, construct $\overline{V}_i$, for $i = 1$ and 2, by (a) blowing up γ_i in V_0; (b) blowing up γ_{3-i} in the result (cf. Fig. 6). Let V_i be the open set in $\overline{V}_i$ of points lying over

$(V_0 - P_{3-i})$. Then by patching V_1 and V_2, we obtain a single *non-projective* variety V_3 in which, over P_1 and P_2 the two curves γ_1 and γ_2 have been blown up in opposite orders! Now let $\sigma_0 : V_0 \to V_0$ be a projective transformation of order 2 which permutes P_1 and P_2, and also γ_1 and γ_2. Then σ_0 induces an automorphism of V_3 of order 2. HIRONAKA has shown that there is no geometric quotient of V_3 by the group $\boldsymbol{Z}_2 = \{1, \sigma\}$, although of course there is an analytic space $[V_3/\text{mod } \boldsymbol{Z}_2]$.

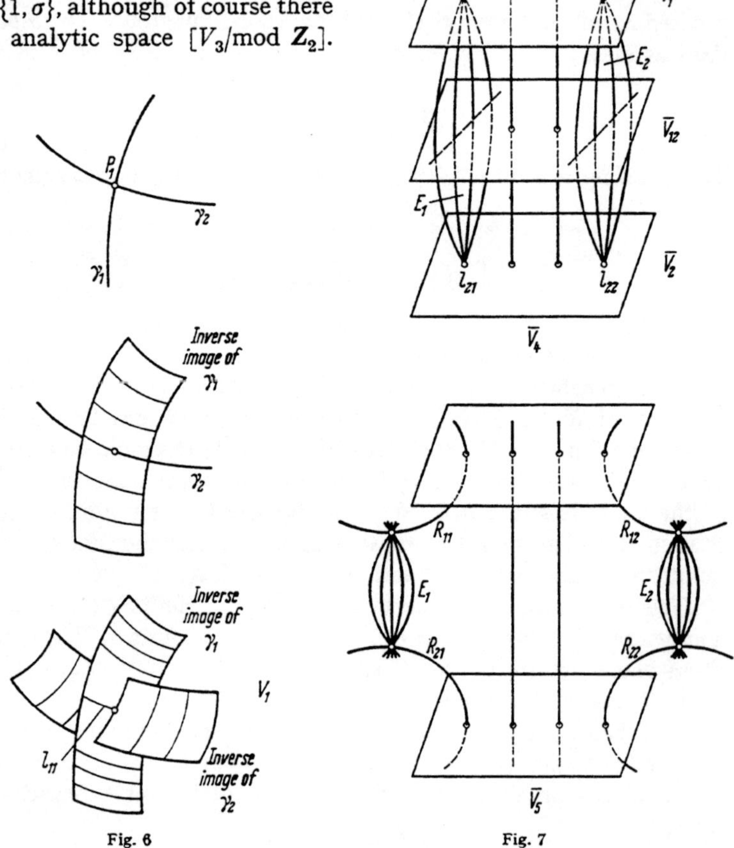

Fig. 6 Fig. 7

This is the first step of the construction. Now fix some projection embeddings of $\overline{V}_1$ and $\overline{V}_2$, say $\overline{V}_1 \subset \boldsymbol{P}_{n_1}$, $\overline{V}_2 \subset \boldsymbol{P}_{n_2}$. Embed $\boldsymbol{P}_{n_1}$ and $\boldsymbol{P}_{n_2}$ as disjoint linear subspaces of $\boldsymbol{P}_{n_1+n_2+1}$ and consider the two subvarieties $\overline{V}_i \subset \boldsymbol{P}_{n_1+n_2+1}$. Let $\overline{V}_4$ be the locus of lines in $\boldsymbol{P}_{n_1+n_2+1}$ joining birationally corresponding points of $\overline{V}_1$ and $\overline{V}_2$ (cf. Fig. 7, where all dimensions have been lowered by one). Now the group $\boldsymbol{G}_m$ operates in a natural way on $\boldsymbol{P}_{n_1+n_2+1}$ so as to leave $\boldsymbol{P}_{n_1}$ and $\boldsymbol{P}_{n_2}$ pointwise fixed. This induces an action of G_m on $\bar{V}_4$ leaving the subvarieties $\bar{V}_1$ and $\bar{V}_2$ pointwise fixed. Now let $\phi: \bar{V}_4 \to V_0$ be the natural projection. Also introduce for convenience a plane $H_0 \subset V_0$ through P_1 and P_2 and left fixed by σ_0 (there are, in fact, exactly two such planes). Let H_4 be its proper inverse image under ϕ, i.e., the irreducible subvariety of $\bar{V}_4$ that dominates H_0 and is a component of $\phi^{-1}(H_0)$. Let $I \subset \mathcal{O}_{\bar{V}_4}$ be the sheaf of ideals vanishing on H_4. Define $\bar{V}_5$ as the I-transform of $\bar{V}_4$ (cf. EGA 2, § 8). We wish to describe explicitly the structure of $\bar{V}_5$. Let $\bar{V}_{12}$ be the birational joint of $\bar{V}_1$ and $\bar{V}_2$, let v_j be the m-adic valuation of the function field $k(V_0)$ centered at P_j, and let l_{ij} be the center of v_j on $\bar{V}_i$. Then one verifies readily that l_{ij} are lines on $\bar{V}_i$ and that $\bar{V}_{12}$ is obtained from V_i by blowing up l_{i1} and l_{i2}. But $\bar{V}_4$ is essentially a (twisted) product of P_1 — (2 points) by $\bar{V}_{12}$, with $\bar{V}_1$ and $\bar{V}_2$ pasted at the two ends. $\bar{V}_5$ differs from $\bar{V}_4$ only at points of H_4 which are singular on $\bar{V}_4$. In fact, the singularities of $\bar{V}_4$ are precisely the four lines $l_{ij} \subset \bar{V}_i \subset \bar{V}_4$, and H_4 passes through all of them. A local computation shows that these singularities are all resolved on $\bar{V}_5$, and each line l_{ij} is blown up into a ruled surface R_{ij}. Moreover, the action of G_m lifts to $\bar{V}_5$, since H_4 is invariant under G_m, and G_m acts non-trivially on the ruled surface R_{ij} with orbit space l_{ij}. Finally, define V_5 to be the open subset of $\bar{V}_5$ obtained by removing:

a) the proper transform of $\bar{V}_1$, $\bar{V}_2$;

b) the proper transforms on $\bar{V}_5$ of the divisors E_1 and E_2 on $\bar{V}_4$; where E_1 is the locus of lines joining l_{11} and l_{21}, and E_2 is the locus of lines joining l_{12} and l_{22};

c) R_{12} and R_{21}.

A rather conceptualized "picture" of V_5 is given in Fig. 7. It is clear that V_5 is invariant under the action of G_m, and that V_3 is a geometric quotient of V_5 by G_m: for V_5 is the union of the lines joining biregularly corresponding points of V_1 and V_2, plus the lines of R_{11} and of R_{22}.

Note also that the automorphism σ_0 can be carried along: i) σ lifts to an isomorphism $\bar{V}_1 \to \bar{V}_2$; ii) if the embeddings $\bar{V}_i \subset P_{n_i}$ are chosen suitably, σ extends to an isomorphism $P_{n_1} \to P_{n_2}$; iii) this isomorphism extends to an automorphism of $P_{n_1+n_2+1}$ of order 2 satisfying $\sigma \cdot \alpha \cdot \sigma^{-1} = \alpha^{-1}$, for $\alpha \in (G_m)$ acting on $P_{n_1+n_2+1}$; iv) σ takes $\bar{V}_4$ into itself, and H_4 into itself, hence σ lifts to a morphism $\sigma: \bar{V}_5 \to \bar{V}_5$; v) this σ carries V_5 into V_5. Let G be the semi-direct product of G_m and Z_2 defined by $\sigma\alpha = \alpha^{-1}\sigma$ for $\alpha \in G_m$, $\sigma \in Z_2$, $\sigma \neq 1$. Then we have now defined an action of G on V_5 such that there is no geometric quotient. Moreover, note that V_5 is quasi-projective (recall that blowing up any sheaf of ideals takes a projective variety into a projective variety).

This completes the second step of the construction. Finally fix some very ample sheaf L on $\bar{V}_5$. By Proposition 1.6, some power L^n of L is

G_m-linearizable. Moreover, it is easy to check that then $\sigma^* L^n \otimes L^n$ admits a G-linearization. Put $M = \sigma^* L^n \otimes L^n$. Since L is very ample, it follows that M is very ample. By Proposition 1.7, there is an action of G on $\boldsymbol{P}_n$, a G-linearization of $\underline{o}(1)$ and a G-linear immersion

$$\phi : V_5 \to \boldsymbol{P}_n$$

such that $\phi^*(\underline{o}(1)) \cong M$. Let $\beta : G \to PGL(n+1)$ be the homomorphism which induces the action of G on $\boldsymbol{P}_n$. It is well-known that the coset space $PGL(n+1)/\beta(G)$ exists (cf. [8], Exposé 8). Define V_6 to be the locally closed reduced subscheme of $\boldsymbol{P}_n \times PGL(n+1)/\beta(G)$ whose geometric point are the pairs

$$(\Sigma(\alpha, \phi x), \alpha \bmod \beta(G))$$

for geometric points α in $PGL(n)$, x in V_5, and where Σ indicates the usual action of $PGL(n+1)$ on $\boldsymbol{P}_n$. Then one sees easily that a geometric quotient of V_6 by $PGL(n+1)$ would also be a geometric quotient of V_5 by G. Moreover, it is not difficult to verify that the action of $PGL(n+1)$ on V_6 is proper.

§ 4. Sequences of linear subspaces

The next example is a generalization of the theory of Chapter 3. This result is the basis of an analysis of vector bundles over an algebraic curve which I shall publish elsewhere. Let $Grass_{k,n}$ stand for the Grassmannian scheme of k-dimensional linear subspaces of $\boldsymbol{P}_n$. Let

$$V = [Grass_{k,n}]^N.$$

We shall study the action of $PGL(n+1)$ on V.

For technical reasons, it is convenient to study the induced action of $SL(n+1)$. Recall that the Grassmannian is a closed subscheme of

$$P = \text{Proj}\, \boldsymbol{Q}\, [\ldots; p_{i_0,\ldots,i_k}; \ldots]$$

where $i_0 < i_1 < \cdots < i_k$, and

$$p_{i_0,\ldots,i_k} = x_{i_0} \wedge \cdots \wedge x_{i_k}$$

are the Plücker coordinates. Since $SL(n+1)$ acts on the affine space whose coordinates are $x_0, \ldots, x_n$, it follows that $SL(n+1)$ acts on the affine space whose coordinates are the p's, i.e. on the affine cone over P. Therefore the invertible sheaf $\underline{o}_P(1)$ is $SL(n+1)$-linearized, and so is the induced ample sheaf on $Grass_{k,n}$. We call this sheaf simply $\underline{o}(1)$. Now suppose that L is a linear subspace of $\bar{\boldsymbol{P}}_n$ of dimension k. Let L also denote the corresponding geometric point of $Grass_{k,n}$. Suppose λ is the 1-PS of $SL(n+1)$ given by the matrices

$$\lambda(\alpha) = [\alpha^{r_i} \cdot \delta_{ij}]_{0 \le i,j \le n}.$$

Then if $p_{i_0,\ldots,i_k}(L)$ are homogeneous coordinates for L, it follows from Proposition 2.3. that:

$$(*)\qquad \mu^{\underline{o}(1)}(L,\lambda) = \max\left\{-r_{i_0} - \cdots - r_{i_k} \,\middle|\, \begin{array}{l}\text{all } i_0 < \cdots < i_k \text{ such} \\ \text{that } p_{i_0,\ldots,i_k}(L) \neq 0\end{array}\right\}.$$

Now, for every subset $I \subset \{0, 1, \ldots, n\}$, let L_I be the linear subspace of $\boldsymbol{P}_n$ defined by the equations $x_i = 0$, for all $i \in I$. If I consists of exactly $n - k$ integers, then the dimension of L_I is k; and, in fact, all the Plücker coordinates of L_I are zero except for one: namely

$$p_{j_0,j_1,\ldots,j_k}(L_I) \neq 0 \quad \text{if} \quad I \cup \{j_0, \ldots, j_k\} = \{0, 1, \ldots, n\}.$$

It follows therefore from (*) that in this case:

$$(**)\qquad \mu^{\underline{o}(1)}(L_I, \lambda) = -\sum_{j \notin I} r_j = \sum_{j \in I} r_j.$$

Now suppose again that L is any linear subspace of $\bar{\boldsymbol{P}}_n$ of dimension k, and assume that the r_i satisfy:

$$r_0 > r_1 > \cdots > r_n.$$

Consider the action of $\lambda(\alpha)$ on $\boldsymbol{P}_n$ as $\alpha \to 0$. One checks that:

(i) points not on the hyperplane L_n specialize to $(0, 0, \ldots, 1)$.

(ii). points on L_n, but not in $L_{(n-1,n)}$ specialize to $(0, 0, \ldots, 1, 0)$.

. . . .

$(n-1)$ points in the line $L_{(2,\ldots,n)}$, not equal to $L_{(1,2,\ldots,n)}$ itself specialize to $(0, 1, \ldots, 0)$.

(n) the point $L_{(1,2,\ldots,n)} = (1, 0, \ldots, 0)$ remains fixed.

Now for each i, $0 \leq i \leq k$, let μ_i be the unique integer μ such that:

$$\dim(L \cap L_{(\mu+1,\ldots,n)}) = i$$
$$\dim(L \cap L_{(\mu,\mu+1,\ldots,n)}) = i - 1.$$

Then it follows that L contains a point in $L_{(\mu_i+1,\ldots,n)} - L_{(\mu_i,\ldots,n)}$ for each i. Therefore, the specialization of L when $\alpha \to 0$ contains the point with coordinates: $x_k = 0\,(k \neq \mu_i)$, $x_{\mu_i} = 1$. Since these points span a linear space of dimension k, it follows that the specialization of L, as $\alpha \to 0$, is equal to L_I, where $I = \{0, 1, \ldots, n\} - \{\mu_0, \ldots, \mu_k\}$. Therefore, by (**), we conclude

$$\begin{aligned}\mu^{\underline{o}(1)}(L,\lambda) &= -\sum_{i=0}^{k} r_{\mu_i} \\ &= -\sum_{i=0}^{n} [\dim L \cap L_{(i+1,\ldots,n)} - \dim L \cap L_{(i,\ldots,n)}] \cdot r_i\end{aligned}$$

where the term in this sum for $i = n$ means $[k - \dim L \cap L_{(n)}] r_n$, and for $i = 0$ means $[\dim L \cap L_{(1,\ldots,n)} - (-1)] r_0$. Rearranging, we find:

$(***)_1$

$$\mu^{\underline{o}(1)}(L,\lambda) = -(k+1)\cdot r_n + \sum_{i=0}^{n-1} [\dim L \cap L_{(i+1,\ldots,n)} + 1] \cdot (r_{i+1} - r_i).$$

By Proposition 2.4, we know that this formula must hold even under the assumptions $r_0 \geq r_1 \geq \cdots \geq r_n$.

Now suppose that $(L^{(1)}, \ldots, L^{(N)})$ is a geometric point of $V = [Grass_{k,n}]^N$. Let $\varrho(1)$ denote the $SL(n+1)$-linearized ample sheaf:

$$\bigotimes_{i=1}^{N} p_i^*(\varrho(1)).$$

Then, as in § 1, we deduce:

$$(***)_N \qquad \mu^{\varrho(1)}((L^{(1)}, \ldots, L^{(N)}), \lambda) = -(k+1) \cdot N \cdot r_n + \sum_{i=0}^{n-1} \left\{ \sum_{j=1}^{N} \dim L^{(j)} \cap L_{(i+1,\ldots,n)} + 1 \right\} \cdot (r_{i+1} - r_i).$$

This formula being a *linear* function of the r_i, its value is positive for all r_i satisfying $r_0 \geq \cdots \geq r_n$ and $\sum r_i = 0$ if and only if it is positive for the extreme sets of r_i's:

$$n - p = r_0 = r_1 = \cdots = r_p; r_{p+1} = \cdots = r_n = -(p+1).$$

Therefore $\mu > 0$ for such r_i if and only if

$$\sum_{i=1}^{N} (\dim L^{(i)} \cap L_{(p+1,\ldots,n)} + 1) < \frac{(p+1)(k+1)N}{n+1}$$

for all p. Since every 1-*PS* λ of $\overline{SL(n+1)}$ is conjugate to a 1-*PS* which is diagonalized and for which $r_0 \geq \cdots \geq r_n$, we deduce from Theorem 2.1 that:

Proposition 4.3. The geometric point $(L^{(1)}, \ldots, L^{(N)})$ of V is properly stable (resp. semi-stable) with respect to $\varrho(1)$ if and only if, for all linear subspaces $L \subset \bar{P}_n$,

$$\frac{\sum_{i=1}^{N} (\dim L^{(i)} \cap L + 1)}{N(k+1)} < \frac{\dim L + 1}{n+1}$$

(resp. $\leq$ in same equation).

§ 5. The projective adjoint action

In this section, we shall merely relate some results of KOSTANT to the present treatment. Suppose that G is a semi-simple group of adjoint type, i.e. acting faithfully on its lie algebra $\mathfrak{g}$. Let $P(\mathfrak{g}^*)$ denote the projective space of 1-dimensional subspaces of $\mathfrak{g}$. We are interested in the action of G on $P(\mathfrak{g}^*)$, which we call the *projective adjoint action*. Note that the sheaf $\varrho(1)$ on $P(\mathfrak{g}^*)$ is G-linearized since G acts on the cone over $P(\mathfrak{g}^*)$. The following is one of Kostant's results:

Proposition 4.4. Let x be a geometric point of $P(\mathfrak{g}^*)$, and let x^* be an element of the lie algebra over x (i.e. of the lie algebra over Ω). Then:

(i) x is stable $\Leftrightarrow x^*$ is a regular semi-simple element, i.e. it lies in a Cartan subalgebra and in the interior of some Weyl chamber.

(ii) x is semi-stable $\Leftrightarrow x^*$ is not nilpotent.

(cf. [20], and use Proposition 2.2).

§ 6. Space curves

In this section, we shall investigate the stability of the chow point of a non-singular space curve. For the sake of simplicity, we work in this section over an algebraically closed field k. Suppose we are dealing in general, with r dimensional subvarieties of $\boldsymbol{P}_n$. As above, $PGL(n+1)$ and its covering $SL(n+1)$ act on $\boldsymbol{P}_n$, and the action of $SL(n+1)$ is induced by its linear representation in

$$V = H^0(\boldsymbol{P}_n, \mathfrak{o}_{\boldsymbol{P}}(1)).$$

Let * denote the operation of taking the dual vector space. Via skew-symmetrization, there is a canonical map

$$\Lambda^{r+1}(V^*) \xrightarrow{\phi_1} \overbrace{V^* \otimes V^* \otimes \cdots \otimes V^*}^{(r+1)\times}.$$

This induces a map

$$S^d[\Lambda^{r+1}(V^*)] \xrightarrow{\phi_d} \overbrace{S^d V^* \otimes \cdots \otimes S^d V^*}^{(r+1)\times}.$$

The image of ϕ_d is a vector space $V_{d,r}$ on which the induced representation of $SL(n+1)$ is irreducible. Chow forms are elements of $V_{d,r}$, and chow points are k-rational points of $\boldsymbol{P}(V^*_{d,r})$, i.e. one-dimensional subspaces of $V_{d,r}$. Let $\check{\boldsymbol{P}}_n = \boldsymbol{P}(V^*)$: the projective space dual to $\boldsymbol{P}_n$. Then since

$$(\check{\boldsymbol{P}}_n)^{r+1} = \operatorname{Proj} \sum_{d=0}^{\infty} \overbrace{(S^d V^*) \otimes \cdots \otimes (S^d V^*)}^{(r+1)\times}$$

it follows that a chow point can also be considered as a divisor on $(\check{\boldsymbol{P}}_n)^{r+1}$ of a suitable type. The exact procedure for determining the chow form of a cycle X of dimension r will not be repeated here (compare, however, § 5.4 and [38]). We will need to know only (i) that formation of the chow form is compatible with specialization, and (ii) the chow form of a cycle of linear subspaces. To compute the latter, let $X_0, \ldots, X_n$ be a basis of V, let $U_0, \ldots, U_n$ be a dual basis of V^*, and let monomials

$$\bigotimes_{j=0}^{r} \prod_{i=0}^{n} [U_i^{(j)}]^{r_{ij}}$$

such that $\left(\sum_{i=0}^{n} r_{ij}\right) = d$, for all j, be a basis of

$$\overbrace{(S^d V^*) \otimes \cdots \otimes (S^d V^*)}^{(r+1)\times}.$$

Let L_I be the linear space defined by the equations $X_i = 0$, $i \in I$. Then:

$$\begin{bmatrix}\text{Chow form of}\\ \sum a_k L_{I_k}\end{bmatrix} = \prod_k \begin{bmatrix}\text{Chow form}\\ \text{of } L_{I_k}\end{bmatrix}^{a_k}$$
$$= \prod_k \phi_1\Big(\bigwedge_{i \notin I_k} U_i\Big)^{a_k}$$
$$= \prod_k \begin{bmatrix}\det\limits_{\substack{0 \le j \le r\\ i \notin I_k}} U_i^{(j)}\end{bmatrix}^{a_k}.$$

In particular, suppose λ is the 1-*PS* of $SL(n+1)$ given by the diagonal matrix $(\alpha^{r_i} \cdot \delta_{ij})_{0 \le i \, j \le n}$. Then the μ for the corresponding chow point and this λ is given by Proposition 2.3 as:

$$\mu\Big(\sum_k a_k L_{I_k}; \lambda\Big) = -\sum_k \sum_{i \notin I_k} a_k \cdot r_i.$$

In particular, if P_i is the point $L_{\{0\,1 \ldots \hat{i} \ldots n\}}$ and l_{ij} is the line joining P_i and P_j, then

$$\mu\Big(\sum_k a_k l_{i_k, j_k}; \lambda\Big) = -\sum_k a_k (r_{i_k} + r_{j_k}).$$

We now consider the second problem: given a non-singular curve γ in $\boldsymbol{P}_n$, find the cycle Z in $\boldsymbol{P}_n$ which is the specialization of the transform of γ by $\lambda(\alpha)$, as $\alpha \to 0$. In almost all cases, Z will be a sum of lines $l_{i,j}$, and then by Property iv), § 2.1, we will be able to compute $\mu(\gamma; \lambda)$. As in § 4, suppose that:

$$r_0 > r_1 > \cdots > r_n,$$

and recall that:

$$\left\| \begin{array}{l}\text{Points on } L_{(i,i+1,\ldots,n)}, \text{ but not on}\\ L_{(i-1,i,\ldots,n)} \text{ spezialize to } P_{i-1}, \text{ when } \alpha \to 0.\end{array}\right.$$

We shall assume that γ is not contained in any hyperplane. Therefore, in particular, $\gamma \not\subset L_{(n)}$, and almost all points of γ specialize to P_n. It is fairly clear that the specialization Z of γ itself depends only on the structure of γ at the finite set of points $\gamma \cap L_{(n)}$. In fact, let $\sigma: PGL(n) \times \boldsymbol{P}_n \to \boldsymbol{P}_n$ be the usual action, and consider the composition F_0:

$$(\boldsymbol{G}_m \times \gamma) \xrightarrow{\lambda \times I} (PGL(n) \times \boldsymbol{P}_n) \xrightarrow{\sigma} \boldsymbol{P}_n.$$

(I denoting the inclusion of γ in $\boldsymbol{P}_n$.) Consider $\boldsymbol{G}_m$ as a subset of $\boldsymbol{A}^1$ by adding the point $\alpha = 0$. Then F_0 defines a rational map F from $\boldsymbol{A}^1 \times \gamma$ to $\boldsymbol{P}_n$ with fundamental points:

$$\{0\} \times Q_i, \text{ where } Q_1 \cup \cdots \cup Q_\nu = I^{-1}(L_{(n)}).$$

In fact, F is a morphism at every point $\{0\} \times Q$, such that $I(Q) \notin L_{(n)}$, and its image is P_n. Suppose, as a rational map, F is given at $\{0\} \times Q_i$ by equations

$X_0 = f_0; X_1 = f_1; \ldots; X_n = f_n; f_i \in \mathfrak{o}_i$, the local ring of $\boldsymbol{A}^1 \times \gamma$ at $\{0\} \times Q_i$.

Let $\mathfrak{A}_i = (f_0, f_1, \ldots, f_n)$. Since $\{0\}\times Q_i$ is an isolated fundamental point of F, $\mathfrak{A}_i$ is primary to the maximal ideal in $\mathfrak{o}_i$. Define V to be the surface obtained by normalizing the result of blowing up all the ideals $\mathfrak{A}_i$ (cf. EGA 2, § 8.1), and let $\bar{\omega}: V \to A^1\times\gamma$ be the projection. An open subset $V_0 \subset V$ is isomorphic to $A^1\times\gamma - \bigcup\{0\}\times Q_i$, and there is a *morphism* F' from V to P_n such that the diagram

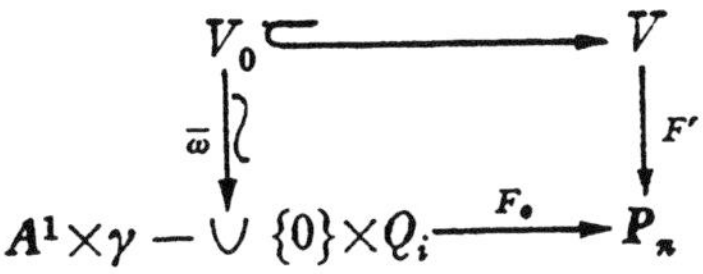

$$\begin{array}{ccc} V_0 & \hookrightarrow & V \\ \bar{\omega}\downarrow\wr & & \downarrow F' \\ A^1\times\gamma - \bigcup\{0\}\times Q_i & \xrightarrow{F_0} & P_n \end{array}$$

commutes. Then the specialization Z of γ can be evaluated via F'. For, on V, the divisors $\{x\}\times\gamma$, $x \in (G_m)_k$ specialize when $x \to 0$ to the Cartier divisor $(\alpha)_V$ given by the zeroes of the function α (α being the coordinate on A^1). Therefore $Z = F'((\alpha)_V)$. In fact, $(\alpha)_V$ is the union of the *proper* transform $\bar{\omega}^{-1}(\{0\}\times\gamma)$ on V, and of divisors $\mathcal{E}_i$ with support on the exceptional locus $\bar{\omega}^{-1}(Q_i)$. Since F takes almost all of $\{0\}\times\gamma$ to a point, in fact:

$$Z = \sum_i F'(\mathcal{E}_i).$$

To describe $F'(\mathcal{E}_i)$ more explicitly, it is convenient to introduce a formal power series representation for I at the points Q_i. In fact, let t_i be a uniformizing parameter at Q_i on γ. Then suppose that I is given by

$$X_0 = (a_0^{(i)} \cdot t_i^{s_{0,i}}) + \text{higher terms}, \quad a_0^{(i)} \neq 0$$

$$X_1 = (a_1^{(i)} \cdot t_i^{s_{1,i}}) + \text{higher terms}, \quad a_1^{(i)} \neq 0$$

$$\vdots$$

$$X_n = (a_n^{(i)} \cdot t_i^{s_{n,i}}) + \text{higher terms}, \quad a_n^{(i)} \neq 0.$$

Note that since I is a *morphism*, for some k, $s_{k,i} = 0$. Therefore, the rational map F is given at $(0)\times Q_i$ by equations:

$$X_k = a_k^{(i)} \cdot \alpha^{r_k} \cdot (t_i^{s_{k,i}} + \text{higher terms})$$

in terms of the uniformizing parameters α, t_i at $(0)\times Q_i$ on $A^1\times\gamma$, and $\mathfrak{A}_i = (\alpha^{r_0} \cdot t_i^{s_{0,i}}, \ldots, \alpha^{r_n} \cdot t_i^{s_{n,i}})$.

We require now a general fact:

Lemma. Let V_0 be a non-singular surface over k, and let P be a closed point of V_0. Let x and y be uniformizing parameters at P on V_0, and let

$$\mathfrak{A} = (x^{r_0} \cdot y^{s_0}, \ldots, x^{r_n} \cdot y^{s_n})$$

where r_i, s_i are integers. For all positive rational numbers $a = p/q$ (p, q relatively prime), let v_a be the discrete, rank 1 valuation centered at P such that

$$v_a(\sum a_{ij}\, x^i \cdot y^j) = \min_{a_{ij}\neq 0} (i \cdot p + j \cdot q).$$

Let V be the normalization of the surface obtained by blowing up $\mathfrak{A}$. Then the exceptional divisors E_a on V are exactly those prime divisors of $k(V_0)$ corresponding to valuations ν_a where

(*) ‖ The least integer in the sequence of integers $r_i p + s_i q$ $(0 \le i \le n)$, occurs at least twice.

Moreover, $x^q \cdot y^{-p}$ generates the function field of E_a.

Proof. In order that the center of a valuation ν on V be positive dimensional, it is necessary and sufficient that its center on one of the surfaces:

$$\operatorname{Spec} \mathfrak{o}_{P,V_0}\left[\frac{x^{r_0} \cdot y^{s_0}}{x^{r_i} \cdot y^{s_i}}, \ldots, \frac{x^{r_n} \cdot y^{s_n}}{x^{r_i} \cdot y^{s_i}}\right]$$

be positive dimensional. Since $\nu_a(x^{r_i} \cdot y^{s_i}) = r_i p + s_i q$, condition (*) is clearly equivalent to whether or not ν_a has a positive dimensional center on one of these surfaces. But I claim that no discrete, rank 1 valuation other than a ν_a could appear on some V: to see this, pass to $\tilde{V}_0 = \operatorname{Spec} \hat{\mathfrak{o}}_{P,V_0}$ ($\hat{}$ representing completion). Then $\mathfrak{A}$ is invariant under the group of automorphisms $(x, y) \to (\alpha x, \beta y)$ of $\hat{\mathfrak{o}}_{P,V_0}$. But it is easy to check that the only discrete, rank 1 valuations invariant under this group are the ν_a. *QED*.

A suggestive way to present this lemma is to graph the points (r_i, s_i) in a plane, and then to form their "Newton polygon". Then there is one exceptional divisor on V for each edge of the Newton polygon.

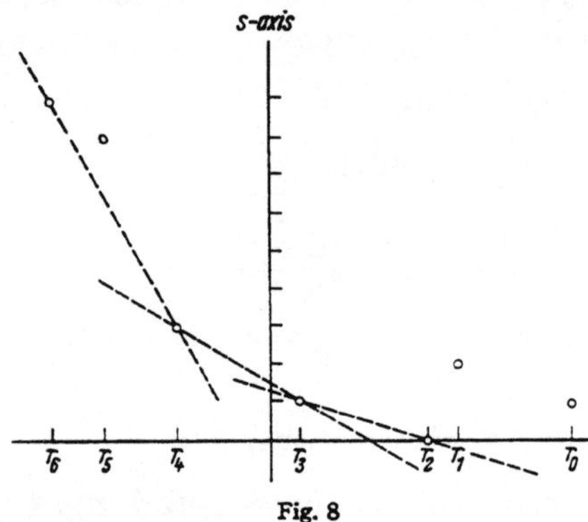

Fig. 8

Newton Polygon of:
$X_0 = t + \cdots$
$X_1 = t^2 + \cdots$
$X_2 = 1$
$X_3 = t + \cdots$
$X_4 = t^3 + \cdots$
$X_5 = t^8 + \cdots$
$X_6 = t^9 + \cdots$

Limit cycle:
$Z = l_{23} + 2l_{34} + 6l_{46}$

Returning to the computation of Z, $F'(\mathcal{E}_i)$ will be determined by forming the Newton polygon of the set of points $(r_k, s_{k,i})$, $0 \le k \le n$. Note that if we exclude the r_k's satisfying some particular linear equations, this polygon will consist of edges through only 2 points $(r_k, s_{k,i})$ at a time. *Assuming this*, divisors E_a appear on V corresponding to ν_a, when $a = p/q$, and for some e, f:

$$(*) \qquad (p r_e + q s_{e,i}) = (p r_f + q s_{f,i}) < (p r_k + q s_{k,i}), \; (k \ne e, f).$$

Moreover:

$$(\alpha)_V = \sum_{\substack{\text{such}\\ a}} v_a(\alpha)\cdot E_a + (\text{proper transf. of } \{0\}\times\gamma)$$

$$= \sum_{\substack{\text{such}\\ a}} p\cdot E_a + (\text{proper transf. of } \{0\}\times\gamma).$$

What is $F'(E_a)$? As a subset of $\boldsymbol{P}_n$, it must be the center of induced valuation v'_a of $F'(V) \subset \boldsymbol{P}_n$. By (*), this satisfies

$$v'_a(X_e/X_f) = v_a(\alpha^{r_e}\cdot t_i^{s_{e,i}}/\alpha^{r_f}\cdot t_i^{s_{f,i}}) = 0$$

$$v'_a(X_k/X_f) = v_a(\alpha^{r_k}\cdot t_i^{s_{e,i}}/\alpha^{r_f}\cdot t_i^{s_{f,i}}) > 0$$

$$(k \neq e, f).$$

Therefore this center is precisely the line $l_{e,f}$ joining P_e and P_f. Moreover

$$(X_e/X_f) = (\alpha^{r_e - r_f})\cdot(t_i^{s_{e,i}-s_{f,i}})$$

generates the function field of $l_{e,f}$, and,

$$\alpha^q\cdot t_i^{-p}$$

generates the function field of E_a. Therefore $F'(E_a)$ equals $l_{e,f}$ with multiplicity:

$$\left|\frac{r_e - r_f}{q}\right| = \left|\frac{s_{e,i} - s_{f,i}}{p}\right|.$$

Putting all this together, we conclude:

$$Z = \sum_{i=1}^{\nu}\left\{\sum_{\substack{\text{pairs } (e,f)\\ \text{satisfying } (*)}} |s_{e,i} - s_{f,i}|\cdot l_{e,f}\right\}.$$

Interpreting (*) via the Newton polygon, and recalling that $r_0 > r_1 > \cdots > r_n$, one finds that the pairs (e, f) satisfying (*) are precisely the consecutive pairs of e's in a sequence $\nu_i = e(1) < e(2) < \cdots < e(\mu) = n$ where ν_i is the largest integer ν such that $s_{\nu,i} = 0$ (cf. Fig. 8). Thus the expression in the above box for each i is a chain of straight lines (with multiplicity) from P_{ν_i} to P_n, where P_{ν_i} is the specialization of Q_i under $\lambda(\alpha)$, when $\alpha \to 0$.

This gives an intuitive picture according to which γ specializes under $\lambda(\alpha)$ to a set of cycles, one for each point Q of $\gamma \cap L_{(n)}$, and that each cycle is a sort of blown up picture of the infinitesimal structure of γ near Q.

Putting together this expression for Z with the previous results on the μ of the Chow point of cycles of lines, we obtain:

$$\mu(\gamma;\lambda) = -\sum_{i=1}^{\nu}\sum_{\substack{\text{pairs } (e,f)\\ \text{satisfying } (*)\\ e>f}} (s_{e,i} - s_{f,i})\cdot(r_e + r_f)$$

(where we have written γ for "the chow point of γ" in μ). But the terms of this sum for each i have a very simple interpretation via the Newton polygon: If this polygon is rotated 90° so that the s-axis is horizontal, this expression is the area under the Newton polygon (with sign, i.e. counted negatively below the s-axis). But the Newton polygon is precisely the polygon which minimizes this area. Therefore $\mu(\gamma; \lambda)$ equals:

$$-\sum_{i=1}^{\nu} \underset{r_i=e(1)<e(2)<\cdots<e(\mu)=n}{\underset{\text{all sequences}}{\operatorname{Min}}} \left[\sum_{j=2}^{\mu} (s_{e(j),i} - s_{e(j-1),i}) \cdot (r_{e(j)} + r_{e(j-1)})\right].$$

Actually, we have proven this formula only when $r_0 > r_1 > \cdots > r_n$, and when the r_i's do not satisfy some finite set of linear equations. But by Proposition 2.14, and the nature of the right hand expression, it follows that it holds for any 1-PS λ such that $r_0 \geq r_1 \geq \cdots \geq r_n$.

The next step is to rearrange this formula to give a *lower* bound for μ. In fact, putting $s'_{k,i} = \min_{e\geq k} (s_{e,i})$:

$$\mu(\gamma; \lambda) = -\sum_{i=1}^{\nu} \underset{0=k(1)<\cdots<k(\mu)=n}{\underset{\text{all sequences}}{\operatorname{Min}}} \sum_{j=2}^{\mu} (s'_{k(j),i} - s'_{k(j-1),i}) \cdot (r_{k(j)} + r_{k(j-1)})$$

$$\geq \underset{0=k(1)<\cdots<k(\mu)=n}{\underset{\text{all sequences}}{\operatorname{Max}}} -\sum_{j=2}^{\mu}\left[\sum_{i=1}^{\nu} s'_{k(j),i} - \sum_{i=1}^{\nu} s'_{k(j-1),i}\right] \cdot (r_{k(j)} + r_{k(j-1)}).$$

For all k, let $e_k = \sum_{i=1}^{\nu} s'_{k,i}$. This has a very simple interpretation:

$s'_{k,i} = I(Q_i; \gamma \cdot H)$ where H is a "generic" hyperplane of the form $\sum_{j=k}^{n} a_j X_j$ hence

> $e_k =$ intersection multiplicity along $L_{(k,k+1,\ldots,n)}$ of γ and a "generic" hyperplane H containing $L_{(k,k+1,\ldots,n)}$.

To show that this lower bound is sometimes positive, we need a combinatorial fact:

Lemma. If the integers e_i satisfy:

i) $e_0 = 0$, $e_n = n + g$, $e_0 \leq e_1 \leq \cdots \leq e_n$,
ii) $e_i \leq i$ for $i = 0, 1, \ldots, n - g$,
iii) $e_i \leq i + g$ for $i = n - g + 1, \ldots, n$,

and if $n \geq 2g$ and $g \geq 1$, then for all real numbers r_i such that $r_0 \geq \cdots \geq r_n$, $\sum_{i=0}^{n} r_i = 0$, and not all $r_i = 0$,

$$\underset{0=k(1)<\cdots<k(\mu)=n}{\underset{\text{all sequences}}{\operatorname{Max}}} \left\{-\sum_{j=2}^{\mu} (e_{k(j)} - e_{k(j-1)}) \cdot (r_{k(j)} + r_{k(j-1)})\right\} > 0.$$

Proof. It is easy to check that this Max only decreases if we replace the e_i by their upper bounds $e_i = i$ or $i + g$. Let $\Sigma_0 \Sigma_1, \ldots, \Sigma_g$ stand for the sums in the above expression for the sequences of $k(j)$:

In Σ_0: $0, 1, \ldots, n$,
In Σ_1: $0, 1, \ldots, n-g-1, n-g+2, \ldots, n$,
In Σ_2: $0, 1, \ldots, n-g-2, n-g+3, \ldots, n$,
. . . .
In Σ_{g-1}: $0, 1, \ldots, n-2g+1, n$,
In Σ_g: $0, n$.

Multiply Σ_i by $\frac{1}{(g+2i)(g+2i+2)}$ for $0 \leq i \leq g-1$, multiply Σ_g by $1/3g(n+g)$ and add these up. The result is found to be

$$\frac{1}{3g}\{r_{n-2g+1} + r_{n-2g+2} + \cdots + r_{n-1} + r_n\}.$$

Under our hypothesis on n and the r_i, this is strictly negative. Therefore one of the Σ_i is strictly negative, and the Max of the lemma is positive. *QED*.

This prepares the way for:

Theorem 4.5. Let γ be a non-singular curve in $\boldsymbol{P}_n$. Then the chow point of γ is stable with respect to the action of $PGL(n+1)$ if

i) γ is not contained in any hyperplane,
ii) the embedding of γ in $\boldsymbol{P}_n$ is defined by a complete linear system,
iii) the genus g of γ is at least 1, and the degree of γ is at least $3g$.

Proof. Since the degree of γ is at least $3g$, the embedding of γ in $\boldsymbol{P}_n$ is defined by a non-special linear system, i.e. by $H^0(\gamma, L)$ where L is an invertible sheaf on γ and $H^1(\gamma, L) = (0)$. Therefore, this degree equals $n+g$ by the Riemann-Roch theorem, and $n \geq 2g$. By the above lemma, the calculation of $\mu(\gamma, \lambda)$, and Theorem 2.1, the theorem follows provided that:

(*) Given any flag $L_0 \subset L_1 \subset \cdots \subset L_{n-1} \subset \boldsymbol{P}_n$, where L_i is a linear space of dimension i, let e_k be the intersection multiplicity along L_{k-1} of γ and a generic hyperplane H containing L_{k-1}. Then these e_k satisfy ii) and iii) in the above lemma.

But (*) is essentially the Weierstrass Gap Theorem. To prove it, let the hyperplanes containing L_{n-k} correspond to the k-dimensional subspace $A_k \subset H^0(\gamma, L)$. Then e_k is the degree of the divisor D_k of base points of A_{n-k+1}, i.e. the subscheme of γ defined by $s=0$ for all $s \in A_{n-k+1}$. This means that $H^0(\gamma, L(-D_k))$ has dimension at least $n-k+1$, while $L(-D_k)$ has degree $n+g-e_k$. Therefore, by the Riemann-Roch theorem:

$$\begin{aligned} n-k+1 &\leq \dim H^0(\gamma, L(-D_k)) \\ &= n-e_k+1+\dim H^1(\gamma, L(-D_k)). \end{aligned}$$

Since $L(-D_k)$ has global sections, $\dim H^1(\gamma, L(-D_k))$ is at most g, and this proves $e_k \leq k + g$ for all k. Moreover, if $L(-D_k)$ has more than $(g+1)$ independent global sections, $\dim H^1(\gamma, L(-D_k)) = 0$; hence if $k \leq n - g$, it follows that $e_k \leq k$. *QED.*

It is an interesting problem to try to give more general conditions under which a non-singular curve in $\boldsymbol{P}_n$ and spanning $\boldsymbol{P}_n$ has a stable chow point. In contrast to Theorem 4.5, this is true also if $n = 2$ or 3, and the genus g is at least 1. However, there are elliptic curves in $\boldsymbol{P}_4$ with unstable chow points.

Chapter 5

The problem of moduli — 1st construction

§ 1. General discussion

What are moduli? Classically Riemann claimed that $3g - 3$ parameters could be defined for every Riemann surface of genus g which would determine its conformal structure. Put differently — and more generally — perhaps the problem is "classify the set of all complex structures on a given differentiable manifold, modulo isomorphisms". Or, in an algebraic setting: given some kind of variety (non-singular? normal?), classify the set of all varieties having something in common with the given one (same numerical invariants of some kind? belonging to a common algebraic family?). We shall be concerned here only with this algebraic form of the problem. This algebraic setting dictates the meaning to be given to the word "classify". Obviously, one will want to classify whatever selected objects one is dealing with by another algebraic object: perhaps a scheme, perhaps a variety, or perhaps a functor of more general type such as a Q-variety. But a choice appears at this point:

(1) if one stays close to the classical intuition, the problem posed is: find a scheme M whose geometric points over an algebraically closed field k are in a natural one-one correspondence with the set of selected objects defined over k. The word natural can be interpreted to mean, e.g. that given any family of the selected objects defined over a scheme S, there is a morphism from S to M mapping every geometric point s of S to that point of M which corresponds to the object in the family over the point s.

(2) if one is tempted by the algebra, one tries to classify also the set of selected objects rational over an arbitrary field, or defined over a ring, or over a general scheme (i.e. a family of the objects over that scheme). Then the problem becomes: find a universal family of the

selected objects defined over a scheme M, such that every other family, say defined over S, is induced from the universal one by a unique morphism from S to M.

The first will be called the *coarse moduli problem*; the second will be called the *fine moduli problem*. In practice, one of the difficulties of the present state of knowledge is that one knows very few ways in which to pose a plausible fine moduli problem.* The fine moduli problem seems otherwize not to be essentially harder (provided one has enough descent machinery!). On the other hand, quite a bit has been done to clarify what "selected objects" are suitable for a coarse moduli problem. A very fundamental discovery is that the natural objects to classify are not "bare" varieties, but *polarized* varieties. For example, suppose a scheme S parametrizes a family of abelian subvarieties of a fixed $\boldsymbol{P}_n$. Then isomorphism between abelian varieties defines an equivalence relation on S. In general, this equivalence relation is extremely bad! For example, it may be an infinite union of algebraic subvarieties of $S \times S$. This pathology is largely eliminated by considering polarized varieties instead.

Definition 5.1. Let X be a scheme, proper over an algebraically closed field k. The subgroup $Pic^\tau(X)$ of $Pic(X)$ is the set of all invertible sheaves L on X such that, for some integer $n \neq 0$, L^n represents a point of the Picard scheme of X in the same component as the identity.

Definition 5.2. Let X be a scheme, proper over an algebraically closed field k. An *inhomogeneous polarization* of X is a coset U of $Pic^\tau(X)$ in $Pic(X)$ consisting of ample invertible sheaves. A *homogeneous polarization* of X is a subset V of $Pic(X)$ containing an inhomogeneous polarization U and consisting exactly of the set of $\lambda \in Pic(X)$ such that

$$n\lambda \in m \cdot U$$

for some positive integers n, m.†

Suppose we wish to classify (coarsely) the set of *all* inhomogeneously polarized non-singular varieties. There are three problems that arise:

(a) it still has a non-separated topology,

(b) it has infinitely many components,

(c) it might happen that some pieces of this set look nice in all

---

* As, for example, a fine moduli scheme for curves is made possible by the concept of a "level n structure" on a curve.

† At least for the moduli of abelian varieties, polarizations seen to be essential to have a reasonable moduli space. This can be seen from the pathologies found by HAYASHIDA-NISHI [135], and NISHII [29] (contrast however NARASIMHAN-NORI [230]). On many varieties, the canonical class defines a canonical polarization (e.g. canonical models of surfaces of general type). An interesting question is whether there is a moduli space for the set of varieties of general type modulo *birational* equivalence.

topological senses, yet fail to be schemes, (i.e. they are not *stable*, as in § 4.3).

Actually, (a) is not too serious: the author and MATSUSAKA have shown ([24]) that if one looks only at non-singular inhomogeneously polarized varieties V, not birational to a variety $\boldsymbol{P}_1 \times V_0$, then the result has a separated topology. This follows from an observation of M. ARTIN. Also (b) is not very frightening: there are never infinitely many components through one point, and it may be that the well-known numerical invariants of varieties divide the set of isomorphism classes into disjoint open pieces each with only a finite number of components. For non-singular *surfaces*, this is also established in [24]. However, in both (a) and (b), the non-singularity hypothesis seems quite unnatural, and its significance quite obscure. Any search for *compact* moduli schemes will certainly require one to drop this condition.

Problem (c) seems to be the hardest. It was to develop techniques for attacking (c) that *I* worked out the theory in this book. For once (a) and (b) have been solved for some selected objects, (c) becomes essentially equivalent to the question of whether an orbit space of some locally closed subset of the Hilbert or Chow schemes by the projective group exists. It may or may not be necessary to impose an extra condition of *stability*, and to get moduli only for those objects which are stable. For the extremely non-separated case of ruled surfaces, it turns out that an appropriate stability condition solves (a), (b), and (c) all in one stroke, (cf. [26], [39]).

§ 2. Moduli as an orbit space

In this section, we will analyze the situation considered in the last section for the case of non-singular curves. We first give the problem of moduli a precise statement.

Definition 5.3. Let S be any noetherian scheme. A *curve* of genus g over S is a morphism $\pi: \Gamma \to S$ which is smooth and proper and whose geometric fibres are irreducible curves of genus g.

Definition 5.4. Let S be any noetherian scheme. Then $\mathcal{M}_g(S)$ is the set of all curves of genus g over S, modulo isomorphism.

Note that the collection of sets $\mathcal{M}_g(S)$ forms a contravariant functor from the category of noetherian schemes to the category of sets: i.e. given any morphism $f: T \to S$, a map

$$\mathcal{M}_g(f): \mathcal{M}_g(S) \to \mathcal{M}_g(T)$$

is defined by associating to $\pi: \Gamma \to S$, the "pull-back" curve $p_2: \Gamma \times_S T \to T$.

Definition 5.5. If $\mathcal{M}_g$ is represented by a scheme M (cf. EGA 0, § 8), then M is called the *fine moduli scheme.*

This definition is, of course, vacuous, and is inserted only to illustrate our general ideas. What is useful is:

Definition 5.6. A scheme M and a morphism ϕ from the functor $\mathcal{M}_g$ to the functor $h_M(S) = \mathrm{Hom}\,(S, M)$ represented by M is called a *coarse moduli scheme* if

(i) for all algebraically closed fields Ω, the map $\phi\,(\mathrm{Spec}\,\Omega)$: $\mathcal{M}_g(\mathrm{Spec}\,\Omega) \to h_M(\mathrm{Spec}\,\Omega)$ is an isomorphism.

(ii) given any scheme N, and any morphism ψ from $\mathcal{M}_g$ to the representable functor h_N, there is a unique morphism $\chi: h_M \to h_N$ such that $\psi = \chi \circ \phi$.

Recall that if $g \geq 2$, then for any curve Γ/S of genus g, the sheaf of relative differential forms $\Omega^1_{\Gamma/S}$ is relatively ample: therefore such curves are endowed with a canonical inhomogeneous polarization, and $\mathcal{M}_g$ is a moduli functor of the type envisioned in § 1. We assume that $g \geq 2$ from now on. In the remainder of this section, we shall describe the essential connection between the functor $\mathcal{M}_g$ and the part of the Hilbert scheme parametrizing pluricanonical curves, solving problems (a) and (b) of § 1, and reducing (c) to a question of stability. In this, $\boldsymbol{P}_n$, $PGL(n+1)$, etc. denote the usual schemes over $\boldsymbol{Z}$. For details on the Hilbert scheme, cf. [13], exposé 221, [40], and Ch. 0, § 5, (c).

Fix a positive integer $\nu \geqq 3$. Put

$$P(x) = (2x \cdot \nu - 1) \cdot (g - 1)$$

$$n = P(1) - 1 = \nu \cdot (2g - 2) - g.$$

Proposition 5.1. There is a unique subscheme H_ν in the Hilbert scheme $Hilb^{P(x)}_{\boldsymbol{P}_n}$ such that, for any morphism $f: S \to Hilb^{P(x)}_{\boldsymbol{P}_n}$, f factors through H_ν if and only if:

i) the induced subscheme Γ in $S \times \boldsymbol{P}_n$ is a curve of genus g over S,

ii) the invertible sheaf on Γ induced by $\underline{o}_{\boldsymbol{P}_n}(1)$ is isomorphic to:

$$(\Omega^1_{\Gamma/S})^\nu \otimes p_1^*(L)$$

for a suitable invertible sheaf L on S,

iii) for every geometric point s of S, the fibre Γ_s of Γ over s spans the projective space $\bar{\boldsymbol{P}}_n$. Hence, by our assumptions on n and $P(x)$, Γ_s is a ν-canonical curve in $\bar{\boldsymbol{P}}_n$.

In the language of representable functors, the Proposition asserts that conditions i), ii), and iii) are represented by the subscheme H_ν.

Proof. Condition i) amounts to asking that Γ be smooth over S, and with connected geometric fibres, (its dimension and arithmetic genus being already specified by the Hilbert polynomial $P(x)$). But this condition is realized precisely on an open subset $U_1 \subset Hilb_{\mathbf{P}_n}^{P(x)}$. For the smoothness of Γ over S, cf. SGA 2, Proposition 1.1, and Theorem 2.1; for the connectedness of the geometric fibres, use the Stein factorization $\Gamma \to S' \to S$ and note that $S' \to S$ is an isomorphism in an open set.

To analyze condition (ii), let L_1 be the restriction of $\varrho_{U_1} \otimes \varrho_{\mathbf{P}_n}(1)$ to the closed subscheme $\Gamma_1 \subset U_1 \times \mathbf{P}_n$ which is the universal curve over U_1. Recall that the Picard scheme $Pic(\Gamma_1/U_1)$ of any curve exists (cf. § 0.5, (d)). Then L_1 and $(\Omega^1_{\Gamma_1/U_1})^\nu$ both define morphisms

$$\lambda_1, \omega^\nu : U_1 \to Pic\,(\Gamma_1/U_1).$$

Condition (ii) is equivalent to $\lambda_1 \circ f = \omega^\nu \circ f$, by the defining property of the Picard scheme. Let U_2 be the fibre product:

$$\begin{array}{ccc} U_2 & \longrightarrow & U_1 \\ \downarrow & & \downarrow{\scriptstyle (\lambda_1, \omega^\nu)} \\ Pic(\Gamma_1/U_1) & \xrightarrow{\Delta} & [Pic(\Gamma_1/U_1) \underset{U_1}{\times} Pic\,(\Gamma_1/U_1)] \end{array}$$

where Δ is the diagonal. Since $Pic(\Gamma_1/U_1)$ is separated over U_1 (cf. § 0.5, (d)), Δ is a closed immersion, and therefore U_2 is a closed subscheme of U_1. Clearly f factors through U_2 if and only if $\lambda_1 \circ f = \omega^\nu \circ f$, i.e. condition (ii) holds. Put

$$\Gamma_2 = \Gamma_1 \underset{U_1}{\times} U_2, \quad \text{and} \quad L_2 = (L_1 \otimes \varrho_{\Gamma_2}) = [\varrho_{U_2} \otimes \varrho_{\mathbf{P}_n}(1)] \otimes \varrho_{\Gamma_2}.$$

To analyze condition (iii), consider the canonical homomorphism:

$$H^0\big(\varrho_{\mathbf{P}_n}(1)\big) \otimes \varrho_{U_2} \xrightarrow{h} \overline{\omega}_{2,*}(L_2)$$

where $\overline{\omega}_2 : \Gamma_2 \to U_2$ is the projection. But $\overline{\omega}_{2,*}(L_2)$ is a locally free sheaf of rank $n+1$ such that, for all geometric points s: Spec $(\Omega) \to U_2$, if Γ_s is the geometric fibre of $\overline{\omega}_2$ over s, then

$$\begin{aligned} \overline{\omega}_{2,*}(L_2) \otimes \Omega &\cong H^0(\Gamma_s, L_2 \otimes \varrho_{\Gamma_s}) \\ &\cong H^0\big(\Gamma_s, (\Omega^1_{\Gamma_s/\Omega})^\nu\big). \end{aligned}$$

This follows because $H^1(\Gamma_s, (\Omega^1_{\Gamma_s/\Omega})^\nu) = 0$ for all s (cf. Ch. 0, § 5, (a)). Therefore, let $\mathcal{F}$ be the cokernel of h. Then, for all geometric points s, $\mathcal{F} \otimes \Omega$ is not zero if and only if the map:

$$H^0\big(\varrho_{\mathbf{P}_n}(1)\big) \otimes \Omega \xrightarrow{h_s} H^0\big(\Gamma_s, (\Omega^1_{\Gamma_s/\Omega})^\nu\big)$$

is not surjective. Since both are vector spaces of dimension $n+1$, this occurs if and only if h_s is not injective, i.e. if and only if Γ_s is contained in a hyperplane in $\bar{\mathbf{P}}_n$. Therefore, if U_3 is the open subset of U_2 where $\mathcal{F} = (0)$, U_3 is precisely the open set where condition (iii) is realized. We put $H_\nu = U_3$. *QED.*

Since conditions i), ii), and iii) are invariant under automorphisms of $\boldsymbol{P}_n$, the action of $PGL(n+1)$ on $Hilb_{\boldsymbol{P}_n}^{P(x)}$ restricts to an action of $PGL(n+1)$ on H_ν. Moreover, by condition i), there is a natural morphism of functors:

$$\pi : h_{H_\nu} \to \mathcal{M}_g.$$

Let $\mathcal{PGL}(n+1)$ be the functor represented by $PGL(n+1)$, i.e. $\mathcal{PGL}(n+1)(S) = \text{Hom}\big(S, PGL(n+1)\big)$. The the action of the group scheme $PGL(n+1)$ on the scheme H_ν induces an action of the functor $\mathcal{PGL}(n+1)$ on the functor h_{H_ν} which we write

$$\sigma : \mathcal{PGL}(n+1) \times h_{H_\nu} \to h_{H_\nu}$$

i.e. a morphism of the product functor on the left to the functor on the right which induces, for every S, an action of the group $\mathcal{PGL}(n+1)$ (S) on the set $h_{H_\nu}(S)$. Then it is obvious that

$$(*) \qquad \pi \circ \sigma = \pi \circ p_2$$

if p_2 is the projection morphism from $\mathcal{PGL}(n+1) \times h_{H_\nu}$ to h_{H_ν}.

Definition 5.7. For all noetherian schemes S, let $\mathcal{M}'_g(S)$ be the quotient of the set Hom (S, H_ν) by the action of the group Hom $\big(S, PGL(n+1)\big)$. Let $\mathcal{M}'_g$ be the functor defined by this collection of sets and by the obvious maps between them.

According to (*), the morphism π factors:

$$h_{H_\nu} \xrightarrow{\pi'} \mathcal{M}'_g \xrightarrow{I} \mathcal{M}_g.$$

Proposition 5.2. I is injective. And for any $\alpha \in \mathcal{M}_g(S)$, there is an open covering $\{U_i\}$ of S, such that the restriction of α to $\mathcal{M}_g(U_i)$, for all i, is in the image of I.

Proof. To prove the first statement, let ϕ_1 and ϕ_2 be morphisms of S to H_ν, and let Γ_1 and Γ_2 be the corresponding induced subschemes of $S \times \boldsymbol{P}_n$. Suppose that Γ_1 and Γ_2 are isomorphic as curves over S:

$$\begin{array}{ccc} \Gamma_1 & \xrightarrow[\psi]{\sim} & \Gamma_2 \\ & {\scriptstyle \bar\omega_1}\searrow \quad \swarrow{\scriptstyle \bar\omega_2} & \\ & S & \end{array}$$

Recall that by condition ii) of Proposition 5.1,

$$\varrho_{\Gamma_1} \otimes \varrho_{\boldsymbol{P}_n}(1) \cong (\Omega^1_{\Gamma_1/S})^\nu \otimes \bar\omega_1^*(L_1),$$

$$\varrho_{\Gamma_2} \otimes \varrho_{\boldsymbol{P}_n}(1) \cong (\Omega^1_{\Gamma_2/S})^\nu \otimes \bar\omega_2^*(L_2),$$

for suitable invertible sheaves L_1, L_2 on S, and where e.g. $\varrho_{\Gamma_1} \otimes \varrho_{\boldsymbol{P}_n}(1)$ is defined from the composition:

$$\Gamma_1 \subset S \times \boldsymbol{P}_n \xrightarrow{p_2} \boldsymbol{P}_n.$$

Recall that the natural homomorphisms:

$$H^0(\mathcal{O}_{\boldsymbol{P}_n}(1)) \otimes \mathcal{O}_S \xrightarrow{h_1} \bar{\omega}_{1,*}(\mathcal{O}_{\Gamma_1} \otimes \mathcal{O}_{\boldsymbol{P}_n}(1))$$

$$H^0(\mathcal{O}_{\boldsymbol{P}_n}(1)) \otimes \mathcal{O}_S \xrightarrow{h_2} \bar{\omega}_{2,*}(\mathcal{O}_{\Gamma_2} \otimes \mathcal{O}_{\boldsymbol{P}_n}(1))$$

induce isomorphisms over every geometric point by condition iii). As the sheaves on both sides are locally free (cf. Proof of Proposition 5.1), h_1 and h_2 are isomorphisms. Noting that ψ induces a compatible isomorphism $\psi^*: (\Omega^1_{\Gamma_2/S})^\nu \to (\Omega^1_{\Gamma_1/S})^\nu$, we obtain:

$$\begin{aligned} S \times \boldsymbol{P}_n &\cong \boldsymbol{P}[H^0(\mathcal{O}_{\boldsymbol{P}_n}(1)) \otimes \mathcal{O}_S] \\ &\cong \boldsymbol{P}[\bar{\omega}_{2,*}(\mathcal{O}_{\Gamma_2} \otimes \mathcal{O}_{\boldsymbol{P}_n}(1))] \\ &\cong \boldsymbol{P}[\bar{\omega}_{2,*}((\Omega^1_{\Gamma_2/S})^\nu)] \\ &\cong \boldsymbol{P}[\bar{\omega}_{1,*}((\Omega^1_{\Gamma_1/S})^\nu)] \\ &\cong \boldsymbol{P}[\bar{\omega}_{1,*}(\mathcal{O}_{\Gamma_1} \otimes \mathcal{O}_{\boldsymbol{P}_n}(1))] \\ &\cong \boldsymbol{P}[H^0(\mathcal{O}_{\boldsymbol{P}_n}(1)) \otimes \mathcal{O}_S] \\ &\cong S \times \boldsymbol{P}_n. \end{aligned}$$

By the functorial characterization of $PGL(n+1)$ (cf. Ch. 0, § 5, (b)), such an isomorphism is defined by an S-valued point β of $PGL(n+1)$. We leave it to the reader to check that

$$\sigma(\beta, \phi_2) = \phi_1.$$

To prove the second statement, let α correspond to the curve $\bar{\omega}: \Gamma \to S$. Then $\bar{\omega}_*((\Omega^1_{\Gamma/S})^\nu)$ is a locally free sheaf $\mathcal{E}_\nu$ on S, and for all geometric points s of S over fields Ω.

$$\mathcal{E}_\nu \underset{\mathcal{O}_S}{\otimes} \Omega \cong H^0(\Gamma_s, (\Omega^1_{\Gamma_s/\Omega})^\nu).$$

For example, this follows since $H^1(\Gamma_s, (\Omega^1_{\Gamma_s/\Omega})^\nu) = (0)$, (cf. Ch. 0, § 5, (a)). Let $\{U_i\}$ be an open covering such that $\mathcal{E}_\nu \mid U_i$ is a *free* sheaf for all i. Then

$$\boldsymbol{P}(\mathcal{E}_\nu \mid U_i) \cong U_i \times \boldsymbol{P}_n.$$

But since $\nu \geq 3$, there is a canonical closed immersion:

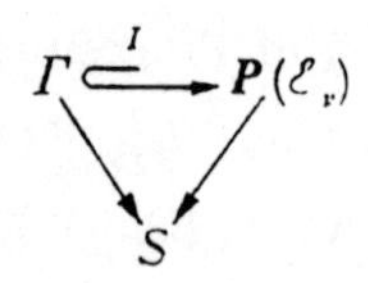

Namely, when S is the Spec of a field, this is a classical result in the theory of curves. In general, note first that

$$\bar{\omega}^*(\mathcal{E}_\nu) \to (\Omega^1_{\Gamma/S})^\nu$$

is surjective, since i) this is true over a field k, and ii) every section of $(\Omega^1_{\Gamma_s/k})^{\nu}$ over Γ_s extends to a section of $\mathcal{E}_\nu$ over some neighborhood of s. Therefore, in the general case, a *morphism* I is defined, and extends the corresponding morphism over a field: then by Proposition 4.6.7, EGA 3, I is actually a closed immersion.

Let $\Gamma_i = \Gamma \underset{S}{\times} U_i$. Then the composition:

$$\Gamma_i \hookrightarrow \boldsymbol{P}(\mathcal{E}_\nu | U_i) \cong U_i \times \boldsymbol{P}_n$$

defines a U_i-valued point β_i of $Hilb_{\boldsymbol{P}_n}$. One checks that this is actually a U_i-valued point of H_ν. And, by definition, the restriction of α to U_i is the curve Γ_i which is the image of β_i under π. *QED.*

Loosely speaking, Proposition 5.2 asserts that $\mathcal{M}_g$ is obtained by a partial "sheafification" of $\mathcal{M}'_g$. This is used in the proof of:

Proposition 5.3. H_ν is irreducible and smooth over Z.

Proof. To prove that H_ν is smooth over Z, we use part (iii) bis of Theorem 3.1, SGA 3: let $Y = \mathrm{Spec}\,(A)$ for some artin local ring of finite type over $\boldsymbol{Z}$, and let $Y_0 = \mathrm{Spec}\,(A/I)$ be a subscheme. We must show that every Y_0-valued point of y_0 of H_ν lifts to a Y-valued point of H_ν. But, according to Theorem 6.3 in the same notes, SGA 3, the element $\pi(y_0)\,\varepsilon\,\mathcal{M}_g(Y_0)$ does lift to an element of $\mathcal{M}_g(Y)$. Moreover, by Proposition 5.2, this element lifts to a Y-valued point $\tilde{y}$ of H_ν. Let $\tilde{y}_0$ be the restriction of $\tilde{y}$ to Y_0. Since $\pi(\tilde{y}_0) = \pi(y_0)$, it follows from Proposition 5.2 that there is a Y_0-valued point α_0 of $PGL(n+1)$ such that

$$\sigma(\alpha_0, \tilde{y}_0) = y_0.$$

Since $PGL(n-1)$ is smooth over $\boldsymbol{Z}$, α_0 lifts to a Y-valued point α of $PGL(n+1)$. Then $\sigma(\alpha, \tilde{y})$ is a Y-valued point of H_ν which lifts y_0.

Since H_ν is smooth over $\boldsymbol{Z}$, H_ν is irreducible if and only if $H_\nu \times \boldsymbol{Q}$ is connected. But, in fact, $H_\nu \times \boldsymbol{C}$ is connected. The only known proofs of this last fact are transcendental. Starting from the connectedness of the Teichmüller space, one can show that any 2 algebraic curves over $\boldsymbol{C}$ are connected by a differentiable family of curves. Therefore any two points of $H_\nu \times \boldsymbol{C}$ are connected by a differentiable line segment. This implies that $H_\nu \times \boldsymbol{C}$ is connected as an algebraic variety. (Compare [4] and [5]). *QED.*

The precise connection between orbit spaces and coarse moduli schemes is an immediate consequence of Proposition 5.2:

Proposition 5.4. A geometric quotient of H_ν by $PGL(n)$ is a coarse moduli scheme for curves of genus g.

Proof. Note first that the set of morphisms ψ from $\mathcal{M}_g$ to representable functors h_N, and the set of morphisms f from H_ν to schemes N such that $f \circ \sigma = f \circ p_2$ (as morphisms from $PGL(n+1) \times H_\nu$ to N; σ denoting the action of $PGL(n+1)$) are canonically isomorphic. Namely, the curve

Γ_ν over H_ν given by the construction of H_ν defines an element $\gamma_\nu \in \mathcal{M}_g(H_\nu)$. Then associate to any morphism $\psi: \mathcal{M}_g \to h_N$, the morphism $\psi(\gamma_\nu)$: $H_\nu \to N$. The fact that this is an isomorphism of the two sets in question is follows formally from Proposition 5.2. Moreover, ψ gives an isomorphism of $\mathcal{M}_g(\mathrm{Spec}\,\Omega)$ and $h_N(\mathrm{Spec}\,\Omega)$ for all algebraically closed fields if and only if f gives an isomorphism of the set of orbits of geometric points of H_ν and the set of geometric points of N. Therefore, if (N, f) is a geometric quotient of H_ν by $PGL(n+1)$, the corresponding ψ makes N into a coarse moduli scheme. *QED.*

§ 3. First chern classes

Before proceeding to the next step — which is to construct a geometric quotient of H_ν by means of the results of § 4.6 — we must insert a digression on some Cartier divisors associated to certain sheaves, and to the images of certain morphisms.* Let X be a noetherian scheme. Then in X there are only finitely many points x such that the depth of $\underline{o}_{x,X}$ is 0, i.e. such that every non-unit of $\underline{o}_{x,X}$ is a 0-divisor. Suppose $\underline{\mathcal{F}}$ is a coherent sheaf on X such that:

i) $\mathrm{Supp}\,(\mathcal{F})$ contains no points of depth 0,

ii) for every $x \in X$, the stalk $\mathcal{F}_x$ is an $\underline{o}_{x,X}$-module of finite Tor-dimension, i.e. admitting a finite projective resolution.

Then I propose to define for such $\mathcal{F}$ an effective Cartier divisor $\mathrm{Div}\,(\mathcal{F})$ on X, whose support is contained in $\mathrm{Supp}\,(\mathcal{F})$.

We require some preliminaries:

Lemma 5.5. Let $0 \to \mathcal{E}_2 \to \mathcal{E}_1 \to \mathcal{E}_0 \to 0$ be any short exact sequence of locally free sheaves on any local ringed space X. Let r_i be the rank of $\mathcal{E}_i$. Then

$$(\Lambda^{r_0}\mathcal{E}_0) \otimes (\Lambda^{r_1}\mathcal{E}_1)^{-1} \otimes (\Lambda^{r_2}\mathcal{E}_2)$$

is canonically isomorphic to $\underline{o}_X$.

Proof. If the short exact sequence splits, the result is clear. Since any such sequence does split locally, the sheaf in question is locally isomorphic to $\underline{o}_X$. As the isomorphism is canonical, it is independent of the splitting and one has such an isomorphism globally. *QED.*

Lemma 5.6. Let $0 \to \mathcal{E}_n \to \mathcal{E}_{n-1} \to \cdots \to \mathcal{E}_0 \to 0$ be any exact sequence of locally free sheaves on any noetherian local ringed space X. Let r_i be the rank of $\mathcal{E}_i$. Then

$$\bigotimes_{i=1}^{n} (\Lambda^{r_i}\mathcal{E}_i)^{(-1)^i}$$

is canonically isomorphic to $\underline{o}_X$.

---

* The theory in this and the next section has been developed and extended by FOGARTY [100] and KNUDSEN [176].

Proof. We use induction on n. If $\tilde{\mathcal{E}}$ is the kernel of the homomorphism $\mathcal{E}_1 \to \mathcal{E}_0$, then we have 2 exact sequences:

$$0 \to \tilde{\mathcal{E}} \to \mathcal{E}_1 \to \mathcal{E}_0 \to 0$$

$$0 \to \mathcal{E}_n \to \cdots \to \mathcal{E}_2 \to \tilde{\mathcal{E}} \to 0.$$

But then $\tilde{\mathcal{E}}$ is also a locally free sheaf: e.g. because, as the kernel of the surjective $\mathcal{E}_1 \to \mathcal{E}_0$, it is flat. Therefore, by the induction assumption:

$$(\Lambda^{r_0} \mathcal{E}_0) \otimes (\Lambda^{r_1} \mathcal{E}_1)^{-1} \otimes (\Lambda^{r_1 - r_0} \tilde{\mathcal{E}}) \cong \underline{o}_X$$

$$(\Lambda^{r_1 - r_0} \tilde{\mathcal{E}}) \otimes (\Lambda^{r_2} \mathcal{E}_2)^{-1} \otimes \cdots \otimes (\Lambda^{r_n} \mathcal{E}_n)^{(-1)^{n-1}} \cong \underline{o}_X .$$

Tensoring the 1st expression by the inverse of the 2nd, we obtain the sought-for isomorphism. *QED.*

Now suppose $\mathcal{F}$ is a coherent sheaf of finite cohomological dimension on a noetherian scheme. Then, for all $x \in X$, there is some neighborhood U of x on which $\mathcal{F}$ admits a resolution:

$$0 \to \mathcal{E}_n \to \mathcal{E}_{n-1} \to \cdots \to \mathcal{E}_0 \to \mathcal{F} \to 0$$

where the $\mathcal{E}_i$ are free coherent sheaves on U. Let $U' = U - \operatorname{Supp}(\mathcal{F})$. Applying lemma 5.6 to the sequence

$$0 \to \mathcal{E}_n \to \mathcal{E}_{n-1} \to \cdots \to \mathcal{E}_0 \to 0$$

which is exact on U', we obtain an isomorphism

$$\underline{o}_X \cong \bigotimes_{i=0}^{n} (\Lambda^{r_i} \mathcal{E}_i)^{(-1)^i}$$

on U'. But since the $\mathcal{E}_i$ are free sheaves *on* U, there is an isomorphism on U

$$\bigotimes_{i=1}^{n} (\Lambda^{r_i} \mathcal{E}_i)^{(-1)^i} \xrightarrow{\sim} \underline{o}_X$$

unique up to a unit. Composing these, we obtain a homomorphism from $\underline{o}_X$ to $\underline{o}_X$ on U', i.e. a non-zero section f of $\Gamma(U', \underline{o}_X)$. Since U' contains all points of U of depth 0 by assumption i) at the beginning of this section, f is not a 0-divisor. Since f is unique up to a unit in U, we have defined a Cartier divisor (f) in U. This (f) is to be $\operatorname{Div}(\mathcal{F})$ in the open set U.

There are two things to be considered: a) is (f) independent of the resolution of $\mathcal{F}$? Only if this is verified can we be sure that the various Cartier divisors (f) in the various open sets U agree on their overlaps and define $\operatorname{Div}(\mathcal{F})$ globally. b) We must check that (f) is actually an *effective* Cartier divisor. But recall that a Cartier divisor is effective if and only if it is effective *at all points of depth* 1. Consequently 2 Cartier divisors are equal if and only if there are equal at all points of depth 1.

Therefore, both questions (a) and (b) need only be considered at points of depth 1.

Suppose $x \in X$ is such that the depth of $\mathfrak{o}_{x,X}$ is 1: we recall a result of AUSLANDER and BUCHSBAUM [3] — if a finitely generated $\mathfrak{o}_{x,X}$ — module has finite cohomological dimension, then it has cohomological dimension 0 or 1. Now suppose

$$(*)_n \qquad 0 \to \mathcal{E}_n \to \mathcal{E}_{n-1} \to \cdots \to \mathcal{E}_0 \to \mathcal{F} \to 0$$

is any free, coherent resolution of $\mathcal{F}$ in a neighborhood U of such an x. Let $\tilde{\mathcal{E}}$ be the kernel of $\mathcal{E}_0 \to \mathcal{F}$. Then by the result just quoted, $\tilde{\mathcal{E}}$ is a free sheaf in some smaller neighborhood $U_0 \subset U$ of x. Therefore we have exact sequences:

$$\left.\begin{array}{r} (*)_1 \qquad 0 \to \tilde{\mathcal{E}} \xrightarrow{h} \mathcal{E}_0 \to \mathcal{F} \to 0 \\ 0 \to \mathcal{E}_n \to \cdots \to \mathcal{E}_2 \to \mathcal{E}_1 \to \tilde{\mathcal{E}} \to 0 \end{array}\right\}$$

on U_0. But this means that the isomorphism

$$\mathfrak{o}_X \xrightarrow{\sim} \bigotimes_{i=0}^{n} (\Lambda^{r_i} \mathcal{E}_i)^{(-1)^i}$$

on $U_0 - \operatorname{Supp}(\mathcal{F})$ defining $\operatorname{Div}(\mathcal{F})$ extends to an isomorphism:

$$\mathfrak{o}_X \xrightarrow{\sim} \Lambda^{r_0} \tilde{\mathcal{E}} \otimes \left\{ \bigotimes_{i=1}^{n} (\Lambda^{r_i} \mathcal{E}_i)^{(-1)^i} \right\}$$

on *all* of U_0; i.e. $\operatorname{Div}(\mathcal{F})$, as defined from the original resolution $(*)_n$ of $\mathcal{F}$, is the same as $\operatorname{Div}(\mathcal{F})$, as defined simply by the resolution $(*)_1$ of $\mathcal{F}$. But this, in concrete terms, works out to be the following: choose bases for the free sheaves $\tilde{\mathcal{E}}$ and $\mathcal{E}_0$ of rank r_0 on U_0, and express h by an $r_0 \times r_0$ matrix h_{ij} of functions on U_0. Then:

$$\operatorname{Div}(\mathcal{F}) = \big(\det(h_{ij})\big),$$

which is clearly effective. As for uniqueness, we need only check uniqueness among all resolution of length 2. But if $\mathfrak{o} = \mathfrak{o}_{x,X}$, $\mathfrak{m}$ is the maximal ideal in $\mathfrak{o}$, and $k = \mathfrak{o}/\mathfrak{m}$, then put $r = \dim_k (\mathcal{F}_x / \mathfrak{m} \cdot \mathcal{F}_x)$. Then there is a minimal resolution: $0 \to \mathcal{E}_1 \xrightarrow{h} \mathcal{E}_0 \to \mathcal{F} \to 0$, where $\mathcal{E}_0$ and $\mathcal{E}_1$ are free of rank r, and *every other resolution* is isomorphic, in some neighborhood of x, to:

$$0 \to \mathcal{E} \oplus \mathcal{E}_1 \xrightarrow{H} \mathcal{E} \oplus \mathcal{E}_0 \to \mathcal{F} \to 0$$

where H is given by a matrix:

$$\left[\begin{array}{c|c} 1_s & * \\ \hline 0 & h \end{array}\right]$$

Therefore, $\det H = \det h$ and the independence is proven.

This completes the definition of Div ($\mathcal{F}$). To illustrate the concept, we give one example: Suppose $\underset{\sim}{o}$ is a regular local ring of dimension 1, $X = \mathrm{Spec}\,(\underset{\sim}{o})$, and $\mathcal{F}$ corresponds to an $\underset{\sim}{o}$-module M of finite length. Let π generate the maximal ideal of $\underset{\sim}{o}$. Then every such M is isomorphic to

$$\sum_{i=1}^{r} \underset{\sim}{o}/(\pi^{n_i})$$

for some sequence of integers $n_1, \ldots, n_r$. Then the matrix h above is $(\pi^{n_i} \cdot \delta_{ij})$. Hence

$$\det(L) = \pi^{\Sigma n_i}.$$

In other words, Div ($\mathcal{F}$) is the closed point of X with multiplicity equal to the length of M as an $\underset{\sim}{o}$-module.

Div ($\mathcal{F}$) has the following functorial property: Let $f: X \to Y$ be a morphism of noetherian schemes, and let $\mathcal{F}$ be a coherent sheaf on Y with properties i) and ii). Suppose that for all $x \in X$, if $y = f(x)$:

$$\mathrm{Tor}_i^{\underset{\sim}{o}_y}(\underset{\sim}{o}_x, \mathcal{F}_y) = (0), \qquad i > 0.$$

Then $f^*(\mathcal{F})$ has finite Tor-dimension since, if we choose locally a finite, free resolution of $\mathcal{F}$ and tensor it with $\underset{\sim}{o}_x$, it gives a finite free resolution of $f^*(\mathcal{F})$. On the other hand, suppose also that for all points $x \in X$ of depth 0, $f(x) \notin \mathrm{Supp}\,(\mathcal{F})$. Then $f^*(\mathcal{F})$ has properties i) and ii). Moreover, one checks immediately that:

$$f^*(\mathrm{Div}\,(\mathcal{F})) = \mathrm{Div}\,(f^*(\mathcal{F})).$$

The concept of the Div of a sheaf can be used to define the Div of certain morphisms. Suppose

$$X \xrightarrow{f} Y$$

is a projective morphism of noetherian schemes. Assume, furthermore:

i) f is of finite Tor-dimension, i.e. for all $x \in X$, the local ring $\underset{\sim}{o}_{x,X}$ is a module of finite Tor-dimension over $\underset{\sim}{o}_{f(x),Y}$, the dimension being bounded for all x.

ii) $f^{-1}(x)$ is empty (resp. finite) if $x \in X$ has depth 0 (resp. Depth 1).

In this situation, I claim that there is a canonical effective Cartier divisor Div (f), whose support is contained in the set $f(X)$. To define this, suppose L is an invertible sheaf on X, such that

$$R^i f_*(L) = (0)$$

for $i > 0$. At least one such L exists because f is a projective morphism. Then we put

$$\mathrm{Div}\,(f) = \mathrm{Div}\,(f_*(L)).$$

First we must check that this makes sense:

Lemma 5.7. $f_*(L)$ has finite Tor-dimension, as $\mathcal{O}_Y$-module.

Proof. The result is local on Y, so it suffices to prove it when $Y = \mathrm{Spec}\,(R)$. Cover X by m affine open subsets U_i, and let

$$M_{i_1,\ldots,i_k} = \Gamma(U_{i_1} \cap \cdots \cap U_{i_k}, L).$$

Consider the Čech co-chain complex:

$$(*) \quad 0 \to \bigoplus_i M_i \to \bigoplus_{i_1<i_2} M_{i_1,i_2} \to \cdots \to M_{1,2,\ldots,m} \to 0.$$

Since the higher direct images of L are (0), and since the affine scheme Y has no higher cohomology for quasi-coherent sheaves,

$$H^i(X, L) \cong H^0(Y, R^i f_*(L)) = (0)$$

for $i > 0$. Therefore $(*)$ is exact, except at the first place, where its cohomology is $f_*(L)$. But now each module $M_{i_1,\ldots,i_k}$ has finite Tor-dimension over R: therefore so does $f_*(L)$. *QED.*

Since, by (ii), $\mathrm{Supp}\,(f_*(L))$ contains no points of depth 0, $\mathrm{Div}\,(f)$ makes sense. Secondly, we must check that it is independent of the choice of L. Let L_1, L_2 be 2 invertible sheaves on X with no higher direct images on Y. *I* claim

$$\mathrm{Div}\,\big(f_*(L_1)\big) = \mathrm{Div}\,\big(f_*(L_2)\big).$$

It suffices to prove this at all points $y \in Y$ of depth 1. But by assumption (ii), $f^{-1}(y)$ is then a finite set. In particular, the sheaf $L_1 \otimes L_2^{-1}$ is isomorphic to $\mathcal{O}_X$ on the set $f^{-1}(y)$, and hence in an open neighborhood U of $f^{-1}(y)$. But since f is a closed map, $U \supset f^{-1}(V)$ for some neighborhood V of y. Therefore, $f_*(L_1) \cong f_*(L_2)$ in V, and their Div's are equal. Thus $\mathrm{Div}\,(f)$ depends only on f.

Finally, $\mathrm{Div}\,(f)$ has some functorial properties. Let

$$\begin{array}{ccc} X' & \xrightarrow{g'} & X \\ {\scriptstyle f'}\downarrow & & \downarrow{\scriptstyle f} \\ Y' & \xrightarrow[g]{} & Y \end{array}$$

be a fibre product. Suppose

i) for all $x \in X$ and $y' \in Y'$ such that $y = f(x) = g(y')$, then

$$\mathrm{Tor}_i^{\mathcal{O}_y}(\mathcal{O}_x, \mathcal{O}_{y'}) = (0), \quad i > 0.$$

(ii) $f'^{-1}(y)$ is empty (resp. finite) if $y \in Y'$ is a point of depth 0 (resp. depth 1).

Then it is not hard to check that f' has finite Tor-dimension and that

$$\mathrm{Div}\,(f') = g^*\big(\mathrm{Div}\,(f)\big).$$

Actually, the situation which will concern us is given by the diagram:

$$\begin{array}{ccc} X' = X \underset{S}{\times} T & \rightarrow & X \\ f' \downarrow & & \downarrow f \\ Y' = Y \underset{S}{\times} T & \rightarrow & Y \\ \downarrow & & \downarrow p \\ T & \rightarrow & S \end{array}$$

where p and $p \circ f$ are flat. Then condition (i) always holds by the change of rings spectral sequence for Tor's and (ii) will hold, *for any* T, if it holds for $T = \mathrm{Spec}\,(\Omega)$, i.e.

(ii)* for all geometric points s of S, let

$$\overline{X} \xrightarrow{f_s} \overline{Y}$$

be the morphism obtained by base extension over s. Then $f_s^{-1}(y)$ is empty (resp. finite) if $y \in \overline{Y}$ is a point of depth 0 (resp. depth 1).

§ 4. Utilization of § 4.6

Consider the chow variety parametrizing cycles of dimension r and degree d in $\boldsymbol{P}_n$. As a set of chow forms, determined up to non-zero multiples, it can be embedded in a projective space of divisors. Namely, let $\check{\boldsymbol{P}}_n$ be the projective space dual to $\boldsymbol{P}_n$, i.e. the linear system $|H|$. Then chow points are divisors in $(\check{\boldsymbol{P}}_n)^{r+1}$ which come from sections of

$$\bigotimes_{i=1}^{r+1} p_i^*\left(\mathcal{O}_{\check{\boldsymbol{P}}_n}(d)\right).$$

Write the projective space of such divisors

$$\mathrm{Div}^{d,d,\ldots,d}[(\check{\boldsymbol{P}}_n)^{r+1}].$$

Then the chow variety is a subvariety here. Now suppose that $P(x)$ is any polynomial of the form:

$$P(x) = d \cdot \frac{x^r}{r!} + \text{lower order terms}.$$

Then I claim that there is a canonical morphism

$$\phi: Hilb^{P(x)}_{\boldsymbol{P}_n} \rightarrow \mathrm{Div}^{d,d,\ldots,d}\,[(\check{\boldsymbol{P}}_n)^{r+1}]$$

which, in particular, maps a geometric point of *Hilb* corresponding to $Z \subset \bar{\boldsymbol{P}}_n$ to the chow form of the cycle of components of dimension r in Z. Moreover, ϕ will be $PGL(n+1)$-linear with respect to the canonical actions of $PGL(n+1)$ on the left and the right. This morphism will then bring the results of § 4.6 to bear on the problems of § 5.2.

What I propose to do is to construct ϕ in case $r = 1$: for one thing, this is the only case we need, and for another the generalization to any r will be entirely mechanical. The method of constructing ϕ will be to assign to every S-valued point of $Hilb_{\mathbf{P}_n}^{P(x)}$ an S-valued point of $\mathrm{Div}^{d,d}[(\check{\mathbf{P}}_n)^2]$. Since it will be clear that this assignment is functorial, it will define a morphism from the functor $Hilb_{\mathbf{P}_n}^{P(x)}$ to the functor represented by $\mathrm{Div}^{d,d}[(\check{\mathbf{P}}_n)^2]$; therefore by EGA 0, Proposition 8.1.7, this defines a unique morphism ϕ of *schemes* as required. But an S-valued point of $Hilb_{\mathbf{P}_n}^{P(x)}$ corresponds functorially to a closed subscheme $Z \subset \mathbf{P}_n \times S$, flat over S, with Hilbert polynomial $P(x)$. And an S-valued point of $\mathrm{Div}^{d,d}(\check{\mathbf{P}}_n^2)$ corresponds functorially to a family of divisors on $(\check{\mathbf{P}}_n)^2$, parametrized by S, i.e. a relative effective Cartier Divisor* $D \subset (\check{\mathbf{P}}_n)^2 \times S$, over S, which induces divisors of type (d, d) on the geometric fibres over S. Therefore the problem is to assign (functorially) a suitable divisor D to every subscheme Z, as above.

To start with, let $H \subset \mathbf{P}_n \times \check{\mathbf{P}}_n$ be the "universal hyperplane", i.e. if

$$\mathbf{P}_n = \mathrm{Proj}\, \mathbf{Z}\, [X_0, \ldots, X_n]$$

$$\check{\mathbf{P}}_n = \mathrm{Proj}\, \mathbf{Z}\, [U_0, \ldots, U_n],$$

then interpret X_i, and U_i as sections of $\mathcal{O}(1)$ on $\mathbf{P}_n$ and $\check{\mathbf{P}}_n$; then H is defined as the zeroes of the section

$$\left[\sum_{i=0}^{n} X_i \otimes U_i\right] \varepsilon\, \Gamma(p_1^*(\mathcal{O}(1)) \otimes p_2^*(\mathcal{O}(1))).$$

Now consider the variety $\mathbf{P}_n \times \check{\mathbf{P}}_n \times \check{\mathbf{P}}_n$. Via p_{12} and p_{13}, this can be mapped onto $\mathbf{P}_n \times \check{\mathbf{P}}_n$. Let H_2 and H_3 be the Cartier divisors $p_{12}^*(H)$ and $p_{13}^*(H)$, and $\mathcal{H}$ be the intersection of H_2 and H_3 (as a subscheme: i.e. if f, g are local equations of H_2, H_3, then $\mathcal{H}$ is defined by the ideal (f, g)). We shall be concerned with the set-up:

$$\begin{array}{ccc} & \mathcal{H} & \\ {}^{p_1}\swarrow & & \searrow^{p_{23}} \\ \mathbf{P}_n & & \check{\mathbf{P}}_n \times \check{\mathbf{P}}_n. \end{array}$$

The important things to remember are:

a) the geometric points of $\mathcal{H}$ consist of triples (x, h_1, h_2) where x is a geometric point of $\mathbf{P}_n$ and h_1, h_2 are hyperplanes, in the corresponding $\bar{\mathbf{P}}_n$, containing x.

---

* This means an effective Cartier divisor which is flat over S; or, equivalently, which does not contain the whole of any of the fibres of $(\check{\mathbf{P}}_n)^2 \times S$ over S. cf. Ch. 0, § 5, (e).

b) $\mathscr{H}$ is a fibre bundle over $\boldsymbol{P}_n$, via p_1, with fibre isomorphic to $\boldsymbol{P}_{n-1}\times\boldsymbol{P}_{n-1}$.

The proof of b) is straightforward and is omitted.

To apply this set-up we make a base extension by an arbitrary noetherian scheme S:

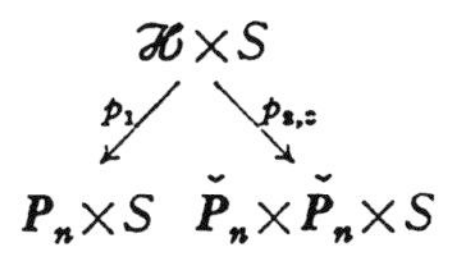

Let the closed subscheme $Z\subset\boldsymbol{P}_n\times S$, flat over S, with Hilbert polynomial $P(x)$ represent a given S-valued point of $Hilb^{P(x)}_{\boldsymbol{P}_n}$. We seek an effective Cartier divisor D in $\check{\boldsymbol{P}}_n\times\check{\boldsymbol{P}}_n\times S$. The method is very simple. Form the fibre product Z^* as follows:

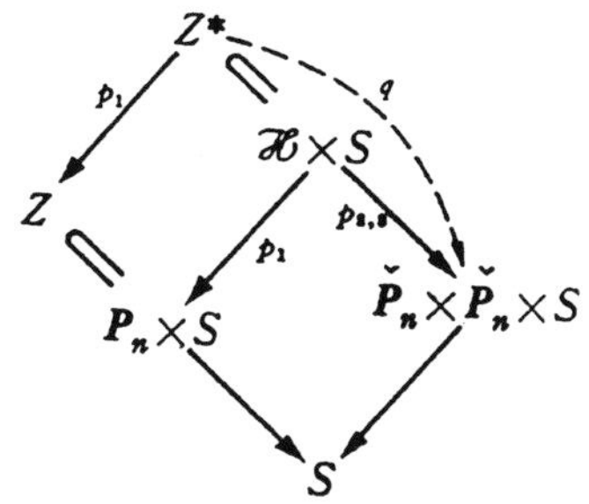

Then D is to be Div (q). We must verify quite a few things: that Div (q) is defined and is a relative Cartier divisor of the right type over S, and that its construction commutes with all base extensions $T\to S$. But note that since $\mathscr{H}\times S$ is flat over $\boldsymbol{P}_n\times S$, therefore Z^* is flat over Z; and Z is flat over S, so Z^* is flat over S. Therefore all the base extensions are of the last type mentioned in § 3. Because of the remarks made there, the only things to check are:

i) q is of finite Tor-dimension,

ii) Div (q) is a *relative* divisor over S,

iii) When $S=\operatorname{Spec}(\Omega)$, the morphism $q\colon Z^*\to\check{\boldsymbol{P}}_n\times\check{\boldsymbol{P}}_n\times\operatorname{Spec}(\Omega)$ has empty fibres (resp. finite fibres) over points of depth 0 (resp. Depth 1).

iv) When $S=\operatorname{Spec}(\Omega)$, the divisor Div $(q)\subset\check{\boldsymbol{P}}_n\times\check{\boldsymbol{P}}_n\times\operatorname{Spec}(\Omega)$ has type (d,d).

First of all, (i) comes from a lemma [generalizing a result of J. P. Serre, and proven by J. Fogarty]

Lemma 5.8. Let

$$\begin{array}{ccc} X & \xrightarrow{\;f\;} & Y \\ & {\scriptstyle p}\searrow \quad \swarrow{\scriptstyle q} & \\ & S & \end{array}$$

be morphisms of finite type between noetherian schemes. Suppose q is a smooth morphism with fibres of dimension n, and that p is flat. Then f has Tor-dimension at most n.

Proof. Look at the diagram:

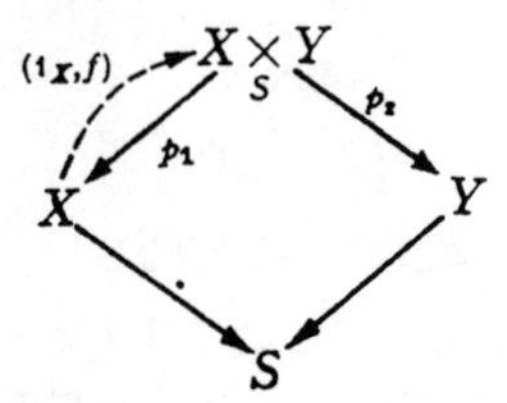

Let $s = (1_X, f)$. This is a section of p_1, and $f = p_2 \circ s$. Since X is flat over S, $X \underset{S}{\times} Y$ is flat over Y. Therefore it suffices to prove that s has Tor-dimension at most n. But since s is a section of p_1, s is an isomorphism of X with a closed subscheme $s(X) \subset X \underset{S}{\times} Y$. Moreover, because p_1 is *smooth*, the subscheme $s(X)$ is locally a complete intersection (i.e. its ideal is generated locally by an $\mathcal{O}_{X \underset{S}{\times} Y}$-sequence): cf. SGA 2, Theorem 4.15. Therefore, the Tor-dimension of the structure sheaf of $s(X)$ is equal to the codimension of $s(X)$, i.e. n. Hence the Tor-dimension of s is n. *QED.*

Now one of the equivalent meanings of a relative divisor D for a flat morphism $f: X \to Y$ is that the induced subscheme on the fibres of f should be a Cartier divisor too. In the case of the projections $\check{P}_n \times \check{P}_n \times S \to S$, this just means that Supp (Div (q)) does not contain completely any of the fibres of $\check{P}_n \times \check{P}_n \times S$ over S. Since Supp (Div (q)) $\subset q(Z^*)$, this follows from (iii), i.e. (ii) follows from (iii).

To prove (iii) and (iv), the simplest thing to say is that this is now just the classical theory of chow forms. Take $S = \operatorname{Spec}(\Omega)$, and write simply $\check{P}_n$ for $\check{P}_n \times \operatorname{Spec}(\Omega)$. Then the closed points of Z^* are just triples (x, h_1, h_2) consisting of closed points $x \in Z$ and of 2 hyperplanes $h_1, h_2 \in \check{P}_n$ containing x. That $q^{-1}(y)$ is empty when y is the generic point of $\check{P}_n \times \check{P}_n$ (the only point of depth 0) means that $h_1 \cap h_2 \cap Z = \Phi$ for *some* set of hyperplanes h_1, h_2: and this is clear since $\dim Z = 1$. That $q^{-1}(y)$ is finite when y is the generic point of a divisor in $\check{P}_n \times \check{P}_n$ (i.e. a point of depth 1), means that the set of hyperplanes h_1, h_2 such that $\dim h_1 \cap h_2 \cap Z = 1$ is the set of closed points on a subset $W \subset \check{P}_n \times \check{P}_n$ of codimension $\geqq 2$. This is also an elementary fact. Finally, to show that Div (q) is of type (d, d), why not show that Div (q) is the "chow form" of the cycle of 1-dimensional components of Z, in which case it is classical? To be precise:

Definition 5.8. Let X be a scheme of finite type over Spec (Ω) of dimension r. Let $X_1, \ldots, X_k$ be the components of X, let Z_i be the underlying set of X_i, let x_i be the generic point of X_i, and let $\mathfrak{o}_i$ be the local ring of X at x_i. Let r_i be the length of $\mathfrak{o}_i$ (which is an Artin local ring) as a module over itself.

$$\text{cycle}(X) = \sum_{\substack{\text{all } i \text{ such that} \\ \dim X_i = r}} r_i \cdot Z_i .$$

Note that if X is an effective Cartier divisor on a normal variety, then cycle (X) is the Weil divisor usually associated to X. Now in the classical definition of the chow form, one also uses the diagram:

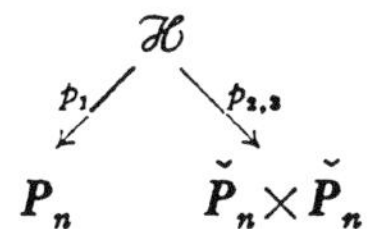

Then one defines:

$$\text{Chow divisor of } Z = (p_{2,3})_* \, [p_1^* (\text{cycle}(Z))]$$

where $(p_{2,3})_*$ and p_1^* are the classical operations on cycles, as given in WEIL's treatise for example (n.b. all 3 schemes here are non-singular varieties, so there is no difficulty). Compare this with our approach: first of all, it is very easy to check:

$$p_1^* (\text{cycle}(Z)) = \text{cycle}(Z^*).$$

The rest follows from:

Lemma 5.9. Let X and Y be non-singular varieties, and let $f: X \rightarrow Y$ be a projective morphism. Let $Z \subset X$ be a closed subscheme such that

$$\dim Z = \dim Y - 1.$$

Let g be the restriction of f to Z. Then:

$$f_* \, [\text{cycle}(Z)] = \text{cycle}\, [\text{Div}(g)].$$

Proof. Let D be an irreducible subset of Y of codimension 1, and let $y \in D$ be its generic point. Since $\dim Z = \dim D$, the set $f^{-1}(y)$ is finite: let $f^{-1}(y) = \{z_1, \ldots, z_k\}$ and let z_i be the generic point of the subset E_i of Z. Then $E_1, \ldots, E_k$ are the components of Z which dominate D, and they all have the same dimension as Z. Let L be an invertible sheaf on X such that

$$R^i g_* (L \otimes \mathfrak{o}_Z) = (0), \quad i > 0.$$

Let L_i be the stalk of $L \otimes \mathfrak{o}_Z$ at z_i, let A_i be the stalk of $\mathfrak{o}_Z$ at z_i, and let A be the stalk of $\mathfrak{o}_Y$ at y. Then A is a discrete, rank 1 valuation ring, A_i is an A-module of finite length, and $L_i \cong A_i$ as A_i-module.

First compute the multiplicity of D in f_* [Cycle (Z)]:

$$\text{Cycle }(Z) = \sum_{i=1}^{k} \begin{Bmatrix}\text{length of } A_i \\ \text{as } A_i\text{-module}\end{Bmatrix} \cdot E_i + \text{other components.}$$

$$f_* [\text{Cycle }(Z)] = \left(\sum_{i=1}^{k} \begin{Bmatrix}\text{length of } A_i \\ \text{as } A_i\text{-module}\end{Bmatrix} [E_i : D]\right) \cdot D + \text{other components}$$

$$\text{Mult. of } D = \sum_{i=1}^{k} \begin{Bmatrix}\text{length of } A_i \\ \text{as } A_i\text{-module}\end{Bmatrix} \cdot [\varkappa(A_i) : \varkappa(A)]$$

$$= \left\{\text{length of } \bigoplus_{i=1}^{k} A_i, \text{ as } A\text{-module}\right\}.$$

Second compute the multiplicity of D in Cycle [Div (g)]:

$$g_* (L \otimes \varrho_Z)_y = \bigoplus_{i=1}^{k} L_i \cong \bigoplus_{i=1}^{k} A_i.$$

$$\text{Div }(g)_y = \text{Div}\left(\bigoplus_{i=1}^{k} A_i\right).$$

$$\text{Mult. of } D \text{ in Cycle } (\text{Div }(g)) = \text{length of Div } \left(\bigoplus_{i=1}^{k} A_i\right), \text{ as } A\text{-module}$$

$$= \text{length of } \bigoplus_{i=1}^{k} A_i, \text{ as } A\text{-module}$$

(according to the computations in § 3 over discrete rank 1 valuation rings.) *QED.*

This completes the proof of:

Conclusion 5.10. There is a canonical morphism

$$\phi : Hilb_{\mathbf{P}_n}^{P(x)} \to \text{Div}^{d,d} (\check{\mathbf{P}}_{n}^{2})$$

which maps a geometric point of $Hilb_{\mathbf{P}_n}^{P(x)}$ corresponding to the 1-dimensional scheme $Z \subset \mathbf{P}_n$ to the chow form of cycle (Z).

Now we can return to the situation of § 2. The morphism ϕ which we have defined restricts to a $PGL(n+1)$-linear morphism:

$$\Phi : H_\nu \to \text{Div}^{2\nu(g-1), 2\nu(g-1)} [\check{\mathbf{P}}_n \times \check{\mathbf{P}}_n].$$

Since H_ν parametrizes smooth curves, and since the chow form of a smooth curve determines that curve, Φ is injective. Write P for the projective space on the right. According to what was proven in § 4.6,

$$\Phi(H_\nu) \subset (P)^s_{(0)} (\varrho_P(1)).$$

Let $L = \Phi^* (\varrho_P(1))$ be the induced $SL(n+1)$-linearized invertible sheaf on H_ν. Then by Proposition 1.18, since an injective morphism of finite type is quasi-affine (SGA 8, Th. 6.2):

$$H_\nu = (H_\nu)^s_{(0)} (L).$$

Therefore we have proven*:

Theorem 5.11. A geometric quotient of H_ν by $PGL(n+1)$ exists. Therefore a coarse moduli scheme for curves over schemes over $\mathbf{Z}$ exists.

Chapter 6

Abelian schemes

The moduli of curves over $\mathbf{Z}$ can also be treated via their Jacobians. For this it is necessary to treat first the moduli of abelian varieties over $\mathbf{Z}$. To do this, we develop first some basic information about *families* of abelian varieties, or, as we shall call them, abelian schemes. We have only included a bare minimum concerning such schemes. This theory is due to GROTHENDIECK, and has been partially sketched by him in the BOURBAKI talks [13] and has been developed by RAYNAUD [341]. For the theory of abelian varieties over fields, see the author's book Abelian Varieties [212] for a systematic exposition.

For the sake of simplicity, all pre-schemes in this chapter will be assumed to be *locally noetherian* and to be *schemes*.

§ 1. Duals

Definition 6.1. Let S be a noetherian scheme. A group scheme $\pi: X \to S$ is called an *abelian scheme* if π is smooth and proper, and the geometric fibres of π are connected.

Although it is known, when $S = \mathrm{Spec}\,(k)$, that abelian schemes are commutative as groups, this does not imply the fact in general. For treating quite a few questions involving passage from the classical to the general case, we will need:

Proposition 6.1. (*Rigidity lemma.*) Given a diagram:

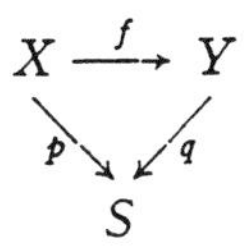

suppose S is connected, p is flat and $H^0(X_s, \underline{o}_{X_s}) \cong \varkappa(s)$, for all points $s \in S$ (X_s denoting the fibre of p over s). Assume that *one* of the following is true:

1) X has a section ε over S, and S consists of one point,

2) X has a section ε over S, and p is a closed map,

3) p is proper.

If, for one point $s \in S$, $f(X_s)$ is set-theoretically a single point, then there is a section $\eta: S \to Y$ of q such that $f = \eta \circ p$.

Proof. We shall prove this successively if 1), 2), and 3) hold. Suppose 1) holds. First define a continuous map $\eta: S \to Y$ as $f \circ \varepsilon$. Then $f = \eta \circ p$

---

* We use Theorem 1.10 to construct the quotient by $PGL(n+1)$. In the text, this is proven only over a field, but it has been extended by SESHADRI to schemes over Spec Z: see Appendix 1G.

as continuous maps. One checks that $p_*(\underline{o}_X) \cong \underline{o}_S$. But f is defined by its underlying map, plus a homomorphism:

$$\underline{o}_Y \to f_*(\underline{o}_X) \cong \eta_*(p_*(\underline{o}_X)) \cong \eta_*(\underline{o}_S).$$

But such a homomorphism is precisely the extra structure required to make η into a morphism of ringed spaces. It follows immediately that this is, in fact, a morphism of local ringed spaces, hence by EGA 1, § 1.8, a morphism of schemes.

Now suppose 2) holds, and set $\eta = f \circ \varepsilon$. To compare f and $\eta \circ p$, let Z be the biggest closed subscheme of X where $f = \eta \circ p$, i.e. if $\Delta \subset Y \times_S Y$ is the diagonal, then $Z = (f, \eta \circ p)^{-1}(\Delta)$. We must show that $Z = X$. But the first part of the proof has already shown that if Z contains $p^{-1}(t)$ set-theoretically, for any $t \in S$, then for all artin subschemes $T \subset S$, concentrated at t, Z contains $p^{-1}(T)$ as a subscheme. But this implies that Z actually contains some open neighborhood U of $p^{-1}(t)$. Since p is a *closed* map, this implies that Z contains an open neighborhood of the form $p^{-1}(U_0)$, for some open neighborhood U_0 of t. In particular, Z contains some open set $p^{-1}(U_0)$ for some open neighborhood U_0 of s. Let U_1 be the maximal open subset of S such that $Z \supset p^{-1}(U_1)$. But now

$$\begin{aligned} t \in U_1 &\Leftrightarrow p^{-1}(t) \subset Z \\ &\Leftrightarrow p^{-1}(t) \text{ is disjoint from } X - Z \\ &\Leftrightarrow t \notin p(X - Z). \end{aligned}$$

But p is flat, hence open, and Z is closed, hence $p(X - Z)$ is open, hence $Y - U_1$ is open. Since S is connected, $Y = U_1$.

Now suppose 3) holds. Then, after a faithfully flat base extension S'/S (e.g. by $S' = X$ itself), we can assume that X'/S' has a section. Then by case 2) we know that $f' : X' \to Y'$ is of the form $\eta' \circ p'$. Since this property determines η' uniquely, and since f' descends to a morphism f, it follows immediately that η' must also descend to $\eta : S \to Y$ (cf. SGA 8, Th. 5.2). Then $f' = \eta' \circ p'$ implies $f = \eta \circ p$. *QED.*

Corollary 6.2. Given a diagram:

$$\begin{array}{ccc} X & \overset{f}{\underset{g}{\rightrightarrows}} & G \\ & {\scriptstyle p}\searrow \quad \swarrow{\scriptstyle q} & \\ & S & \end{array}$$

assume that G is a group scheme over S, S is connected, p is flat and proper, and $H^0(X_s, \underline{o}_{X_s}) \cong \kappa(s)$, for all points $s \in S$. If, for some point $s \in S$, the morphisms f_s and g_s from Y_s to G_s are equal, then there is a section $\eta : S \to G$ such that

$$f = (\eta \circ p) \cdot (g)$$

(where the dot denotes multiplication of X-valued points of G).

Proof. This reduces to Proposition 6.1 for $f \cdot g^{-1}$.

Corollary 6.3. Given a diagram:

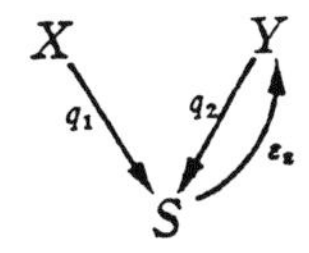

where ε_2 is a section for q_2, assume that q_1 is proper and flat, Y is connected, and $H^0(X_s, \varrho_{X_s}) \cong \varkappa(s)$, for all $s \in S$. Suppose that G is a group scheme over S, and that $f: X \underset{S}{\times} Y \to G$ is an S-morphism. Then there exist S-morphisms $g: X \to G$ and $h: Y \to G$ such that

$$f = (g \circ p_1) \cdot (h \circ p_2).$$

Proof. This reduces to Corollary 6.2, for the Y-morphisms (f, p_2) and $[f \circ (1_X, \varepsilon_2 \circ q_1)] \times 1_Y$ from $X \underset{S}{\times} Y$ to $G \underset{S}{\times} Y$

Corollary 6.4. Let X be an abelian scheme and G any group scheme over a scheme S. If $f: X \to G$ is an S-morphism taking the identity for X to the identity for G, then f is a homomorphism.

Proof. Apply Corollary 6.3 to $f \circ \mu: X \underset{S}{\times} X \to G$, where μ is the group law for X.

Corollary 6.5. If X is an abelian scheme over a scheme S, then X is a commutative group scheme.

Proof. Apply Corollary 6.4 to the inverse morphism from X to X.

Corollary 6.6. If X is an abelian scheme over a scheme S, then X has only one structure of group scheme over S with the given identity $\varepsilon: S \to G$ as the identity.

Proof. Apply Corollary 6.4 to 1_X, with 2 different group laws considered on domain and image.

Our next goal is to define a dual abelian scheme, for any ***projective*** abelian scheme $\pi: X \to S$. First of all, it is known that *Pic* (X/S) exists in this case (cf. Ch. 0, § 5, (d)). Let $Pic^\tau(X/S)$ be the open subgroup pre-scheme whose geometric points correspond to the invertible sheaves some power of which are algebraically equivalent to zero. Then, since π is smooth and projective, it is known that $Pic^\tau(X/S)$ is projective over S (Ch. 0, § 5, (d)). Also, the geometric fibres of $Pic^\tau(X/S)$ are reduced and connected, since a) there is no torsion on abelian varieties A, and since b) dim $H^1(A, \varrho_A)$ equals the dimension of the Picard variety of A.

Proposition 6.7. In this situation, $Pic^\tau(X/S)$ is smooth over S.

Proof. We use criterion iii) bis, Theorem 3.1 of SGA 3. Namely, suppose $s \in S$, and A is an artin local ring which is a finite $\varrho_{s,S}$-algebra. Suppose $I \subset A$ is an ideal such that $\mathfrak{m} \cdot I = (0)$, $\mathfrak{m}$ being the maximal ideal in A. Then we must show that any Spec (A/I)-valued point of $Pic^\tau(X/S)$ can be lifted to a Spec (A)-valued point. But the former (resp. the latter) corresponds to an invertible sheaf L_0 (resp. L) on

$$X \underset{S}{\times} \mathrm{Spec}\,(A/I), \left(\text{resp. } X \underset{S}{\times} \mathrm{Spec}\,(A)\right).$$

Now, according to Proposition 7.1 of SGA 3, there is an obstruction to obtaining an L extending L_0, and it is an element of

$$H^2(\overline{X}, \varrho_{\overline{X}}) \underset{k}{\otimes} I,$$

where $k = A/\mathfrak{m}$, and $\overline{X} = X \underset{S}{\times} \mathrm{Spec}\,(k)$.

However, let $\mu: X \underset{S}{\times} X \to X$ be the multiplication. Then the 3 morphisms μ, p_1, p_2 define 3 homomorphisms

$$\mu^*, p_1^*, p_2^*: Pic\,(X/S) \to Pic\,(X \underset{S}{\times} X/S).$$

Let $\varrho = \mu^* - p_1^* - p_2^*$. According to a Theorem of WEIL, when $S = \mathrm{Spec}\,(k)$, k being a field, $Pic^\tau(X/S)$ is contained in the kernel of ϱ: this is the theorem of the square (cf. LANG [21], Ch. III). But then by Proposition 6.1, (condition (1)), it follows that $Pic^\tau(X/S)$ is in the kernel of ϱ whenever S consists of one point. Therefore L_0, which is a Spec (A/I)-valued point of $Pic^\tau(X/S)$, is in the kernel of ϱ. Therefore $\varrho(L_0)$ can be trivially lifted to a Spec (A)-valued point of $Pic\,(X \underset{S}{\times} X/S)$. But how are the obstructions to lifting L_0 and to lifting $\varrho(L_0)$ related? It is readily checked that they are related by the homomorphism:

$$H^2(\overline{X}, \varrho_{\overline{X}}) \underset{k}{\otimes} I \xrightarrow{\mu^*-p_1^*-p_2^*} H^2(\overline{X} \times \overline{X}, \varrho_{\overline{X} \times \overline{X}}) \underset{k}{\otimes} I.$$

But using the known structure of the cohomology of abelian varieties, and the Kunneth formula, it follows readily that this homomorphism is *injective*. Therefore the obstruction to lifting L_0 must also vanish. This proves that L exists, hence $Pic^\tau(X/S)$ is smooth over S. *QED.*

Corollary 6.8. *$Pic^\tau(X/S)$ is a projective abelian scheme, and it is written $\hat{X}$.*

To illustrate the concepts of abelian scheme and dual abelian scheme, we apply them to jacobians:

Proposition 6.9. *Let $\pi: \Gamma \to S$ be a curve of genus g, where $g \geq 2$. Let $J = Pic^\tau(\Gamma/S)$. Then J is a projective abelian scheme over S, and there is a canonical isomorphism:*

$$\theta: J \to \hat{J}.$$

Proof. Since Γ is smooth and projective over S, it follows that $J = Pic^\tau(\Gamma/S)$ is projective over S (Ch. 0, § 5, (d)). To prove that J is smooth over S, we use again the criterion, Theorem 3.1 of SGA 3.

This time the obstructions vanish automatically since the dimension of the fibres of π is 1, and the obstruction is in H^2 (compare with the proof of Proposition 6.7). Moreover, over a field, curves have no torsion, so their Pic^τ is connected. Therefore J is an abelian scheme.

Recall that $Pic\,(\Gamma/S)$ is a disjoint union of open subsets $J^{(d)}$ where $J = J^{(0)}$ and $J^{(d)} + J^{(e)} \subset J^{(d+e)}$: this is a special case of the decomposition of $Pic\,(X/S)$ by Hilbert polynomials (cf. Ch. 0, § 5, (c)). Here $J^{(d)}$ is to have geometric points representing invertible sheaves whose c_1 has degree d. Moreover there is a canonical S-morphism

$$\phi : \Gamma \to J^{(1)}.$$

To define ϕ, let $\Delta \subset \Gamma \times_S \Gamma$ be the diagonal. Then Δ is a Cartier divisor, and even a relative Cartier divisor over Γ via p_1 or p_2: i.e. being the image of a section of p_i, it is flat over Γ via p_i, and since π is simple of dimension 1, Δ intersects the fibres of p_i in Cartier divisors. Therefore, it is a relative Cartier divisor (Ch. 0, § 5, (e)). Then the invertible sheaf $\mathcal{O}_{\Gamma \times_S \Gamma}(\Delta)$ defines a morphism ϕ from Γ to $Pic\,(\Gamma/S)$; and as geometric points of Γ are mapped to points of $Pic\,(\Gamma/S)$ associated to invertible sheaves of degree 1, $\phi(\Gamma) \subset J^{(1)}$.

To define θ, suppose first that π admits a section $\varepsilon : S \to \Gamma$. Then the difference $(\phi - \phi \circ \varepsilon \circ \pi)$ defines an S-morphism ψ from Γ to J. Associated to ψ is a contravariant morphism:

$$\begin{array}{ccc} \hat{\psi} : Pic^\tau(J/S) & \to & Pic^\tau(\Gamma/S) \\ \| & & \| \\ \hat{J} & & J \end{array}$$

Now it is a classical result that $\hat{\psi}$ is an isomorphism when $S = \operatorname{Spec}(k)$, k a field: therefore by EGA 3, 4.6.7, it is always an isomorphism. Put $\theta = \hat{\psi}^{-1}$. Is this θ independent of the section ε, if there are 2 sections $\varepsilon_1, \varepsilon_2$ of Γ over S? Then ψ_1 will differ from ψ_2 by the translation T_{12} with respect to the S-valued point of $\phi \circ \varepsilon_1 - \phi \circ \varepsilon_2$ of J, i.e. $\psi_2 = T_{12} \circ \psi_1$. Hence $\hat{\psi}_2 = \hat{\psi}_1 \circ \hat{T}_{12}$. But $\hat{T}_{12}$ is trivial by the theorem of the square when $S = \operatorname{Spec}(k)$, k a field; and it is trivial in general by Proposition 6.1.

Now suppose Γ/S has no section. It certainly acquires a section after a suitable faithfully flat base extension S'/S of finite type, e.g. take $S' = \Gamma$. Therefore, a morphism $\theta' : J' \to \hat{J}'$ has been defined. But by the independence of θ' from the choice of section, the two morphisms

$$\theta_1', \theta_2' : (J' \times_{S'} (S' \times_S S')) \to (\hat{J}' \times_{S'} (S' \times_S S'))$$

induced from the projections $p_1, p_2 : S' \times_S S' \to S'$ must be equal. Therefore by the theory of descent (SGA 8, Theorem 5.2), θ' is induced by a morphism $\theta : J \to \hat{J}$. *QED.*

§ 2. Polarizations

In this section, we shall study invertible sheaves L on abelian schemes $\pi: X \to S$ which are relatively ample for the morphism π. First of all, we introduce a fundamental construction: let $\mu: X \times_S X \to X$ be the group law, and consider at first for *any* L on X the invertible sheaf

$$\mu^*(L) \otimes p_1^*(L)^{-1} \otimes p_2^*(L)^{-1}$$

on $X \times_S X$. Regarding $X \times_S X$ as a scheme over X via p_1, this sheaf defines an X-valued point:

$$\psi: X \to Pic\,(X/S).$$

If $\varepsilon: S \to X$ is the identity, then, by construction, the induced S-valued point, $\psi \circ \varepsilon$, is the identity of $Pic\,(X/S)$. Since the geometric fibres of X over S are connected, it follows that ψ is an X-valued point of $Pic^{\tau}(X/S)$, i.e. an S-morphism from X to $\hat{X}$. Moreover, by Corollary 6.4, it follows that this morphism is a homomorphism.

Definition 6.2. This homomorphism will be denoted

$$\Lambda(L): X \to \hat{X}.$$

It is clear that $\Lambda(L_1 \otimes L_2^{\pm 1}) = \Lambda(L_1) \pm \Lambda(L_2)$. Moreover, as remarked in the course of the proof of Proposition 6.7, it is a Theorem of Weil that when $S = \mathrm{Spec}\,(k)$, k being a field, and when L is algebraically equivalent to 0, then $\Lambda(L) = 0$. Therefore, by the rigidity lemma, in any case $\Lambda(L) = 0$ when L is algebraically equivalent to 0 on the fibres X_s of π. Therefore, when L is relatively ample for $\pi, \Lambda(L)$ is an invariant of the "polarization" defined by L. This motivates:

Definition 6.3. Let $\pi: X \to S$ be a projective abelian scheme. A *polarization* of X is an S-homomorphism:

$$\lambda: X \to \hat{X}$$

such that, for all geometric points of S, the induced $\bar{\lambda}: \bar{X} \to \hat{\bar{X}}$ is of the form $\Lambda(\bar{L})$, for some *ample* invertible sheaf $\bar{L}$ on $\bar{X}$.

Note that such a homomorphism λ must be finite and surjective (since this is true over a field, cf. Lang [21], Cor. 2,p. 96). As an example, let Γ/S be a curve of genus $g \geq 2$, and let $\theta: J \to \hat{J}$ be the homomorphism defined in Proposition 6.9. $-\theta$ is, in fact, a polarization since it is a classical fact, when $S = \mathrm{Spec}\,(k), k$ a field, that $-\theta = \Lambda(\varrho_J(\Theta)), \Theta$ being the "Θ-divisor". Actually this would be an awkward concept were it not possible somehow to lift polarizations to invertible sheaves globally. To do this, let $\mathcal{L}$ be the universal invertible sheaf on $X \times_S \hat{X}$, i.e. that sheaf defining the embedding $\hat{X} \subset Pic\,(X/S)$ via the functorial definition of *Pic*. $\mathcal{L}$ is only determined up to tensoring with $p_2^*(L)$, any L on $\hat{X}$. But if $\varepsilon: S \to X$ is the identity for X, then $\mathcal{L}$ can be uniquely normalized so that the sheaf induced on $\hat{X}$ via:

$$\hat{X} \cong S \underset{S}{\times} \hat{X} \xrightarrow{\varepsilon \times 1_{\hat{X}}} X \underset{S}{\times} \hat{X}$$

is $\mathfrak{o}_{\hat{X}}$, (compare Ch. 0, § 5, (d)). Then set:

$$L^{\Delta}(\lambda) = (1_X, \lambda)^*(\mathcal{L})$$

(here $(1_X, \lambda)$ maps X to $X \underset{S}{\times} \hat{X}$). The pleasing result is:

Proposition 6.10. If λ is a polarization, then $\Lambda(L^{\Delta}(\lambda)) = 2\lambda$.

Proof. Suppose we first prove that $\Lambda(L^{\Delta}(\lambda)) = 2\lambda$ whenever $S = \mathrm{Spec}\,(k)$, k a field. Then it will follow in general by virtue of Corollary 6.2. To prove it over k, we may as well take k to be algebraically closed. Then the original λ is induced by an invertible sheaf L on the abelian variety X, and it suffices to prove that $L^{\Delta}(\lambda) \otimes L^{-2}$ is algebraically equivalent to 0. But note that by definition of $\Lambda(L)$, and the universal mapping property of Picard schemes,

$$\mu^*(L) \otimes p_1^*(L)^{-1} \otimes p_2^*(L)^{-1} \cong (1_X \times \lambda)^*(\mathcal{L}).$$

If $\Delta: X \to X \underset{S}{\times} X$ is the diagonal, then $(1_X, \lambda) = (1_X \times \lambda) \circ \Delta$, hence

$$\begin{aligned} L^{\Delta}(\lambda) &\cong \Delta^*\{\mu^*(L) \otimes p_1^*(L)^{-1} \otimes p_2^*(L)^{-1}\} \\ &= \psi_2^*(L) \otimes L^{-2} \end{aligned}$$

where $\psi_2: X \to X$ is multiplication by 2. But it is a classical result that $\psi_2^*(L)$ is algebraically equivalent to L^4 (cf. LANG [21], Prop. 2, p. 92), hence the result. *QED*.

This shows that there is a global way of passing back from polarizations to sheaves. The final point which we must investigate in this connexion is *which* sheaves arise from polarizations in this way. Note first of all that since $\mathcal{L}$ was normalized on the identity section of $X \underset{S}{\times} \hat{X}$ over S, it follows that $L^{\Delta}(\lambda)$ is always normalized along the identity section $\varepsilon: S \to X$, i.e. $\varepsilon^*(L^{\Delta}(\lambda)) \cong \mathfrak{o}_S$. This is a canonical way to normalize invertible sheaves on X/S, i.e. so that they correspond in a 1—1 way with sections of *Pic* (X/S). We will require the following result:

Proposition 6.11. Let $\pi: X \to S$ be a projective abelian scheme, and let L be an invertible sheaf on X, ample over S, such that $\varepsilon^*(L) \cong \mathfrak{o}_S$, where ε is the identity section. If $f: T \to S$ is any base extension, let $X_T = X \underset{S}{\times} T$ and let $L_T = p_1^*(L)$ be the induced sheaf on X_T. Then for any k, there is at most one homomorphism $\lambda: X_T \to \hat{X}_T$ such that $L_T = L^{\Delta}(k\lambda)$. Moreover, there is a closed subscheme $S_0 \subset S$ such that one such λ exists if and only if f factors through S_0.

Proof. To prove the uniqueness of λ, suppose $L_T = L^{\Delta}(k\lambda_1) = L^{\Delta}(k\lambda_2)$. By Proposition 6.10, $2k\lambda_1 = 2k\lambda_2$. Therefore, for all geometric points t in T, $\bar{\lambda}_1 - \bar{\lambda}_2$ maps the *connected* abelian variety $\bar{X}_T$ into the *finite* set of points of order $2k$ on $\hat{\bar{X}}_T$; therefore $\bar{\lambda}_1 = \bar{\lambda}_2$. Then by Corollary 6.2, $\lambda_1 = \lambda_2$.

On the other hand, to construct S_0, first ask after which base extensions will $\Lambda(L_T)$ equal $2k\mu$ for one (and, by the above, only one) $\mu: X_T \to \hat{X}_T$. Let ψ_{2k} be the morphism from X to X which is multiplication by $2k$. Note that ψ_{2k} is flat by virtue of:

Lemma 6.12. Let X and Y be group schemes over S. Assume X is flat over S and Y is smooth over S. Then a surjective homomorphism $f: X \to Y$ is flat.

Proof of lemma. Using the assumption that X and Y are flat over S, by § 17, EGA 4, it suffices to prove that the morphism $f_s: X_s \to Y_s$ of the fibres is flat, for every $s \in S$. But since f_s is surjective, and Y_s is regular, it follows from Corollary 6.11, SGA 4, that f_s is flat over a non-empty set $V \subset Y_s$. Since f_s commutes with translations, f_s is also flat over every translate of V in the group scheme Y_s. Therefore f_s is everywhere flat. *QED.*

The question of whether or not $\Lambda(L)$ is of the form $\mu \circ \psi_{2k}$ is a problem of descent. We use Th. 5.2, SGA 8. Let $X \underset{X}{\times} X$ be the fibre product with respect to ψ_{2k} and ψ_{2k}. Then μ exists if and only if $\Lambda(L) \circ p_1 = \Lambda(L) \circ p_2$:

$$\begin{array}{ccccc} X \underset{X}{\times} X & \overset{p_1}{\underset{p_2}{\rightrightarrows}} & X & \xrightarrow{\psi_{2k}} & X \\ & & \downarrow{\scriptstyle \Lambda(L)} & \swarrow{\scriptstyle \mu ?} & \\ & & \hat{X} & & \end{array}$$

(Similarly, after any base extension $T \to S$). But let $\ker(\psi_{2k})$ be the fibre product in:

$$\begin{array}{ccc} \ker(\psi_{2k}) & \xrightarrow{I} & X \\ \downarrow{\scriptstyle \pi} & & \downarrow{\scriptstyle \psi_{2k}} \\ S & \xrightarrow[\varepsilon_X]{} & X \end{array}$$

This is a closed subgroup scheme of X over S whose geometric points are the points of order $2k$. Then via p_1 and $p_2 - p_1$, one finds:

$$X \underset{X}{\times} X \xrightarrow{\sim} X \underset{S}{\times} \ker(\psi_{2k}).$$

Therefore $\Lambda(L) \circ p_1 = \Lambda(L) \circ p_2$ if and only if $\Lambda(L) \circ I$ is the trivial morphism from $\ker(\psi_{2k})$ to $\hat{X}$, i.e. $\varepsilon_{\hat{X}} \circ \pi$. The conclusion is that, after base extension $T \to S$, a μ exists if and only if the induced morphisms $\Lambda(L) \circ I$ and $\varepsilon_{\hat{X}} \circ \pi$ are equal after this base extension.

But after which base extensions will $\Lambda(L) \circ I = \varepsilon_{\hat{X}} \circ \pi$? Let $Z \subset \mathrm{Ker}\,(\psi_{2k})$ be the maximal closed subscheme where $\Lambda(L) \circ I = \varepsilon_{\hat{X}} \circ \pi$. Let $\mathcal{A} = \pi_*(\varrho_{\mathrm{Ker}(\psi_{2k})})$, and let the sheaf of $\mathcal{A}$-ideals $\mathcal{J}$ define Z. By the lemma, noting that $\mathrm{Ker}\,(\psi_{2k})$ is finite over S, it follows that $\mathcal{A}$ is a locally free coherent sheaf. For every $s \in S$, let $j_1, \ldots, j_n$ span $\mathcal{J}$ locally, and let S_1 be the subscheme defined by the vanishing of sections j_i of the locally free sheaf $\mathcal{A}$. Then $\Lambda(L) \circ I = \varepsilon_{\hat{X}} \circ \pi$ after a base extension $f: T \to S$ if and only if the induced morphism from $T \underset{S}{\times} \mathrm{Ker}\,(\psi_{2k})$ to $\mathrm{Ker}\,(\psi_{2k})$ factors through Z, i.e. if and only if $f^*(\mathcal{J})$ is the zero-subsheaf of $f^*(\mathcal{A})$, i.e. if and only if f factors through S_1.

Replacing S by S_1, we may assume that $\Lambda(L) = 2k\mu$. After which base extensions will $L = L^{\Delta}(k\mu)$? But L and $L^{\Delta}(k\mu)$ define 2 sections λ and λ' of *Pic* (X/S). The set of all $f: T \to S$ such that $\lambda \circ f = \lambda' \circ f$ is the set of f factoring through a closed subscheme $S_0 \subset S_1$, since *Pic* (X/S) is separated over S. Finally, as both L and $L^{\Delta}(k\mu)$ are normalized along the identity section, $\lambda \circ f = \lambda' \circ f$ if and only if $L_T = L^{\Delta}(k\mu_T)$. *QED.*

The following result carries over from fields to general base schemes without difficulty:

Proposition 6.13. Let $\pi: X \to S$ be a projective abelian scheme. Let L be an invertible sheaf on X, relatively ample for π. Then

(i) $R^i\pi_*(L) = (0)$, if $i > 0$.

(ii) $\pi_*(L)$ is a locally free sheaf on S. Let r be its rank.

(iii) Let $\Lambda(L): X \to \hat{X}$ be the finite flat morphism defined by L. Then the degree of $\Lambda(L)$, i.e. the rank of $\Lambda(L)_*(\varrho_X)$, is r^2.

(iv) If $n \geq 2$, then the sections of $\pi_*(L^n)$ have no common zeroes in X. Therefore there is a morphism

$$\phi_n: X \to \boldsymbol{P}(\pi_* L^n).$$

If $n \geq 3$, ϕ_n is a closed immersion.

Proof.* Over a field, (ii) is vacuous, and (iii) and (iv) have been proven by NISHI ([29] and [30]). When (i) is proven for the fibres of a flat and proper morphism π, (i) and (ii) globally follow automatically (cf. Ch. 0, § 5, (a)). Of course (iii) globalizes, and (iv) globalizes by Proposition 4.6.7, EGA 3, and the fact that when (i) is true along the fibres, then $\pi_*(L)$ generates $H^0(L_s)$, $s \in S$ (cf. Ch. 0, § 5, (a), or Corollary 4.6.2, EGA 3). The only remaining point is to prove that (i) holds on the fibres on π: i.e. suppose A is an abelian variety over a field k, and L is an ample invertible sheaf on A. Then we must prove that $H^i(A, L) = (0)$, $i > 0$. Also, we can assume that k is algebraically

* Over a field, (i), (iii), and (iv) are also proven in [29], § 16, 17.

closed for simplicity. But since L is ample, there is an n_0 such that

$$H^i(A, L^n) = (0), \quad n \geq n_0,\ i > 0.$$

Moreover, suppose that an invertible sheaf L' is algebraically equivalent to L^n. Then L' is isomorphic to some translate of L^n, with respect to an element $x \varepsilon A_k$, since $\Lambda(L^n)$ is surjective. Since translations are automorphisms of A, $H^i(A, L') = (0)$, if $i > 0$, $n \geq n_0$. Now let $\psi_k: A \to A$ be multiplication by k. Then $\psi_k^*(L)$ is algebraically equivalent to L^{k^2} (cf. LANG [21], p. 92). Therefore if $k \geq \sqrt{n_0}$,

$$\begin{aligned} H^i[A, \psi_{k,*}(\psi_k^*(L))] &\cong H^i(A, \psi_k^*(L)) \\ &\cong (0), \quad i > 0. \end{aligned}$$

On the other hand, if the characteristic does not divide k, then L is a direct summand of $\psi_{k,*}(\psi_k^*(L))$: i.e. via

$$L \underset{i}{\overset{k^{-2g}Tr}{\leftrightarrows}} \psi_{k,*}(\psi_k^*(L))$$

where i is the canonical homomorphism, g is the dimension of A, and Tr signifies the trace. Therefore $H^i(A, L) = (0)$ for $i > 0$. *QED.*

§ 3. Deformations

We prove in this section a result which has been announced by GROTHENDIECK, but for which no proof has appeared:

Theorem 6.14. Let S be a connected, locally noetherian scheme. Let $\pi: X \to S$ be a smooth projective morphism, and let $\varepsilon: S \to X$ be a section of π. Assume that for one geometric point s of S, the fibre X_s of π is an abelian variety with identity $\varepsilon(s)$. Then X is an abelian scheme over S with identity ε.

This will be proven in 3 steps. The first is:

Proposition 6.15. Let $S = \operatorname{Spec}(A)$, where A is an Artin local ring. Let $\mathfrak{m} \subset A$ be the maximal ideal, and let $I \subset A$ be an ideal such that $\mathfrak{m} \cdot I = (0)$. Let $\pi: X \to S$ be a smooth proper morphism, and let $\varepsilon: S \to X$ be a section. Let $S_0 = \operatorname{Spec}(A/I)$ and let $X_0 = X \times_S S_0$. Assume that X_0 is an abelian scheme over S_0 with identity $\varepsilon \mid S_0$. Then X is an abelian scheme over S with identity ε.

Proof. Let $k = A/\mathfrak{m}$, and let $\bar{X} = X \times_S \operatorname{Spec}(k)$. Let $\mu_0: X_0 \times_{S_0} X_0 \to X_0$ be the morphism $\mu_0(x, y) = x - y$, (for T-valued points x, y of X_0, any scheme T/S). Let $\bar{\mu}$ be the restriction of μ_0 to $\bar{X} \times \bar{X}$. The first problem is to extend μ_0 to some morphism $\mu: X \times_S X \to X$. By Corollary 5.2, SGA 3, if $\mathcal{J}$ is the tangent sheaf to $\bar{X}$, then there is an obstruction

$$\beta \,\varepsilon\, H^1(\bar{X} \times \bar{X}, \bar{\mu}^*(\mathcal{J}) \underset{k}{\otimes} I)$$

whose vanishing is necessary and sufficient for the existence of an extension μ. But consider $g_1, g_2: X_0 \to X_0 \underset{S_0}{\times} X_0$ defined by $g_1(x) = (x, e)$, $g_2(x) = (x, x)$ (for a T-valued point x of X_0, for any scheme T/S). Let $\bar{g}_i = g_i \mid \bar{X}$. Then $\mu_0 \circ g_1 = 1_{X_0}$, and $\mu_0 \circ g_2 = (\varepsilon \circ \pi) \mid X_0$. Therefore both $\mu_0 \circ g_1$ and $\mu_0 \circ g_2$ do extend to morphisms from X to X, and the corresponding obstructions

$$\beta_1, \beta_2 \,\varepsilon\, H^1(\bar{X}, (\bar{\mu} \circ \bar{g}_i)^* (\mathcal{J}) \underset{k}{\otimes} I)$$

to extending these morphisms are zero. But in any case, one must have $\beta_i = \bar{g}_i^* (\beta)$ by a simple functorial property of the obstruction*. Therefore $g_1^* \beta = g_2^* \beta = 0$.

Now recall that $\mathcal{J}$ is a trivial vector bundle: in fact

$$\mathcal{J} \cong \mathcal{O}_{\bar{X}} \otimes H^0(\bar{X}, \mathcal{J}).$$

Therefore, by the Kunneth formula:

$$\begin{aligned} H^1(\bar{X} \times \bar{X}, \bar{\mu}^* (\mathcal{J}) \underset{k}{\otimes} I) &\cong H^1(\bar{X} \times \bar{X}, \mathcal{O}_{\bar{X} \times \bar{X}}) \otimes H^0(\bar{X}, \mathcal{J}) \otimes I. \\ &= \{p_1^* H^1(\bar{X}, \mathcal{O}_{\bar{X}}) \oplus p_2^* H^1(\bar{X}, \mathcal{O}_{\bar{X}})\} \otimes H^0(\bar{X}, \mathcal{J}) \otimes I. \end{aligned}$$

It follows from this that if $g_1^* \alpha = g_2^* \alpha = 0$, for any α, then $\alpha = 0$. In particular $\beta = 0$, and μ exists.

However the extension μ of μ_0 is not unique. By Corollary 5.2, SGA 3 again, the set of all extensions is a principal homogeneous space under

$$H^0(\bar{X} \times \bar{X}, \bar{\mu}^* \mathcal{J} \otimes I)$$

which is isomorphic to $H^0(\bar{X}, \mathcal{J}) \otimes I$. By the same result, the set of all extensions of μ_0, restricted to the identity $(\varepsilon_0, \varepsilon_0)(S) \subset X_0 \underset{S_0}{\times} X_0$, is a principal homogeneous space under $H^0(\bar{X}, \mathcal{J}) \otimes I$. Therefore, we conclude that there is a unique extension of μ_0 for every extension of μ_0 restricted to the identity. We require that the extension μ take the identity of $X \underset{S}{\times} X$ to X, i.e. $\mu_0(\varepsilon, \varepsilon) = \varepsilon$, and this determines μ uniquely.

It remains to prove that μ defines a true group law on X. Of course, one can formally define both multiplication and inverse in terms of μ, and it is only necessary to prove that various identities are satisfied by the resulting morphisms. But all these identities take the form $h_1 = h_2$, where

$$h_i : \overbrace{X \underset{S}{\times} \cdots \underset{S}{\times} X}^{l \times} \to X$$

---

* In general, if one has $X \xrightarrow{f} Y \xrightarrow{g} G$, there is a "chain rule" for computing the obstruction to extending $g \circ f$ in terms of the obstructions to extending g and f separately.

and h_i is a certain composition of μ's, projections, diagonals, and identity morphisms. Also, in all cases (as the reader can check), $h_1 \circ (\varepsilon, \varepsilon, \ldots, \varepsilon) = h_2 \circ (\varepsilon, \varepsilon, \ldots, \varepsilon) = \varepsilon$. Moreover, since μ_0 defines a group law on X_0, $h_1 = h_2$ on the subscheme $X_0 \underset{S_0}{\times} X_0$. By Corollary 6.2, it follows that $h_1 = h_2$. *QED.*

Proposition 6.16. Let F be the functor on the category of locally noetherian S-schemes defined by:

$F(T) = \{$set of all structures of abelian scheme on $X \underset{S}{\times} T$ over T with identity $(\varepsilon \circ f, 1_T) : T \to X \underset{S}{\times} T$, where $f: T \to S$ is the given morphism$\}$.

Then F is represented by an open set $U \subset S$.

Proof. A structure of abelian scheme on any scheme Y/T (where Y is simple and projective over T) such that an identity $\eta: T \to Y$ is fixed, is the same thing as a morphism $\mu: Y \underset{T}{\times} Y \to Y$ satisfying various identities, such as $\mu \circ (\eta, \eta) = \eta$. Here μ is to be the map $x, y \to x - y$ as above. As a corollary of the existence of Hilbert schemes, it follows that there exists a scheme $Hom_T(Y \underset{T}{\times} Y, Y)$ representing the family of all morphisms from $Y \underset{T}{\times} Y$ to Y (cf. exposé 221, [13]; Ch. 0, § 5, (c)). Therefore, a structure of abelian scheme is a section of $Hom_T(Y \underset{T}{\times} Y, Y)$ of a particular sort. If $Y = X \underset{S}{\times} T$, then the set of such sections is isomorphic to the set of all S-morphisms:

$$f: T \to Hom_S(X \underset{S}{\times} X, X).$$

I claim that for the identities in question to be satisfied, it is necessary and sufficient that f factor through a certain closed subscheme $Z \subset Hom_S(X \underset{S}{\times} X, X)$. But in each case, the identities amount to relations of the form $\gamma_1 \circ f = \gamma_2 \circ f$ where

$$\gamma_i : Hom_S(X \underset{S}{\times} X, X) \to Hom_S\Big(\overbrace{X \underset{S}{\times} \cdots \underset{S}{\times} X}^{l\times}, X\Big)$$

where γ_i is some kind of canonical morphism representing the process of taking some μ, and composing it with itself, with projections, diagonals, etc. [Such processes always give morphisms of the functor represented by the 1st scheme to the functor represented by the 2nd scheme; by EGA 0, § 8, any such morphism will come from a morphism of schemes.] Then take Z to be the scheme-theoretic intersection of the subschemes $(\gamma_1, \gamma_2)^{-1}(\Delta_l)$, Δ_l being the diagonal in $\Big[Hom_S(\overbrace{X \underset{S}{\times} \cdots \underset{S}{\times} X}^{l\times}, X)\Big]^2$.

Therefore Proposition 6.16 is equivalent to the assertion that the projection $\overline{\omega}: Z \to S$ is an open immersion. But Proposition 6.15 is precisely the criterion for $\overline{\omega}$ to be smooth (Theorem 3.1, SGA 3). By Corollary 6.6, $\overline{\omega}$ is geometrically injective, i.e. $F(\operatorname{Spec} \Omega)$ has at most one element in it, for any algebraically closed field Ω. Therefore, by Theorem 5.1, SGA 1, and Corollary 1.4, SGA 2, $\overline{\omega}$ is an open immersion.

The third step is a Theorem of KOIZUMI ([19] p. 377) to the effect that when, in the above situation, $S = \operatorname{Spec}(R)$, R a discrete, rank 1 valuation ring, and the generic fibre of X is an abelian variety, then X is an abelian scheme over S. This, combined with Proposition 6.16 implies that the U of that Proposition is closed. Hence the Theorem follows. *QED.*

Chapter 7

The method of covariants — 2nd construction

In the classical invariant theory of the 19th century, it was found to be very hard to construct the "ring of invariants" by itself. Instead, it was embedded in the "ring of covariants", or the "ring of mixed concomitants" in which stronger operations were available for constructing and combining such forms. When G is acting on X, we follow a similar approach: instead of seeking a direct construction of invariants on X, we seek a single "covariant", i.e., a G-linear morphism ϕ from X to Y. If the action of G on Y is easier to handle than the action on X, we may be able to circumvent the original difficulties. In this chapter, we construct moduli schemes over $\mathbf{Z}$ in this way. As in Chapter 6, all pre-schemes will be assumed to be schemes and to be locally noetherian, without explicit mention of this hypothesis.

§ 1. The technique

The technique rests on:

Proposition 7.1. Let G be a group scheme, flat and of finite type over S. Let X and Y be schemes of finite type over S, let σ and τ be actions of G on X and Y, and let $\phi: X \to Y$ be a G-linear morphism. Assume that Y is a principal fibre bundle over an S-scheme Q, with group G, and with projection $\pi: Y \to Q$ (cf. § 0.4). Assume that there exists an $L \in \mathit{Pic}^G(X)$ which is relatively ample for ϕ, and that Q is quasi-projective over S. Then there is a scheme P, quasi-projective over S, and an S-morphism $\overline{\omega}: X \to P$ such that X becomes a principal fibre bundle over P with group G, and projection $\overline{\omega}$.

Proof. Consider the diagram:

$$\begin{array}{ccccc} G\underset{S}{\times}X & \overset{\sigma}{\underset{p_2}{\rightrightarrows}} & X & & \\ \downarrow{\scriptstyle 1_G\times\phi} & & \downarrow{\scriptstyle\phi} & & \\ G\underset{S}{\times}Y & \overset{\tau}{\underset{p_2}{\rightrightarrows}} & Y & \xrightarrow{\pi} & Q \end{array}$$

Note that π is faithfully flat and $G\underset{S}{\times}Y \cong Y\underset{Q}{\times}Y$: the set-up for descent theory. Since $G\underset{S}{\times}X$ is the fibre product of X and $G\underset{S}{\times}Y$ over Y with respect to both of the pairs of morphisms (ϕ, τ) and (ϕ, p_2), we see that X, as a Y-scheme, comes with descent data for the morphism π (the co-cycle condition coming from the fact that σ is a group *action*). Moreover the G-linearization of L is exactly descent data for L for the morphism π. Since L is ϕ-ample, this descent data for X and L together is effective: this is SGA 8, Proposition 7.8. Therefore the diagram can be completed to:

$$\begin{array}{ccc} X & \xrightarrow{\overline{\omega}} & P \\ \downarrow{\scriptstyle\phi} & & \downarrow{\scriptstyle\psi} \\ Y & \xrightarrow{\pi} & Q \end{array}$$

where $X = Y\underset{Q}{\times}P$; and there is an invertible sheaf M on P which is ψ-ample and such that $L \cong \overline{\omega}^*(M)$. Also ψ is of finite type since ϕ is (SGA 8, Proposition 3.3). Therefore, P is quasi-projective over S. Moreover $\overline{\omega}$ is flat and surjective since π is so; and $G\underset{S}{\times}X \cong X\underset{P}{\times}X$ since this is true for Y and Q. Therefore X is a principal fibre bundle over P with group G. *QED.*

Note the analogy between this Proposition and Proposition 2.18: the latter is basically the same result except that while $S = \text{Spec}\,(k)$, k a field, and G is reductive, the action τ is *only assumed proper*, and (Q, π) is *only assumed to be a geometric quotient* of Y by G; the conclusion is simply that a geometric quotient $(P, \overline{\omega})$ of X by G exists, where P is quasi-projective. There certainly should be a common generalization of these 2 results.

Amplification 7.2. Let G be a group scheme, flat and of finite type over S. Let σ be an action of G on a scheme Y of finite type over S. Assume that Y is a principal fibre bundle over an S-scheme Q, with group G and with projection $\pi: Y \to Q$. For $i = 1$ and 2, let $\psi_i: P_i \to Q$ be morphisms of finite type. Let $X_i = Y\underset{Q}{\times}P_i$, and let G act on X_i

by the product of σ and trivial action on P_i. Then if $f: X_1 \to X_2$ is a Y-morphism,

$$[f \text{ is } G\text{-linear}] \Leftrightarrow \begin{bmatrix} f = 1_Y \times g \text{ for some } Q\text{-morphism} \\ g \text{ from } P_1 \text{ to } P_2. \end{bmatrix}$$

Proof. This follows immediately from Theorem 5.2, SGA 8, using the fact that $Y \underset{Q}{\times} Y \cong G \underset{S}{\times} Y$. *QED.*

§ 2. Moduli as an orbit space

In this section we propose to define the functors involved in moduli problems for abelian varieties and to construct, for these functors, the analogs of the schemes H_ν of § 5.2.

Definition 7.1. Let $\pi: X \to S$ be an abelian scheme whose fibres have dimension g. Assume that the characteristics of the residue fields of all $s \in S$ do not divide n. Then if $n \geq 2$, a *level n structure* on X/S consists of $2g$ sections $\sigma_1, \sigma_2, \ldots, \sigma_{2g}$ of X over S, such that i) for all geometric points s of S, the images $\sigma_i(s)$ form a basis for the group of points of order n on the fibre $\overline{X}_s$, and ii) $\psi_n \circ \sigma_i = \varepsilon$, where $\psi_n: X \to X$ is multiplication by n, and ε is the identity. In order to state our theorems uniformly, without special cases, it is convenient to call X/S by itself a *level 1 structure.*

Definition 7.2. If S is any locally noetherian scheme, let $\mathcal{A}_{g,d,n}(S)$ be the set of triples:

i) an abelian scheme X over S of dimension g,

ii) a polarization $\overline{\omega}: X \to \hat{X}$ of degree d^2, i.e., $\overline{\omega}_*(\varrho_X)$ is locally free of rank d^2, (cf. lemma 6.12),

iii) a level n structure $\sigma_1, \ldots, \sigma_{2g}$ of X over S, all up to isomorphism.

Note that the collection of sets $\mathcal{A}_{g,d,n}(S)$ forms a contravariant functor from the category of locally noetherian schemes to the category of sets in the obvious way: this functor is written $\mathcal{A}_{g,d,n}$.

Definition 7.3. If $\mathcal{A}_{g,d,n}$ is represented by a scheme $A_{g,d,n}$, then $A_{g,d,n}$ will be called the *fine moduli scheme,* of level n, for g-dimensional abelian varieties, with polarizations of degree d^2.

Definition 7.4. Suppose A is a scheme, and ϕ is a morphism from $\mathcal{A}_{g,d,n}$ to the functor h_A (where $h_A(S) = \mathrm{Hom}(S, A)$) represented by A. Then A is called a *coarse moduli scheme* if

i) for all algebraically closed fields Ω, $\phi(\mathrm{Spec}\,\Omega): \mathcal{A}_{g,d,n}(\mathrm{Spec}\,\Omega) \to h_A(\mathrm{Spec}\,\Omega)$ is an isomorphism,

ii) for all morphisms ψ from $\mathcal{A}_{g,d,n}$ to representable functors h_B, there is a unique morphism $\chi: h_A \to h_B$ such that $\psi = \chi \circ \phi$.

Instead of defining the scheme corresponding to H_ν directly, it is convenient to first define the functor which it represents.

Definition 7.5. Let $\pi: X \to S$ be an abelian scheme of dimension g, and let $\bar{\omega}: X \to \hat{X}$ be a polarization of X of degree d^2. Consider:

$$\mathcal{E} = \pi_* \left(L^\Delta(\bar{\omega})^3\right).$$

According to Propositions 6.10 and 6.13, this is a locally free sheaf on X of rank $6^g \cdot d$*. Put

$$m = 6^g \cdot d - 1.$$

Then a *linear rigidification* of X/S is an S-isomorphism

$$\phi: \boldsymbol{P}(\mathcal{E}) \xrightarrow{\sim} \boldsymbol{P}_m \times S.$$

It is not quite trivial that a linear rigidification of a polarized abelian scheme X/S induces a linear rigidification of every polarized abelian scheme $X \underset{S}{\times} T/T$ obtained by base extension. If $\bar{\omega}: X \to \hat{X}$ is the polarization of X, then $\bar{\omega} \times 1_T: X \underset{S}{\times} T \to \hat{X} \underset{S}{\times} T = \widehat{X \underset{S}{\times} T}$ is the polarization of $X \underset{S}{\times} T$. And

$$L^\Delta(\omega)^3 \underset{\mathcal{O}_S}{\otimes} \mathcal{O}_T \cong L^\Delta(\bar{\omega} \times 1_T)^3.$$

However, we must prove that π_* commutes with base extension in this case, i.e.

$$\pi_* \left[L^\Delta(\bar{\omega})^3 \underset{\mathcal{O}_S}{\otimes} \mathcal{O}_T\right] \cong \pi_* \left[L^\Delta(\bar{\omega})^3\right] \underset{\mathcal{O}_S}{\otimes} \mathcal{O}_T.$$

This follows from Proposition 6.13, part (i) and the criterion that we have used before (Ch. 0, § 5, (a)). Consequently, it does follow that

$$\boldsymbol{P}[\pi_*(L^\Delta(\bar{\omega})^3)] \underset{S}{\times} T = \boldsymbol{P}[\pi_*(L^\Delta(\bar{\omega} \times 1_T)^3)],$$

hence linear rigidifications are functorial.

Definition 7.6. For any locally noetherian scheme S, let $\mathcal{H}_{g,d,n}(S)$ be the set of all linearly rigidified abelian schemes X over S, with level n structure and polarization of degree d^2, up to isomorphism. Let $\mathcal{H}_{g,d,n}$ be the functor defined by these sets.

Suppose $\pi: X \to S$, $\{\sigma_i\}: S \to X$, $\bar{\omega}: X \to \hat{X}$, and ϕ gives an element α of $\mathcal{H}_{g,d,n}(S)$. Let $\varepsilon: S \to X$ be the identity morphism. Via ϕ, and

* i.e. By Prop. 6.13, the square of its rank is the degree of the homomorphism $\Lambda(L^\Delta(\bar{\omega})^3)$. But $\Lambda(L^\Delta(\bar{\omega})^3) = 6\bar{\omega}$ by Prop. 6.10. Since $\bar{\omega}$ has degree d^2, and multiplication by n has degree n^{2g} in an abelian scheme of relative dimension n, we obtain the result.

taking into account Proposition 6.13, we can embed X canonically in the projective space:

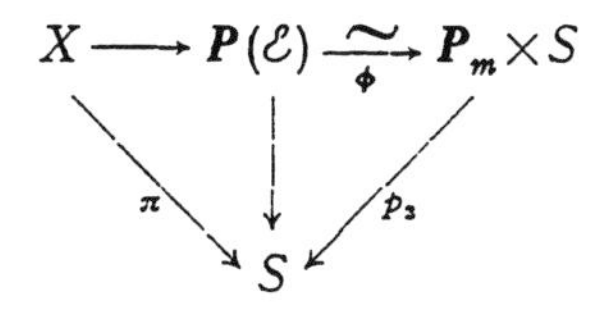

I claim that the whole structure defining $\alpha \in \mathcal{H}_{g,d,n}(S)$ is determined only by

(a) the embedding $\qquad I: X \hookrightarrow (\boldsymbol{P}_m \times S)$,

(b) the $2g+1$-sections of X/S: $\varepsilon, \sigma_1, \sigma_2, \ldots, \sigma_{2g}$.

Namely, by Corollary 6.6, ε determines the group law on X. The σ_i determine the level n-structure. On the other hand,

(*) $\qquad (p_1 \circ I)^* \left(\varrho_{\boldsymbol{P}_m}(1)\right) \cong L^\Delta(\overline{\omega})^3 \otimes \pi^* \varepsilon^* [(p_1 \circ I)^* \left(\varrho_{\boldsymbol{P}_m}(1)\right)]$,

since $\varepsilon^*\{L^\Delta(\overline{\omega})\} \cong \varrho_S$ (cf. § 6.2). Therefore, $L^\Delta(\overline{\omega})^3$ is determined by (a) and (b), hence

$$6\overline{\omega} = \Lambda\left(L^\Delta(\overline{\omega})^3\right)$$

is determined. But the group of S-homomorphisms from X to $\hat{X}$ has no torsion, hence $\overline{\omega}$ is determined. Putting $M = \varepsilon^* [(p_1 \circ I)^* \left(\varrho_{\boldsymbol{P}_m}(1)\right)]$, it follows that the embedding I gives a homomorphism:

$$\begin{array}{rcl} H^0\left(\boldsymbol{P}_m, \varrho_{\boldsymbol{P}_m}(1)\right) \otimes M^{-1} & \to & \pi_* \{(p_1 \circ I)^* \left(\varrho_{\boldsymbol{P}_m}(1)\right) \otimes \pi^*(M^{-1})\} \\ & & \wr\| \text{ via } (*) \\ & & \pi_* \left(L^\Delta(\overline{\omega})^3\right) \\ & & \| \\ & & \mathcal{E} \end{array}$$

which induces the isomorphism ϕ:

$$\boldsymbol{P}(\mathcal{E}) \xrightarrow[\phi]{\sim} \boldsymbol{P}[H^0\left(\boldsymbol{P}_m, \varrho_{\boldsymbol{P}_m}(1)\right) \otimes M^{-1}] = \boldsymbol{P}_m \times S.$$

Now the subscheme $I(X) \subset \boldsymbol{P}_m \times S$ is flat over S, and by Proposition 6.13, its Hilbert polynomial is easily computed to be:

$$P(X) = 6^g \cdot d \cdot X^g.$$

Therefore $I(X)$ defines an S-valued point of $Hilb_{\boldsymbol{P}_m}^{P(X)}$, i.e. the unique morphism $f: S \to Hilb_{\boldsymbol{P}_m}^{P(X)}$ such that

$$\begin{array}{ccc} I(X) & = & Z \times_{Hilb} S \\ \cap & & \cap \\ \boldsymbol{P}_m \times S & = & (\boldsymbol{P}_m \times Hilb) \times_{Hilb} S \end{array}$$

where $Z \subset \mathbf{P}_m \times Hilb_{\mathbf{P}_m}^{P(X)}$ is the universal flat family of subschemes with Hilbert polynomial $P(X)$. Now $I(X)$ comes with $2g+1$ sections over S also given.

For all k, put:

$$Hilb_{\mathbf{P}_m}^{P(X),k} = \overbrace{Z \underset{Hilb}{\times} \cdots \underset{Hilb}{\times} Z}^{k\text{-factors}}.$$

Note that there are k canonical sections in the diagram:

$$\begin{array}{c} Z \underset{Hilb}{\times} Hilb_{\mathbf{P}_m}^{P(X),k} = Z'^{k)} \\ \tau_i \uparrow \downarrow p_2 \\ Hilb_{\mathbf{P}_m}^{P(X),k} \end{array}$$

defined via the k-projections of $Hilb_{\mathbf{P}_m}^{P(X),k}$ onto Z. As in Ch. 0, § 5, (c), the closed subscheme

$$Z^{(k)} \subset \mathbf{P}_m \times Hilb_{\mathbf{P}_m}^{P(X),k}$$

and its k sections $\{\tau_i\}$ are the universal family of closed subschemes with k sections. Applying this to $I(X)$, we get not only an S-valued point of $Hilb_{\mathbf{P}_m}^{P(X)}$, but also one of $Hilb_{\mathbf{P}_m}^{P(X),2g+1}$.This is obviously functorial, so it defines a morphism of functors:

$$\mathcal{H}_{g,d,n} \xrightarrow{\Phi} \mathcal{H}ilb_{\mathbf{P}_m}^{P(X),2g+1}$$

(where the script letters indicate that the Hilb here means the functor rather than the scheme). By our comments above, it follows that this morphism is *injective*, i.e. $\Phi(\alpha)$ determines a subscheme $X \subset \mathbf{P}_m \times S$, flat over S, with Hilbert polynomial $P(X)$ and with $2g+1$ sections; and if this comes from any abelian scheme with level n structure, etc., all this extra structure is *uniquely determined*.

Proposition 7.3. There is a locally closed subscheme

$$H_{g,d,n} \subset Hilb_{\mathbf{P}_m}^{P(X),2g+1}$$

such that an S-valued point of $Hilb_{\mathbf{P}_m}^{P(X),2g+1}$ is in the image of Φ if and only if it is an S-valued point of $H_{g,d,n}$. Therefore $H_{g,d,n}$ represents $\mathcal{H}_{g,d,n}$.

Proof. For ease of notation put $H_0 = Hilb_{\mathbf{P}_m}^{P(X),2g+1}$, and let $Z_0 \subset \mathbf{P}_m \times H_0$, with $\tau_i : H_0 \to Z_0$, be the universal closed subscheme over H_0 with sections τ_i. We must analyze the conditions under which an induced subscheme $Z_0 \underset{H_0}{\times} S$ with sections:

$$(\tau_i \times 1_S) : S = H_0 \underset{H_0}{\times} S \to Z_0 \underset{H_0}{\times} S.$$

defined via $f : S \to H_0$, comes from an abelian scheme with level n structure, etc. We do this in a series of steps:

(I) Exactly as in the proof of Proposition 5.1, there is an open subscheme $H_1 \subset H_0$ such that the geometric fibres of Z_0 over H_0 are smooth and connected if and only if they are fibres over H_1. Any f inducing an abelian scheme must factor through H_1. Let $Z_1 = Z_0 \underset{H_0}{\times} H_1$.

(II) By Theorem 6.14, there is an open and closed subset $H_2 \subset H_1$ such that the geometric fibre over a point s of H_1 is an abelian variety if and only if s is a point of H_2. Let $Z_2 = Z_1 \underset{H_1}{\times} H_2$, and let τ_i still represent the sections of Z_2 over H_2. Put $\varepsilon = \tau_{2g+1}$. Then by Theorem 6.14, Z_2 has the structure of an abelian scheme over H_2 with identity ε; and, if $Z_2 \underset{H_2}{\times} S$ ever has the structure of abelian scheme over S, with identity induced by ε, it follows from Corollary 6.6 that this group structure is the same as that induced from the group structure on Z_2/H_2.

(III) Let $H_3 \subset H_2$ be the closed subscheme where

$$\begin{array}{ccc} Z_2 & \xrightarrow{\psi_n} & Z_2 \\ \uparrow{\scriptstyle \tau_i} & & \uparrow{\scriptstyle \varepsilon} \\ H_2 & = & H_2 \end{array}$$

commutes, for all $1 \leq i \leq 2g$, and where ψ_n is multiplication by n. Let $Z_3 = Z_2 \underset{H_2}{\times} H_3$, and let τ_i still denote the induced sections of Z_3/H_3. If $\tau_i, \ldots, \tau_{2g}$ are to induce a level n structure on some $Z_2 \underset{H_2}{\times} S$, it is certainly necessary that $f: S \to H_2$ factor through H_3.

(IV) Let $H_4 \subset H_3$ be the open subset where a) the residue field characteristics do not divide n, and b) the images of $\tau_1, \ldots, \tau_{2g}$ in each geometric fibre of Z_3/H_3 form a basis of the points of order n on that abelian variety. This certainly exists, because to form a basis means that there is no equation:

$$a_1\tau_1 + a_2\tau_2 + \cdots + a_{2g}\tau_{2g} = \varepsilon, \qquad 0 \leq a_i < n$$

relating the image points, unless all $a_i = 0$. Therefore, H_4 is the intersection over the finite set of sequences $\{a_i \mid 0 \leq a_i < n, \text{ some } a_i \neq 0\}$, of open sets $U\{a_i\}$; where $U\{a_i\}$ is the set of points where the 2 sections $\sum a_i\tau_i$ and ε of Z_3/H_3 differ. If $Z_4 = Z_3 \underset{H_3}{\times} H_4$, and if τ_i still denotes the restriction of τ_i to H_4, then the τ_i define a level n structure of Z_4/H_4; and whenever the induced sections of $Z_3 \underset{H_3}{\times} S/S$ form a level n structure, the morphism $f: S \to H_3$ factors through H_4.

(V) Finally, we look at the polarization: by definition Z_0 was a subscheme of $\boldsymbol{P}_m \times H_0$. Therefore Z_4 is a subscheme of $\boldsymbol{P}_m \times H_4$. Let L_4 be the invertible sheaf on Z_4 equal to

$$\underline{o}_{Z_4} \otimes \{p_1^*(\underline{o}_{\boldsymbol{P}_m}(1))\}.$$

Let π be the projection from Z_4 to H_4 and let $L'_4 = L_4 \otimes \pi^*(\varepsilon^* L_4)^{-1}$; then L'_4 is trivial along the identity section. If the induced abelian scheme $Z_4 \times_{H_4} S$ together with its projective embedding in $\boldsymbol{P}_m \times S$ comes from a polarized and linearly rigidified abelian scheme, then that polarization $\overline{\omega}_S$ must satisfy:

$$L'_4 \otimes_{\varrho_H} \varrho_S \cong L^\Delta(\overline{\omega}_S)^3.$$

According to Proposition 6.11, there is a closed subscheme $H_5 \subset H_4$ such that $L'_4 \otimes_{\varrho_H} \varrho_S$ is of the form $L^\Delta(\overline{\omega}_S)^3$ if and only if $f: S \to H_4$ factors through H_5. Let $Z_5 = Z_4 \times_{H_4} H_5$. Then Z_5 has a polarization

$$\overline{\omega}_5 : Z_5 \to \hat{Z}_5$$

such that if $L'_5 = L'_4 \otimes \varrho_{Z_5}$, then

(*) $$L'_5 \cong L^\Delta(\overline{\omega}_5)^3.$$

(VI) We have seen that Z_5/H_5 has both a level n-structure and a polarization. Moreover, this polarization is related to the projective embedding via (*). From this, we deduce a homomorphism

($\tilde{*}$) $$H^0(\varrho_{\boldsymbol{P}_m}(1)) \otimes M^{-1} \to \pi_*(L^\Delta(\overline{\omega}_5)^3)$$

where $M = \varepsilon^*[L_5] = \varepsilon^*[\varrho_{Z_5} \otimes p_1^*(\varrho_{\boldsymbol{P}_m}(1))]$ is an invertible sheaf on H_5. Now if, after any base extension $f: S \to H_5$, $\boldsymbol{P}[\pi_*(L^\Delta(\overline{\omega}_5)^3)]$ admits a linear rigidification which defines the *given* embedding of Z_5 in $\boldsymbol{P}_m \times H_5$, it follows that ($\tilde{*}$) becomes an isomorphism after that base extension and that ($\tilde{*}$) defines this linear rigidification by taking $\boldsymbol{P}$ on both sides. Put H_6 equal to the open subset of H_5 where the kernel and cokernel of ($\tilde{*}$) vanish. Let $Z_6 = Z_5 \times_{H_4} H_6$. Then over H_6, ($\tilde{*}$) induces the commutative diagram:

$$\begin{array}{ccc} Z_6 & \xrightarrow{I} & \boldsymbol{P}(\pi_*(L^\Delta(\overline{\omega}_6)^3)) \\ \downarrow J & & \wr\downarrow \boldsymbol{P}((\tilde{*})) \\ \boldsymbol{P}_m \times H_6 & = & \boldsymbol{P}(H^0(\varrho_{\boldsymbol{P}_m}(1)) \otimes M^{-1}) \end{array}$$

where J is the given embedding with which we started, and I is the canonical embedding associated to the very ample $L^\Delta(\overline{\omega}_6)^3$. Therefore Z_6/H_6 is linearly rigidified in such a way as to induce the original projective embedding.

This proves that $H_{g,d,n} = H_6$. *QED.*

We fix the notation:

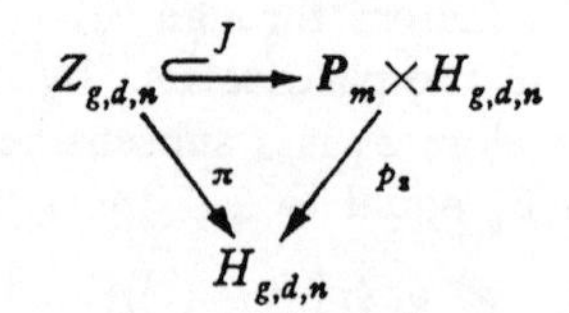

$$\begin{array}{ccc} Z_{g,d,n} & \xhookrightarrow{J} & \boldsymbol{P}_m \times H_{g,d,n} \\ & \searrow^{\pi} \quad {}^{p_2}\swarrow & \\ & H_{g,d,n} & \end{array}$$

for the abelian scheme Z_6 and its projective embedding which occurred in this proof: $Z_{g,d,n}$ is the universal linearly rigidified abelian scheme.

Now note that the group scheme $PGL(m+1)$ acts on all these schemes. In fact, it acts on the functor $\mathscr{H}_{g,d,n}$ as follows; let $\phi: \boldsymbol{P}(\mathscr{E}) \xrightarrow{\sim} \boldsymbol{P}_m \times S$ be a linear rigidification of a polarized abelian scheme X/S, and let $\alpha: \boldsymbol{P}_m \times S \xrightarrow{\sim} \boldsymbol{P}_m \times S$ be the action of an S-valued point of $PGL(m+1)$ on $\boldsymbol{P}_m$. Then the action of α does not affect the polarized abelian scheme, but it transforms the linear rigidification to $\alpha \circ \phi$. Note that with respect to this action on $\mathscr{H}_{g,d,n}$ and the obvious action of $PGL(m+1)$ on the functor $\mathscr{H}ilb_{\boldsymbol{P}_m}^{P(X),2g+1}$, the morphism Φ of functors if $PGL(m+1)$-linear. It follows that there is an action of $PGL(m+1)$ on the scheme $H_{g,d,n}$ and that this action is induced from the action on the larger scheme $Hilb_{\boldsymbol{P}_m}^{P(X),2g+1}$. Finally $PGL(m+1)$ acts on the product $\boldsymbol{P}_m \times H_{g,d,n}$; and it acts on $Z_{g,d,n}$ since this represents the functor of linearly rigidified polarized abelian schemes with level n structure, and with 1 extra section. Moreover J is readily checked to be $PGL(m+1)$-linear.

It is important to check:

Proposition 7.4. The schemes $Z_{g,d,n}$ and $H_{g,d,n}$ are quasi-projective over Spec $(\boldsymbol{Z})$ and carry $PGL(m+1)$ linearized ample invertible sheaves

Proof. Z and H are embedded respectively in $\boldsymbol{P}_m \times Hilb_{\boldsymbol{P}_m}^{P(X),2g+1}$ and $Hilb_{\boldsymbol{P}_m}^{P(X),2g+1}$; and both of these are embedded $PGL(m+1)$-linearly in schemes of the type

$$(\boldsymbol{P}_m)^k \times Hilb_{\boldsymbol{P}_m}^{P(X)}$$

for some k. The result is obviously true for $\boldsymbol{P}_m$; hence if we prove it for $Hilb_{\boldsymbol{P}_m}^{P(X)}$, it follows for all the other subschemes. But GROTHENDIECK has exhibited embeddings of the Hilbert schemes in certain Grassmannian spaces parametrizing the subspaces of suitable dimension of $H^0(\boldsymbol{P}_m, o_{\boldsymbol{P}_m}(N))$ for a large N, (cf. [13], exposé 221). These embeddings are also $PGL(m+1)$-linear, and since these Grassmannians obviously prosess $PGL(m+1)$-linearized ample invertible sheaves, so does $Hilb_{\boldsymbol{P}m}^{P(x)}$. *QED.*

Another important point concerning Z/H is this: suppose X/S is an arbitrary polarized abelian scheme with level n structure and linear rigidification: $\phi: \boldsymbol{P}(\mathscr{E}) \xrightarrow{\sim} \boldsymbol{P}_m \times S$. By Proposition 7.3, there is a *unique* morphism $f: S \to H$ such that X/S and all its structure is isomorphic to $Z \times S$ with all of its structure. But more than that is true! There is even a *unique* S-isomorphism F of X and $Z \times_H S$ such that all the structures correspond. One need only look at the 2 linear rigidifications and note that the following must commute:

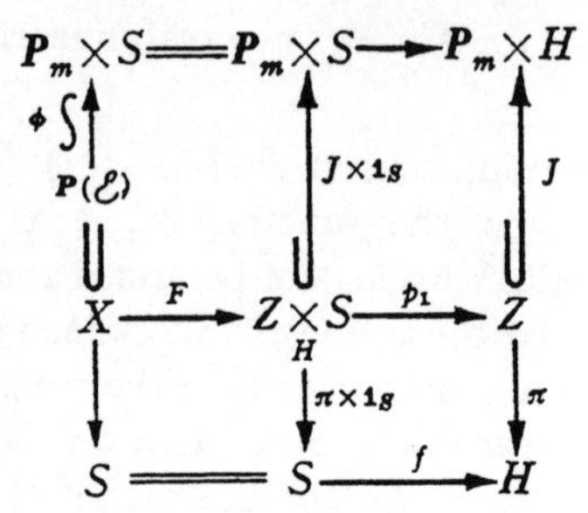

There is only one possible isomorphism F of the 2 subschemes X and $Z \times_H S$ of $\mathbf{P}_m \times S$.

The analog of Proposition 5.2 is:

Proposition 7.5. Let $\pi : X \to S$, $\bar{\omega} : X \to \hat{X}$, and $\{\sigma_i\} : S \to X$ be a polarized abelian scheme with level n structure over S. Then (a) if $\phi_1, \phi_2 : \mathbf{P}(\mathcal{E}) \cong \mathbf{P}_m \times S$ are two linear rigidifications of X/S, then ϕ_1 and ϕ_2 differ by the action of an S-valued point of $PGL(m+1)$; (b) there is an open covering $\{U_i\}$ of S such that the restriction of X/S to U_i admits a linear rigidification.

Proof. (b) is obvious; (a) follows from the fact that any S-automorphism of $\mathbf{P}_m \times S$ is the action of some S-valued point of $PGL(m+1)$ (cf. Ch. 0, § 5, (b)). *QED.*

The analog of Proposition 5.4 is:

Proposition 7.6. If $H_{g,d,n}$ is a principal fibre bundle over a scheme A, quasi-projective over $\mathbf{Z}$, with group $PGL(m+1)$, then A is a fine moduli scheme for polarized abelian varieties with level n structure.

Proof. For the sake of simplicity, we drop the subscripts g, d, n in the course of this proof: First of all, by Proposition 7.1 and Proposition 7.4, there is a scheme Z_0 over A and a morphism from Z to Z_0 such that i) $Z \cong Z_0 \times_A H$ and ii) Z is a principal fibre bundle over Z_0 with group $PGL(m+1)$. Next I claim that the group law on Z/H, the polarization, and the level n structure all descend to the scheme Z_0/A. All these are applications of Amplification 7.2 once one verifies that the group law $\mu : Z \times_H Z \to Z$, the polarization $\bar{\omega} : Z \to \hat{Z}$ and the level n structure $\sigma_i : H \to Z$ are all $PGL(m+1)$-linear morphisms. But the various schemes $Z, Z \times_H Z, \hat{Z}$, etc. all represent functors such as the set of linearly rigidified abelian schemes with level n structure *and* with 0,1, or 2 extra sections or with a given invertible sheaf of some type. And the action of $PGL(m+1)$ on each scheme comes from an action of $PGL(m+1)$ on each functor which alters *only* the linear rigidification. Therefore operations intrinsic to the abelian schemes such as multiplication, applying a polarization to a rational point, etc., all commute with the action of $PGL(m+1)$ on these functors. This proves that all this structure descends to Z_0/A.

To prove the Proposition, we must show that for every abelian scheme $\pi: X \to S$, with polarization and level n structure, there is a unique $f: S \to A$ such that X/S, with its structure, is isomorphic to $Z_0 \times_A S/S$, with its structure. We shall show that there is also a unique $F: X \to Z_0$ which induces an isomorphism of X and $Z_0 \times_A S$, with their structures. If we show both of these, it follows that we need only prove them *locally on* S: for then the unique f_i's and F_i's defined, say, on $U_i \subset S$ and $X_i = X \times_S U_i \subset X$, must patch together to give a global f and F. This being the case, we can assume by Proposition 7.5 that X/S admits a linear rigidification: i.e. it is induced together with all its structure from Z/H via $g: S \to H$ and $G: X \to Z$. Then X/S is clearly induced from Z_0/A by *some* base extension.

To prove uniqueness, we first deduce the following: if $g: S \to H$ is a morphism, and $\pi: X \to S$ is the induced polarized abelian scheme with level n structure and linear rigidification ϕ_0, then there is a unique $G: X \to Z$ setting up an isomorphism $(G, \pi): X \to Z \times_H S$ of polarized abelian schemes with level n structure. We saw above that there was only one G, say G_0, under which also the linear rigidifications of X and $Z \times_H S$ correspond. But suppose that G is any such morphism. Then there is a unique morphism ϕ completing the diagram:

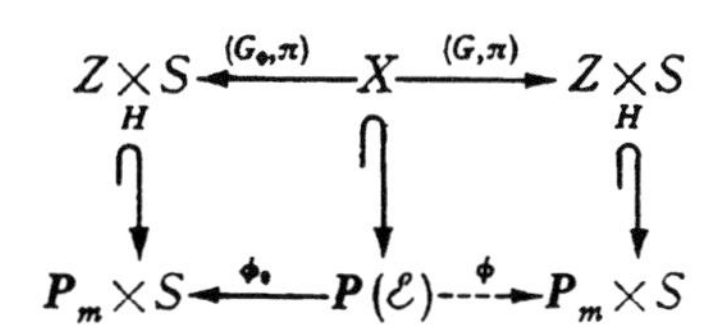

and ϕ is another linear rigidification. But let $\phi \circ \phi_0^{-1}$ be the action of the S-valued point α of $PGL(m+1)$. Then the diagram means that the action of α transforms the subscheme $Z \times_H S$ (and its $2g+1$ sections) into itself. By the functorial definition of H, this means that when the S-valued point α of $PGL(m+1)$ acts on the S-valued point g of H, the result is still the point g. But since H is a principal homogeneous space over A, $PGL(m+1)$ acts freely on H. Therefore $\alpha = \varepsilon$, hence $\phi = \phi_0$, hence $G = G_0$. This proves that G is unique.

Now suppose that $f_i: S \to A$ and $F_i: X \to Z_0$, for $i = 1, 2$, both set up isomorphisms of X and $Z_0 \times_A S$ as polarized abelian schemes with level n structure. As a first step, suppose that the f_i factor through H

via $g_i : S \to H$. Then it follows that the 2 abelian schemes $Z \underset{H}{\times} S$ (via g_1 and g_2) are isomorphic with all their structure except their linear rigidification. Therefore, by Proposition 7.5, g_1 and g_2 differ only by the action of an S-valued point of $PGL(m)$. Therefore $f_1 = f_2$. Now we may assume that $f = f_1 = f_2$ is factored by $g : S \to H$. But the two morphisms

$$G_i = (F_i, g) : X \to Z_0 \underset{A}{\times} H = Z$$

set up isomorphisms of X and $Z \underset{H}{\times} S$: therefore, by what we just proved $G_1 = G_2$. Therefore $F_1 = F_2$.

Finally, suppose the f_i do not factor through H. But after a suitable faithfully flat base extension S'/S, the morphisms $f'_i : S' \to A$ will factor through H: e.g. let $S' = H \underset{A}{\times} S$. It then follows that the extended morphisms are equal: $f'_1 = f'_2$ and $F'_1 = F'_2$. Therefore $f_1 = f_2$ and $F_1 = F_2$ by Theorem 5.2, SGA 8. *QED.*

§ 3. The covariant

We now construct the covariant.

Proposition 7.7. Let A be an abelian variety over an algebraically closed field k. Let $\phi : A \to \mathbf{P}_m$ be a closed immersion, let g be the dimension of A, and let r be the degree of $\phi(A)$ in $\mathbf{P}_m$. Assume that $\phi(A)$ is not contained in any hyperplane. Then for any n such that char $(k) \nmid n$, and $n > \sqrt{(m+1) \cdot r}$, the set of points of order n on A, arbitrarily ordered, is a point of

$$(\mathbf{P}_m)^{n^{2g}}_{\text{stable}}.$$

Proof. We use the calculus of the chow ring to prove this. It will certainly suffice if we show that less than $n^{2g}/m + 1$ of the n^{2g} points of order n are contained in any hyperplane H. To estimate this number, let γ be a 1-cycle on A through the identity in A obtained by intersecting A with a suitable linear subspace in $\mathbf{P}_m$ of codimension $g - 1$. Let $\lambda_n : A \to A$ be multiplication by n. Therefore if $h \subset A$ is a hyperplane section $\phi^{-1}(H)$

$$\{\text{number of points of order } n \text{ on } H\} \leq \left(\lambda^{-1}(\gamma) \cdot h\right).$$

But $(\lambda^{-1}(\gamma) \cdot h) = (\lambda^*(\bar{\gamma}) \cdot \bar{h})$, where $\bar{\gamma}$ and $\bar{h}$ are the elements of the chow ring of A representing γ and h. Calculating, we find:

$$\begin{aligned} n^2(\lambda^* \bar{\gamma} \cdot \bar{h}) &= (\lambda^* \bar{\gamma} \cdot \lambda^* \bar{h}) \\ &= \lambda^*(\bar{\gamma} \cdot \bar{h}) \\ &= n^{2g} \cdot (\bar{\gamma} \cdot \bar{h}) \\ &= n^{2g} \cdot r \end{aligned}$$

since $\lambda^* \bar{h} = n^2 \bar{h}$ (cf. LANG [21], p. 92), and $\deg \lambda = n^{2g}$. Therefore,

$$\frac{\text{number of points of order } n \text{ on } H}{n^{2g}} \leq \left(\frac{r}{n^2}\right) < \left(\frac{1}{m+1}\right). \quad QED.$$

Corollary 7.8. If $n > 6^g \cdot d \cdot \sqrt{g!}$, there is a $PGL(m)$-linear morphism:

$$H_{g,d,n} \to (\boldsymbol{P}_m)^{n^{2g}}_{\text{stable}}.$$

Proof. Let $H = H_{g,d,n}$ for simplicity. Let $\pi: Z \to H$, $\{\sigma_i\}: H \to Z$, and $I: Z \to \boldsymbol{P}_m \times H$ be the universal abelian scheme over H, its level n structure, and the projective embedding determined by its linear rigidification respectively. Number the sequences of $2g$ integers between 0 and $n-1$ in some way, and let:

$$i^{\text{th}} \text{ sequence} = (k_1^{(i)}, k_2^{(i)}, \ldots, k_{2g}^{(i)}), \qquad 0 \leq k_j^{(i)} \leq n-1.$$

Define ϕ_i to be the composition:

$$H \xrightarrow{\sum_{j=1}^{2g} k_j^{(i)} \sigma_j} Z \xrightarrow{I} \boldsymbol{P}_m \times H \xrightarrow{p_1} \boldsymbol{P}_m.$$

Since there are n^{2g} sequences of k's, this defines

$$\prod_{i=1}^{n^{2g}} \phi_i : H \to (\boldsymbol{P}_m)^{n^{2g}}.$$

By Proposition 7.7, the image is stable. *QED.*

Theorem 7.9. If $n > 6^g \cdot d \cdot \sqrt{g!}$, the fine moduli scheme $A_{g,d,n}$ for abelian varieties of dimension g, with level n structure and polarization of degree d exists. It is quasi-projective over Spec $(\boldsymbol{Z})$.

Proof. This follows from Theorem 3.8, Proposition 7.1, Proposition 7.4, Proposition 7.6, and Corollary 7.8. *QED.*

Actually, using the so-called "lemma of Serre" (cf. [5], 17—18), it can be shown that Theorem 7.9 is true even if $n \geq 3$. We omit this result and go on to prove:

Theorem 7.10. For all g, d, n, the coarse moduli scheme $A_{g,d,n}$ exists. It is quasi-projective over every open set Spec $(\boldsymbol{Z}) - (p)$ in Spec $(\boldsymbol{Z})$.

Proof. By an argument identical to that in Proposition 5.4, it follows that if A is a geometric quotient of $H_{g,d,n}$ by the action of $PGL(m+1)$, then A is a coarse moduli scheme. This geometric quotient will be constructed in 2 pieces. For every integer k, let

$$H^{(k)}_{g,d,n} = H_{g,d,n} \times \left\{\text{Spec}\,(\boldsymbol{Z}) - \bigcup_{p|k} (p)\right\}.$$

We shall show that, for any k, the geometric quotient $H^{(k)}_{g,d,n}/PGL(m+1)$ exists. Taking these quotients for 2 relatively prime k's, we can glue them together and obtain $H_{g,d,n}/PGL(m+1)$. Moreover, we shall show that each piece $H^{(k)}_{g,d,n}/PGL(m+1)$ is quasi-projective.

Now suppose $\sigma_1, \ldots, \sigma_{2g}$ is a level nk-structure on an abelian scheme X/S. Then $k\sigma_1, \ldots, k\sigma_{2g}$ is a level n structure on X/S. This defines a morphism $p_n^{(k)}$ from $\mathcal{H}_{g,d,nk}$ to $\mathcal{H}_{g,d,n}$, hence from $H_{g,d,nk}$ to $H_{g,d,n}$. Moreover suppose

$$T = (a_{ij}) \in \Gamma_n$$

$$\Gamma_n = GL(2g; \mathbf{Z}/(n))$$

(i.e. each $a_{ij} \in \mathbf{Z}/(n)$, and det (a_{ij}) is a unit in $\mathbf{Z}/(n)$). Then if $\sigma_1, \ldots, \sigma_{2g}$ is a level n structure, the set of sections $\sum_{j=1}^{2g} a_{1j}\sigma_j, \ldots, \sum_{j=1}^{2g} a_{2g,j}\sigma_j$ is also a level n structure. Therefore Γ_n acts on $\mathcal{H}_{g,d,n}$, hence on $H_{g,d,n}$; and it acts on $\mathcal{A}_{g,d,n}$, hence if a fine moduli scheme $A_{g,d,n}$ exists, it acts on $A_{g,d,n}$. Finally, there is a canonical morphism from Γ_{nk} to Γ_n. Let $\Gamma_n^{(k)}$ be its kernel. Then we have:

Lemma 7.11. Let n and k be positive integers. Then $p_n^{(k)}(H_{g,d,nk}) \subset H_{g,d,n}^{(k)}$, and, taking $H_{g,d,n}^{(k)}$ as image, $p_n^{(k)}$ is a finite étale morphism. Moreover, with respect to the action of $\Gamma_n^{(k)}$ on $H_{g,d,nk}$, we have:

$$\Gamma_n^{(k)} \times H_{g,d,nk} \cong H_{g,d,nk} \underset{H_{g,d,n}}{\times} H_{g,d,nk}.$$

In other words, $H_{g,d,nk}$ is a principal covering of $H_{g,d,n}$ with Galois group $\Gamma_n^{(k)}$ (cf. SGA 5, Def. 2.7).

Proof. For the sake of simplicity, we drop the subscripts g and d. To prove the lemma, let Z_n/H_n be the universal linearly rigidified abelian scheme. Assume that $n \geq 2$: if $n = 1$, the same argument will be valid with slight modifications. Let $\sigma_1, \ldots, \sigma_{2g}$ be the level n structure on $Z_n^{(k)}/H_n^{(k)}$. Let $\psi_k : Z_n^{(k)} \to Z_n^{(k)}$ be multiplication by k. Recall that ψ_k is flat (lemma 6.12) and finite, (since Z_n is proper over H_n and the fibres of ψ_k consist of finite sets of points). Therefore $\psi_{k,*}(\underline{o}_Z)$ is a locally free sheaf, and, in fact, its rank is k^{2g} (cf. LANG [21], Th. 6, p. 109). But since no residue field characteristics divide k, the geometric fibres of ψ_k consist of exactly k^{2g} points; therefore the scheme-theoretic fibres are reduced. This proves that ψ_k is étale. Define G_i as the fibre product:

$$\begin{array}{ccc} G_i & \longrightarrow & Z_n^{(k)} \\ \downarrow & & \downarrow \psi_k \\ H_n^{(k)} & \xrightarrow{\sigma_i} & Z_n^{(k)} \end{array}$$

and define $G = [G_1 \underset{H_n}{\times} \cdots \underset{H_n}{\times} G_{2g}]$. Then G is a finite étale covering of $H_n^{(k)}$. Let $Y = G \underset{H_n}{\times} Z_n^{(k)}$, and let σ_i extend to $\sigma_i' : G \to Y$. Then the projections of G to G_i, and from G_i to $Z_n^{(k)}$ define $2g$ sections $\tau_1, \ldots, \tau_{2g}$ of Y over G such that $k\tau_i = \sigma_i$. Let G_0 be the open subset in G where no relation

$$\sum_{i=1}^{2g} a_i \tau_i = \varepsilon, \quad \begin{cases} 0 \leq a_i < nk - 1 \\ \text{some } a_i > 0 \end{cases}$$

between these sections holds. Then I claim that $H_{nk} = G_0$, $Y \underset{G}{\times} G_0$ being the universal linearly rigidified abelian scheme over G_0 with level nk-structure $\{\tau_i\}$. The proof of this is easy, and is left to the reader. This shows that H_{nk} is étale over $H_n^{(k)}$.

Now the action of $\Gamma_n^{(k)}$ on an abelian scheme with level nk structure preserves the underlying level n structure. Therefore, there is a canonical morphism, (over H_{nk}):

$$\Gamma_n^{(k)} \times H_{nk} \xrightarrow{\pi} H_{nk} \underset{H}{\times} H_{nk}.$$

Since the projection from $\Gamma_n^{(k)} \times H_{nk}$ to H_{nk} is proper, it follows that π is proper. Therefore, to prove that π is an isomorphism, it suffices to prove that π induces an isomorphism on the geometric fibres over H_{nk} (EGA 3, 4.6.7). Or that the geometric fibres of H_{nk} over $H_n^{(k)}$ are finite sets of reduced points, on which $\Gamma_n^{(k)}$ is acting freely and transitively. But we have just proven that these fibres are reduced; and on an abelian variety, it is clear that two level nk structures inducing the same level n structure are transformed into each other by a unique element of $\Gamma_n^{(k)}$. Therefore π is an isomorphism.

Moreover, the image of $p_n^{(k)}$ is certainly $H_n^{(k)}$: for an abelian variety with level n structure admits a level nk structure refining its level n structure if and only if its characteristic does not divide k. Therefore H_{nk} is faithfully flat over $H_n^{(k)}$, and by SGA 8, Cor. 4.8, it is finite if and only if the projection from $H_{nk} \underset{H_n}{\times} H_{nk}$ to H_{nk} is finite. But since π is an isomorphism of schemes over H_{nk}, this is proven too. *QED.*

We now return to the construction of $H_{g,d,n}^{(k)}/PGL(m+1)$. Replacing k by any of its powers does not change $H_{g,d,n}^{(k)}$; therefore we may assume that k is large enough so that

$$nk > 6^g \cdot d \cdot \sqrt{g!}.$$

Then a fine moduli scheme $A_{g,d,nk}$ exists by Theorem 7.9, it is quasi-projective over $\mathbf{Z}$, and the finite group $\Gamma_n^{(k)}$ acts on it. But it is well-known that whenever a finite group acts on a quasi-projective scheme, then a geometric quotient exists. Therefore, we have the diagram:

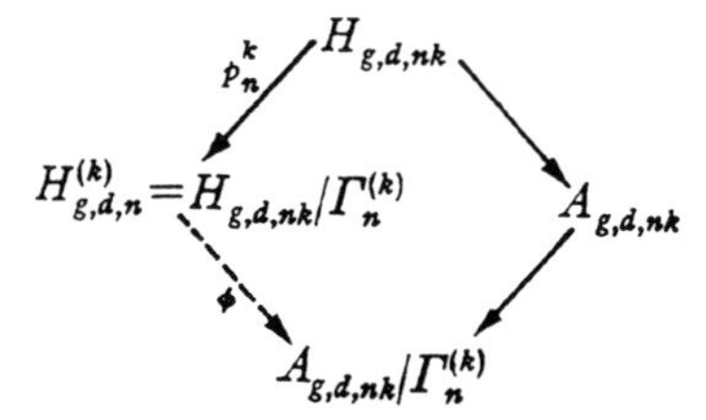

According to the following lemma, there is a unique ϕ such that this diagram commutes, and $(A_{g,d,nk}/\Gamma_n^{(k)}, \phi)$ is then a geometric quotient of

$H^{(k)}_{g,d,n}$ by $PGL(m+1)$, as required:

Lemma 7.12. Suppose we have a diagram:

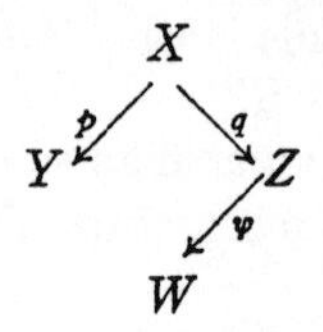

of schemes over a fixed scheme S. Suppose G and H are group schemes over S. Suppose that $G \underset{S}{\times} H$ acts on X, H acts on Y, G acts on Z, p is H-linear, q is G-linear, (Y, p) is a geometric quotient of X by G, (Z, q) is a geometric quotient of X by H, and (W, ψ) is a geometric quotient of Z by G. Then there is a unique morphism $\phi: Y \to W$ such that $\phi \circ p = \psi \circ q$; and (W, ϕ) is a geometric quotient of Y by H.

Proof. Immediate. *QED.*

It can actually be proved that $A_{g,d,n}$ is quasi-projective *globally* over Spec $(\mathbf{Z})$. This result belongs to the general theory of the Picard group of the moduli problem which I shall consider elsewhere.

§ 4. Application to curves

We conclude this chapter by constructing the coarse moduli schemes for curves. There are two quite different ways to obtain this result from the results of the previous sections: one is quite functorial, using the existence of fine moduli schemes for high level abelian varieties, and avoiding Torelli's theorem; the other, which I follow, depends more on our precise knowledge of the relation between curves and their jacobians, and is throughout a "coarse" treatment.

We begin with the jacobian functor

$$\mathcal{M}_g \xrightarrow{j} \mathcal{A}_{g,1,1}$$

defined as follows. Let $\pi: \Gamma \to S$ be a curve representing an element of $\mathcal{M}_g(S)$. According to Proposition 6.9, $Pic^{\tau}(\Gamma/S)$ is an abelian scheme over S, which we will write J. And there is a canonical isomorphism:

$$\theta: J \to \hat{J}.$$

Moreover, it is well-known over a field that $-\theta$ is induced by the Θ-divisor class, and that this divisor class is ample. Therefore, by Definition 6.3, $-\theta$ is a polarization. J and $-\theta$ define an element of $\mathcal{A}_{g,1,1}(S)$, and this is obviously a functor.

A great deal is known about j, thanks to Torelli's theorem (cf. [2], [22], [35], [341]), and to recent work of MATSUSAKA ([23]), and AHLFORS ([1]).

(*) (Torelli) If Ω is an algebraically closed field, then

$$j(\operatorname{Spec}\Omega) : \mathcal{M}_g(\operatorname{Spec}\Omega) \to \mathcal{A}_{g,1,1}(\operatorname{Spec}\Omega)$$

is *injective.*

Theorem 7.13. If $\nu \geq 3$, and H_ν is the scheme defined in § 5.2, then a geometric quotient of H_ν by $PGL(n+1)$ exists. Moreover, it is quasi-projective over every open subset Spec $(\mathbf{Z}) - (p)$ in Spec $\mathbf{Z}$.

Proof. Let Γ_ν be the curve over H_ν induced by the embedding $H_\nu \subset Hilb_{\mathbf{P}_n}^{P(X)}$ of Proposition 5.1. Let Γ_ν/H_ν define the element $\alpha_\nu \in \mathcal{M}_g(H_\nu)$. Let $A_{g,1,1}$ be the coarse moduli scheme for level 1, degree abelian schemes. Then $j(\alpha_\nu)$ defines a morphism:

$$j_\nu : H_\nu \to A_{g,1,1}.$$

But, if Ω is any algebraically closed field, the set of geometric points of $A_{g,1,1}$ is isomorphic to $\mathcal{A}_{g,1,1}(\operatorname{Spec}\Omega)$; on the other hand $\mathcal{M}_g(\operatorname{Spec}\Omega)$ is isomorphic by Proposition 5.2 to $\mathcal{M}'_g(\operatorname{Spec}\Omega)$, hence by definition of $\mathcal{M}'_g$ to the set of orbits under $PGL(n+1)$ in the set of geometric points of H_ν over Ω. Therefore, by Torelli's theorem, j_ν sets up an injection from the set of orbits of geometric points of H_ν to the set of geometric points of $A_{g,1,1}$.

Now let Z be the closure of $j_\nu(H_\nu)$, and let Z' be the normalization of Z. By a theorem of Nagata, Z' is finite over Z (cf. EGA 4, § 7.7); therefore Z' is still quasi-projective. Moreover, by Proposition 5.3, j_ν factors:

$$H_\nu \xrightarrow{j'_\nu} Z' \longrightarrow Z \subset A_{g,1,1}.$$

Moreover, if $\sigma_\nu : PGL(n+1) \times H_\nu \to H_\nu$ is the action of $PGL(n+1)$ no H_ν, it is still true that $j'_j \circ \sigma_\nu = j'_\nu \circ p_2$. For $PGL(n+1) \times H_\nu$ is a reduced and irreducible scheme, and as Z' and Z are generically isomorphic, $j'_\nu \circ \sigma_\nu$ equals $j'_\nu \circ p_2$ at the generic point of $PGL(n+1) \times H_\nu$; therefore they are everywhere equal. This shows that all the conditions of Proposition 0.2 are fulfilled. Consequently $j'_\nu(H_\nu)$ is an open subset of Z', and it is a geometric quotient of H_ν by $PGL(n+1)$. *QED.*

As a matter of fact, it follows from the results of Matsusaka cited above and Hoyt [151] that if $Z'_0 = j'_\nu(H_\nu)$,

i) the image Z_0 of Z'_0 in Z is open,

ii) the morphism $Z'_0 \to Z_0$ is finite, birational and radicial.

Corollary 7.14. The coarse moduli scheme over $\mathbf{Z}$ for curves of genus g, $(g \geq 2)$ exists, and is quasi-projective over every open subset Spec $(\mathbf{Z}) - (p)$ in Spec $(\mathbf{Z})$.

Proof. Proposition 5.4 and 7.13. *QED.*

Chapter 8

The moment map

The second edition of this book noted[27] a recent observation of Mumford, Guillemin and Sternberg, that geometric invariant theory for complex projective varieties is linked to the concept of the moment map[28] in symplectic geometry. In the last decade the relationship between geometric invariant theory and the moment map has been studied and exploited extensively. In this chapter we shall describe some of these developments. We shall also discuss some important applications of Yang-Mills theory (which arose in theoretical physics) to the study of moduli spaces of bundles and related objects. The link between the Yang-Mills functional and holomorphic vector bundles can be explained using symplectic geometry and the moment map (though infinite-dimensional manifolds are involved).

Throughout this chapter $\boldsymbol{P}_n$ will denote complex projective space $\boldsymbol{P}(\boldsymbol{C}^{n+1})$, $GL(n+1)$ will denote the complex general linear group and in general standard notations will refer to schemes over $\boldsymbol{C}$.

§ 1. Symplectic geometry

First let us recall the definition of a moment map for the action of a group on a symplectic manifold.

A symplectic manifold is a C^∞ manifold X equipped with a non-degenerate closed 2-form ω. (In particular any Kähler manifold is a symplectic manifold.) The Darboux lemma ([592] Theorem 22.2) tells us that near any point of X there exist local coordinates $(q_1, \dots, q_n, p_1, \dots, p_n)$ with respect to which ω is given by the standard symplectic form

$$\omega = \sum_{j=1}^{n} dp_j \wedge dq_j\,. \tag{1}$$

Suppose that a Lie group K with Lie algebra $\mathfrak{k}$ acts smoothly on X and preserves the symplectic form ω. We shall assume throughout this chapter that K is compact and connected[29] although this is not necessary for the definition of a moment map. The infinitesimal action of K is a

[27] See Appendix 2C.

[28] The terminology "momentum map" is also used for this concept and is perhaps more appropriate (cf. Example 8.1(i)).

[29] The assumption that K is connected involves very little loss of generality but simplifies some results.

Lie algebra homomorphism from $\mathfrak{k}$ to the Lie algebra of smooth vector fields on X: let us denote the vector field on X associated to $a \in \mathfrak{k}$ by

$$x \mapsto a_x \, .$$

A moment map (or momentum map) for the action of K on X is a map

$$\mu : X \to \mathfrak{k}^*$$

which is K-equivariant with respect to the given action of K on X and the coadjoint action of K on $\mathfrak{k}^*$, and satisfies

$$d\mu(x)(\xi).a = \omega_x(\xi, a_x)$$

for all $x \in X$, $\xi \in T_x X$ and $a \in \mathfrak{k}$. In other words the derivative $d\mu_a$ of the component[30] μ_a of μ along $a \in \mathfrak{k}$ is the 1-form on X which corresponds under the duality defined by ω to the vector field on X induced by a. In the language of symplectic geometry μ_a is a Hamiltonian function for this vector field.

If a moment map exists then it is uniquely determined by these conditions up to the addition of a constant in $\mathfrak{k}^*$ fixed by the coadjoint action. If the compact group K is semisimple then a unique moment map always exists; another condition which ensures the existence of a moment map is the vanishing of $H^1(X; \boldsymbol{R})$ [862], [723], §26 of [592].

Examples 8.1. (i). The cotangent bundle T^*Y of any n-dimensional manifold Y carries a canonical symplectic form ω. This is given by the standard symplectic form (1) with respect to any local coordinates $(q_1, \ldots, q_n)$ on Y and the induced coordinates $(p_1, \ldots, p_n)$ on its cotangent spaces. If Y is the configuration space of a classical mechanical system then T^*Y is the phase space of the system and the coordinates $p = (p_1, \ldots, p_n) \in T^*_q Y$ are traditionally called the momenta of the system.

When a Lie group K acts on Y the induced action on T^*Y preserves ω and there is a moment map

$$\mu : T^*Y \to \mathfrak{k}^*$$

whose components μ_a along $a \in \mathfrak{k}$ are given by pairing the momenta p with the vector fields on X induced by the infinitesimal action of K:

$$\mu_a(p, q) = p.a_q$$

---

[30] The component μ_a of μ along $a \in \mathfrak{k}$ is defined by $\mu_a(x) = \mu(x).a$ for all $x \in X$, where . denotes the natural pairing between $\mathfrak{k}^*$ and $\mathfrak{k}$.

(see e.g. [723]). In particular when $K = SO(3)$ acts by rotations on $Y = \boldsymbol{R}^3$ then μ is the angular momentum, or moment of momentum, about the origin.

(ii). Now let X be a nonsingular complex projective variety embedded in complex projective space $\boldsymbol{P}_n$, and let K be a compact Lie group acting on X via a complex linear representation $\rho : K \to GL(n+1)$. By an appropriate choice of coordinates on $\boldsymbol{P}_n$ we may assume that ρ maps K into the unitary group $U(n+1)$. Then the action of K preserves the Fubini-Study form ω on $\boldsymbol{P}_n$, which restricts to a Kähler form on X. It is not difficult to check (cf. e.g. [659] Lemma 2.5) that there is a moment map[31] $\mu : X \to \mathfrak{k}^*$ defined by

$$\mu(x).a = \frac{\overline{\hat{x}}^t \rho_*(a)\hat{x}}{2\pi i \|\hat{x}\|^2} \tag{2}$$

for all $a \in \mathfrak{k}$, where $\hat{x} \in \boldsymbol{C}^{n+1} - \{0\}$ is a representative vector for $x \in \boldsymbol{P}_n$. Equivalently if we identify the Lie algebra $\boldsymbol{u}(n+1)$ with its dual using the pairing

$$a.b = -tr(ab)$$

then

$$\mu(x) = \rho^*(\frac{i\hat{x}\overline{\hat{x}}^t}{2\pi\|\hat{x}\|^2}).$$

Up to a scalar factor[32] and appropriate identifications this agrees with the formula of Appendix 2C.

It follows immediately from the definition of a moment map $\mu : X \to \mathfrak{k}^*$ that the kernel of its derivative $d\mu(x)$ at a point $x \in X$ is the annihilator with respect to ω_x of the tangent space to the K-orbit through x. It also follows that $\zeta \in \mathfrak{k}^*$ is a regular value of μ if and only if $a_x \neq 0$ for all $x \in \mu^{-1}(\zeta)$ and nonzero $a \in \mathfrak{k}$; in other words if and only if the stabilizer in K of every $x \in \mu^{-1}(\zeta)$ is finite. These two facts imply that if the stabilizer K_ζ of any $\zeta \in \mathfrak{k}^*$ acts freely on $\mu^{-1}(\zeta)$ then $\mu^{-1}(\zeta)$ is a submanifold of X and the symplectic form ω induces a symplectic structure on the quotient $\mu^{-1}(\zeta)/K_\zeta$ [723]. With this symplectic structure the quotient $\mu^{-1}(\zeta)/K_\zeta$ is called the Marsden-Weinstein reduction, or symplectic quotient, at ζ of the action of K on X. (We can of course also consider the quotient $\mu^{-1}(\zeta)/K_\zeta$ when the

[31] The moment map defined here depends on the linearization of the action. Conversely a moment map defines a lift to an appropriate line bundle of the infinitesimal action of K on X (see [508] §6.5.1).

[32] The scalar factor is due to differences in convention on the normalization of the Fubini-Study form.

action of K_ζ on $\mu^{-1}(\zeta)$ is not free, but in this case it is likely to have singularities).

Examples 8.2. (i). Let

$$E : T^*Y \to \boldsymbol{R}$$

be the total energy function for a classical mechanical system with phase space T^*Y (cf. Example 8.1(i)). The evolution of the system is described by the Hamiltonian flow of E, given in local coordinates $(q_1, \dots, q_n, p_1, \dots, p_n)$ by

$$\frac{dq_j}{dt} = \frac{\partial E}{\partial p_j}$$

and

$$\frac{dp_j}{dt} = -\frac{\partial E}{\partial q_j}.$$

Equivalently E is a Hamiltonian function for the vector field defining the flow. If E is invariant under the action of a compact group K on T^*Y induced from an action on Y then the Hamiltonian flow of E preserves $\mu^{-1}(\zeta)$ and maps K_ζ-orbits to K_ζ-orbits for each $\zeta \in \mathfrak{k}^*$. Thus it induces a flow on the quotient $\mu^{-1}(\zeta)/K_\zeta$ which is, of course, the Hamiltonian flow of the function induced by E. Thus the reduced mechanical system describes the original flow on $\mu^{-1}(\zeta)$ modulo the action of K_ζ. In the case of the motion of a rigid body this idea goes back to Euler.

(ii). Now let K be the circle S^1 acting on $\boldsymbol{P}_n$ via the representation

$$\rho : S^1 \to U(n+1)$$

where $\rho(t)$ is the diagonal matrix with entries $(t^{-1}, t, \dots, t)$. If the dual of the Lie algebra of S^1 is identified appropriately with $\boldsymbol{R}$ then we obtain from Example 8.1(ii) that

$$\mu(x) = \frac{-|x_0|^2 + |x_1|^2 + \cdots + |x_n|^2}{|x_0|^2 + |x_1|^2 + \cdots + |x_n|^2}$$

where x has homogeneous coordinates $\hat{x} = (x_0, \dots, x_n)$. Thus $x \in \mu^{-1}(\zeta)$ if and only if

$$(1+\zeta)|x_0|^2 = (1-\zeta)(|x_1|^2 + \cdots + |x_n|^2).$$

Therefore

1. $\mu^{-1}(\zeta)$ is empty if $|\zeta| > 1$.
2. ± 1 are not regular values of μ.

3. If $|\zeta| < 1$ then $\mu^{-1}(\zeta)$ can be identified by taking $x_0 = 1$ with the sphere of radius $(1+\zeta)/(1-\zeta)$ in $\boldsymbol{C}^n$, so the quotient $\mu^{-1}(\zeta)/K_\zeta$ is $\boldsymbol{P}_{n-1}$.

The induced symplectic form on $\boldsymbol{P}_{n-1}$ is invariant under the action of $U(n)$ and therefore must agree up to multiplication by a constant scalar with the Fubini-Study form.

Notice that $\boldsymbol{P}_{n-1}$ is also the quotient associated by geometric invariant theory to the linear action of $GL(1)$ on $\boldsymbol{P}_n$ obtained by complexifying ρ.

§ 2. Symplectic quotients and geometric invariant theory

In the last example we saw a symplectic quotient which could be identified with the quotient variety associated by geometric invariant theory to the complexified group action. This example is not a special case: the phenomenon occurs quite generally, as we shall see.

Recall that a complex Lie group G is reductive (equivalently linearly reductive or geometrically reductive) if and only if it is the complexification of any maximal compact subgroup K (see for example [629] XVII.5). If G acts linearly on a nonsingular complex projective variety $X \subseteq \boldsymbol{P}_n$ via a representation

$$\rho : G \to GL(n+1)$$

then we can choose coordinates so that ρ restricts to a unitary representation of K. Then there is a moment map $\mu : X \to \mathfrak{k}^*$ defined by equation (2) in §1 of this Chapter, and a symplectic quotient (or Marsden-Weinstein reduction) $\mu^{-1}(0)/K$. We also have the categorical quotient of X^{ss} by the action of G, which is a projective variety containing as an open subset the geometric quotient $X^s_{(0)}/G$. Throughout this chapter we shall denote this categorical quotient by $X//G$.

Theorem 8.3. (i). Any $x \in X$ is semistable if and only if the closure of its orbit meets $\mu^{-1}(0)$; i.e. if and only if

$$\overline{O_G(x)} \cap \mu^{-1}(0) \neq \emptyset .$$

(ii). The inclusion of $\mu^{-1}(0)$ into X^{ss} induces a homeomorphism

$$\mu^{-1}(0)/K \to X//G .$$

Remark. The proof we shall give follows from the work of KEMPF and NESS [175] as described in Appendix 2C. Alternative arguments which apply more generally when X is a Kähler manifold can be found in [659] 7.5 and [508 §6.5.2.

Proof. For any $v \in \mathbf{C}^{n+1}$ let G_v be the stabilizer of v in in G. In [175] KEMPF and NESS consider the function p_v on G defined by

$$p_v(g) = \|g \cdot v\|^2$$

and show that it has the following properties.

(a) All critical points of p_v are minima.
(b) If p_v attains a minimum it does so on exactly one double coset $KgG_v \in K \setminus G / G_v$.
(c) p_v attains a minimum if and only if the orbit $O_G(v)$ is closed in $\mathbf{C}^{n+1}$.

It also follows easily from equation (2) in §1 that if $\hat{x} \in \mathbf{C}^{n+1} - \{0\}$ represents $x \in X$ then

(d) $dp_{\hat{x}}(g) = 0$ if and only $\mu(gx) = 0$.

At this point it is useful to note in addition the following fact.

(e) If $\hat{x} \in \mathbf{C}^{n+1} - \{0\}$ represents $x \in X$ and the orbit $O_G(\hat{x})$ is closed in $\mathbf{C}^{n+1}$ then $x \in X^{ss}$ and the orbit $O_G(x)$ is closed in X^{ss}.

The implication in (e) that $x \in X^{ss}$ comes from Proposition 2.2. Moreover if $y \in \overline{O_G(x)} \cap X^{ss}$ is represented by $\hat{y} \in \mathbf{C}^{n+1} - \{0\}$ then there is a nonconstant invariant homogeneous polynomial f such that $f(\hat{y}) \neq 0$, i.e. $y \in X_f$, and hence (since X_f is open and G-invariant) $x \in X_f$. Therefore without loss of generality we may suppose that $f(\hat{x}) = 1 = f(\hat{y})$. Then since $y \in \overline{O_G(x)}$ it follows by considering the restriction to $f^{-1}(1)$ of the natural map from $\mathbf{C}^{n+1}$ to $\mathbf{P}_n$ that after multiplying $\hat{y}$ by a suitable root of unity we may assume that $\hat{y} \in \overline{O_G(\hat{x})}$. But by assumption $O_G(\hat{x})$ is closed in $\mathbf{C}^{n+1}$, so $\hat{y} \in O_G(\hat{x})$ and therefore $y \in O_G(x)$.

Now we can prove (i). For if $x \in X^{ss}$ and $\hat{x} \in \mathbf{C}^{n+1} - \{0\}$ represents x then by Proposition 2.2 the closure of the orbit $O_G(\hat{x})$ in $\mathbf{C}^{n+1}$ does not contain 0, and hence this closure contains a nonzero closed orbit whose image in X is contained in the closure of $O_G(x)$. This together with (c) and (d) shows that

$$\overline{O_G(x)} \cap \mu^{-1}(0) \neq \emptyset .$$

Conversely (a), (c), (d) and Proposition 2.2 imply that

$$\mu^{-1}(0) \subseteq X^{ss} ,$$

and therefore $x \in X^{ss}$ whenever

$$\overline{O_G(x)} \cap \mu^{-1}(0) \neq \emptyset$$

because X^{ss} is G-invariant and open in X. This proves (i).

To prove (ii) we observe that since the quotient map $\phi : X^{ss} \to X//G$ is continuous and G-invariant the inclusion of $\mu^{-1}(0)$ in X^{ss} induces a continuous map $\psi : \mu^{-1}(0)/K \to X//G$. As $\phi(x) = \phi(y)$ if and only if

$$\overline{O_G(x)} \cap \overline{O_G(x)} \cap X^{ss} \neq \emptyset$$

when $x, y \in X^{ss}$, it follows from (i) that ψ is surjective. Moreover if

$$\mu(x) = 0 = \mu(y)$$

and $x \notin O_K(y)$, then $x \notin O_G(y)$ by (a), (b) and (d), while by (a), (c), (d) and (e) the orbits $O_G(x)$ and $O_G(y)$ are closed in X^{ss} so $\phi(x) \neq \phi(y)$.

Therefore the induced map $\psi : \mu^{-1}(0)/K \to X//G$ is a continuous bijection from a compact space to a Hausdorff space, and hence is a homeomorphism.

Remark 8.4. We have already observed[33] that the derivative $d\mu(x) : T_xX \to \mathfrak{k}^*$ of μ at x is surjective if and only if the stabilizer K_x of x in K is finite. Moreover if $x \in \mu^{-1}(0)$ it follows from the proof of [659] 7.2 that the stabilizer of x in K is finite if and only if the stabilizer of x in G is finite. The argument runs as follows.

Suppose that $g \cdot x = x$. We can write $g = k \exp(ia)$ for some $k \in K$ and $a \in \mathfrak{k}$. Then the map $h : \boldsymbol{R} \to \boldsymbol{R}$ defined by

$$h(t) = \mu(\exp(ita) \cdot x).a$$

vanishes at 0 and 1 because x and $\exp(ia) \cdot x = k^{-1} \cdot x$ both lie in $\mu^{-1}(0)$. Thus there is a value of $t \in (0,1)$ such that

$$0 = h'(t) = d\mu(y)(ia_y).a = \omega_y(ia_y, a_y) = < a_y, a_y >$$

where $y = \exp(ita)x$ and $< \cdot, \cdot >$denotes the Riemannian metric induced by the Kähler structure. This means that $a_y = 0$, so that $\exp(\boldsymbol{R}a)$ and $\exp(i\boldsymbol{R}a)$ fix y and hence also x. We conclude that if the stabilizer of x in K is finite then $a = 0$ so $g \in K$ and therefore the stabilizer of x in G is finite.

From this it follows easily that if $x \in \mu^{-1}(0)$ then $x \in X^s_{(0)}$ if and only if $d\mu(x)$ is surjective. For if $\hat{x} \in \boldsymbol{C}^{n+1}$ represents x then as in the proof of Theorem 8.3 we know that the orbit $O_G(\hat{x})$ is closed

[33] In §1, just before Examples 8.2.

in C^{n+1}. Therefore by Appendix 1B we have $x \in X^s_{(0)}$ if and only if $\dim O_G(x) = \dim G$, and we have just noted that this happens if and only if $d\mu(x)$ is surjective.

Thus the homeomorphism $\psi : \mu^{-1}(0)/K \to X//G$ induced by the inclusion of $\mu^{-1}(0)$ in X^{ss} restricts to a homeomorphism

$$\mu^{-1}_{\text{reg}}(0)/K \to X^s_{(0)}/G$$

where $\mu^{-1}_{\text{reg}}(0)$ denotes the set of $x \in \mu^{-1}(0)$ such that $d\mu(x)$ is surjective.

Example 8.5. For fixed $n \geq 1$ consider the diagonal action of the special linear group $G = SL(2)$ on $X = (\boldsymbol{P}_1)^n$ (cf. Chapter 3). As in Chapter 3 for $1 \leq j \leq n$ let $L_j = p_j^*(\mathcal{O}_{\boldsymbol{P}_1}(1))$ where p_j is the projection of $(\boldsymbol{P}_1)^n$ onto its jth factor. Then the tensor product $L_1 \otimes \ldots \otimes L_n$ admits a natural $SL(2)$-linearization. Equivalently the nth tensor power of the standard representation of $SL(2)$ on $\boldsymbol{C}^2$ defines a linearization of the action of $SL(2)$ on $X = (\boldsymbol{P}_1)^n$ with respect to the Segre embedding.

It follows from Proposition 3.4 in §2 of Chapter 3 that $X^s_{(0)}$ consists of those n-tuples $x = (x^{(1)}, \ldots, x^{(n)}) \in (\boldsymbol{P}_1)^n$ which contain no point of $\boldsymbol{P}_1$ with multiplicity at least $n/2$. Similarly X^{ss} consists of those n-tuples $x = (x^{(1)}, \ldots, x^{(n)}) \in (\boldsymbol{P}_1)^n$ which contain no point of $\boldsymbol{P}_1$ with multiplicity strictly greater than $n/2$ (cf. [659] 16.1). In particular $X^{ss} = X^s_{(0)}$ if and only if n is odd.

The dual $\mathfrak{k}^*$ of the Lie algebra of the maximal compact subgroup $K = SU(2)$ of G is a real vector space of dimension three, and $\boldsymbol{P}_1$ can be identified K-equivariantly with the unit sphere in $\boldsymbol{R}^3$. Using example 8.1(ii) it is not difficult to check that after appropriate identifications the moment map

$$\mu : (\boldsymbol{P}_1)^n \to \mathfrak{k}^*$$

sends an n-tuple of points on the unit sphere to its sum in $\boldsymbol{R}^3$ (modulo a constant scalar factor depending on conventions). Thus $\mu^{-1}(0)$ consists of those n-tuples of points with center of gravity at the origin. It is easy to see that $d\mu$ is surjective at such an n-tuple unless it contains two antipodal points on the sphere, each occuring with multiplicity $n/2$. From this it is clear that

$$\mu^{-1}(0) \subseteq X^{ss}$$

and

$$\mu^{-1}_{\text{reg}}(0) = \mu^{-1}(0) \cap X^s_{(0)},$$

as expected from Theorem 8.3(i) and Remark 8.4. It is also easy to see that if $x \in X$ is semistable but not properly stable, so that some point of $\boldsymbol{P}_1$ occurs with multiplicity exactly $n/2$ in x, then the closure of the

G-orbit of x in X^{ss} meets $\mu^{-1}(0)$ in precisely one K-orbit, which is one of those in $\mu^{-1}(0)\setminus\mu_{\text{reg}}^{-1}(0)$. Without using Theorem 8.3(i) it is not quite so easy to see that if x is properly stable then its G-orbit is closed near $\mu^{-1}(0)$ and meets $\mu^{-1}(0)$ in precisely one K-orbit.

To illustrate Theorem 8.3(ii) let us compare $X//G$ and $\mu^{-1}(0)/K$ in the simplest nontrivial case, which is when $n=4$. In this case $X//G$ has dimension one, so it must be isomorphic to $\boldsymbol{P}_1$ [173]. This is easy to see directly, since $X^s_{(0)}$ consists of 4-tuples of distinct points in $\boldsymbol{P}_1$ and it is well known from the theory of the cross-ratio that the quotient of this subset of $(\boldsymbol{P}_1)^4$ by $SL(2)$ is $\boldsymbol{P}_1$ with three points removed. There are nine orbits in X which are semistable but not properly stable[34] and three of these map to each of the three missing points in $\boldsymbol{P}_1$.

Now consider $\mu^{-1}(0)/K$; this consists of all 4-tuples $(x^{(1)},\dots,x^{(4)})$ of points on the unit sphere with center of gravity at the origin, modulo rotations of the sphere. Let y be the center of gravity of $x^{(1)}$ and $x^{(2)}$; then $-y$ is the center of gravity of $x^{(3)}$ and $x^{(4)}$. By rotating we can assume that y lies on the north-south axis and that $x^{(4)}$ lies on a fixed line of longitude. Then $x^{(1)}$ can lie anywhere on the sphere, and its position determines the positions of $x^{(2)}$ and y, and hence the positions of $-y$, $x^{(4)}$ and finally $x^{(3)}$. So we see that $\mu^{-1}(0)/K$ can be identified with the unit sphere, and hence with $\boldsymbol{P}_1$, as expected from Theorem 8.3(ii) and the paragraph before this.

§ 3. Kähler and hyperkähler quotients

There are other quotient constructions closely related to symplectic quotients and to geometric invariant theory. Let us suppose now that X is a compact Kähler manifold and that G is a complex reductive group acting holomorphically on X, with a maximal compact subgroup K which preserves the Kähler form[35]. If $\mu: X\to\mathfrak{k}^*$ is a moment map for the action of K on X whose image contains 0 then we can consider the symplectic quotient (or Marsden-Weinstein reduction) $\mu^{-1}(0)/K$. It turns out that away from its singular points this quotient $\mu^{-1}(0)/K$ inherits a Kähler structure from that of X.

This can be proved in different ways. The crucial ingredient in all the proofs is the observation that if $x\in\mu_{\text{reg}}^{-1}(0)$ then the tangent space to $\mu^{-1}(0)/K$ at the orbit $O_K(x)$ can be identified in a natural way with

[34] Three of these orbits contain 4-tuples with two points each of multiplicity two – these are closed in X^{ss} – and six contain 4-tuples with one point of multiplicity two and two points of multiplicity one – these are not closed in X^{ss}.

[35] This can always be achieved by averaging the form.

the quotient of T_xX by $T_xO_G(x)$. This is because the Kähler form ω is the imaginary part of a Hermitian metric on X whose real part is a Riemannian metric, and the kernel of the derivative $d\mu(x)$ consists of those $\xi \in T_xX$ which are orthogonal with respect to this Riemannian metric to ia_x for all $a \in \mathfrak{k}$. In [624] it is checked directly that the almost complex structure thus induced on $\mu_{\mathrm{reg}}^{-1}(0)/K$ (or rather the open subset where the action of K is free) is integrable, and that the quotient symplectic form is in fact Kähler for this complex structure.

Another method of proof (see [659] §7) is to show that the inclusion of $\mu_{\mathrm{reg}}^{-1}(0)$ in a certain open subset $X_{(0)}^{\min}$ of X induces a diffeomorphism

$$\mu_{\mathrm{reg}}^{-1}(0)/K \to X_{(0)}^{\min}/G,$$

and the symplectic structure on $\mu_{\mathrm{reg}}^{-1}(0)/K$ combines with the complex structure on $X_{(0)}^{\min}/G$ to give a Kähler structure. To define this open set $X_{(0)}^{\min}$ one considers the gradient flow (with respect to the Kähler metric) of the function $f : X \to \boldsymbol{R}$ given by

$$f(x) = \|\mu(x)\|^2,$$

where the norm is obtained from an invariant inner product on $\mathfrak{k}$, such as the Killing form. Let $X^{\min}$ denote the set of all points whose paths of steepest descent under f have limit points in the set $\mu^{-1}(0)$ where f attains its minimum value. Then $X^{\min}$ is an open subset of X which coincides with the set X^{ss} of semistable points in the situation of geometric invariant theory considered in §2 ([659] Theorem 8.10; cf. Example 8.10 in §6 below). Moreover, using the fact ([659] Lemma 6.6) that the gradient flow of f is given by

$$\operatorname{grad} f(x) = 2i\mu(x)_x$$

which means that paths of steepest descent are contained in G-orbits, it can be shown that the analogue of Theorem 8.3 holds with X^{ss} replaced by $X^{\min}$ and $X//G$ interpreted as the quotient of $X^{\min}$ by the equivalence relation

$$x \sim y \Longleftrightarrow \overline{O_G(x)} \cap \overline{O_G(x)} \cap X^{\min} \neq \emptyset$$

([659] Theorem 7.4 and Remark 7.5). In addition the analogue of Remark 8.4 holds with $X_{(0)}^{s}$ replaced by the open subset $X_{(0)}^{\min}$ of $X^{\min}$ consisting of all points in X whose paths of steepest descent have limit points in $\mu_{\mathrm{reg}}^{-1}(0)$.

A rarer, but nonetheless important, quotient construction can be used when X is a hyperkähler manifold; that is, when X has complex structures i, j and k satisfying the quaternionic relations

$$ij = k = -ji$$

and a Riemannian metric which is Kähler for each of i, j and k separately. The basic example of a hyperkähler manifold is $\boldsymbol{H}^n$ where $\boldsymbol{H}$ is the space of quaternions with the standard flat metric. Compact examples are not common but it can be shown that these include all $K3$ surfaces [381], [382]. (Quaternionic projective space $\boldsymbol{P}_n(\boldsymbol{H})$ is not a hyperkähler manifold, although it is a quaternionic Kähler manifold in the terminology of [828]). A source of interesting examples of hyperkähler manifolds is the hyperkähler quotient construction of HITCHIN, KARLHEDE, LINDSTRÖM and ROČEK [627], which we shall now describe.

Let us suppose that a compact Lie group K acts on the hyperkähler manifold X and that the action is holomorphic with respect to each of the complex structures i, j and k and preserves the metric. Then K acts symplectically with respect to each of the three corresponding Kähler forms ω_1, ω_2 and ω_3. If moment maps μ_1, μ_2 and $\mu_3 : X \to \mathfrak{k}^*$ exist for each of these symplectic forms then the quotient

$$X///K = (\mu_1^{-1}(0) \cap \mu_2^{-1}(0) \cap \mu_3^{-1}(0))/K$$

has an induced hyperkähler structure away from its singularities. This is because the map

$$\mu_2 + i\mu_3 : X \to \mathfrak{k}^* \otimes_{\boldsymbol{R}} \boldsymbol{C}$$

is holomorphic with respect to the complex structure i on X (this follows from the definition of a moment map and the quaternionic relations)[36], so that away from its singularities

$$\mu_2^{-1}(0) \cap \mu_3^{-1}(0) = (\mu_2 + i\mu_3)^{-1}(0)$$

is a complex submanifold of X and

$$(\mu_1^{-1}(0) \cap \mu_2^{-1}(0) \cap \mu_3^{-1}(0))/K$$

is its Kähler quotient. Thus the nonsingular part of

$$(\mu_1^{-1}(0) \cap \mu_2^{-1}(0) \cap \mu_3^{-1}(0))/K$$

---

[36] It is in fact a holomorphic moment map for the action of K on X with respect to the holomorphic symplectic structure $\omega_2 + i\omega_3$.

has a Kähler structure induced from the Kähler structure on X given by i, and the same is true for j and k. It is easy to check that these fit together to give a hyperkähler structure on the nonsingular part of

$$(\mu_1^{-1}(0) \cap \mu_2^{-1}(0) \cap \mu_3^{-1}(0))/K.$$

There are important examples of hyperkähler manifolds which are moduli spaces (see for example [366], [367], [497], [498], [623–625], [627], [695], [696]). One of these is the moduli space $\mathcal{M}(k,l)$ of holomorphic $SL(l)$ bundles on $\boldsymbol{P}_2$ with a fixed holomorphic trivialization on the line at infinity and second Chern class k. This can be identified with the moduli space of based gauge equivalence classes of anti-self-dual connections (instantons) on a principal $SU(l)$ bundle over the sphere S^4 with $c_2 = k$ (see [497]). Each of these moduli spaces has a description in terms of monads known as the ADHM construction (see [365] or [49]) which is based on the Horrocks construction of vector bundles [149]. The ADHM construction can be interpreted as a hyperkähler quotient of an open subset of $\boldsymbol{H}^{k(k+l)}$ by an action of the unitary group $U(k)$ (this observation is due to Hitchin and Donaldson). More precisely (see [497]) the moduli space of instantons is the quotient of the set of sequences of complex matrices $(\alpha_1, \alpha_2, a, b)$ of sizes $k \times k$, $k \times k$, $l \times k$ and $k \times l$, satisfying

8.6(i) $[\alpha_1, \alpha_2] + ba = 0$;

(ii) $[\alpha_1, \alpha_1^*] + [\alpha_2, \alpha_2^*] + bb^* - a^*a = 0$ where * denotes the transposed complex conjugate;

(iii) $\begin{pmatrix} \alpha_1 + \lambda \\ \alpha_2 + \mu \\ a \end{pmatrix}$ is injective and $(\lambda - \alpha_1, \alpha_2 - \mu, b)$ is surjective for all complex numbers λ and μ;

by the natural (free) action of $U(k)$

$$\left.\begin{array}{l} \alpha_i \mapsto p\alpha_i p^{-1} \\ a \mapsto ap^{-1} \\ b \mapsto pb \end{array}\right\} p \in U(k).$$

In [497] Donaldson showed that the equation 8.6(ii) is the vanishing of a moment map for the action of $U(k)$ on $\boldsymbol{C}^{2k(k+l)}$. He also showed that the equation 8.6(i) defines a complex subvariety V of $\boldsymbol{C}^{2k(k+l)}$, and the open condition 8.6(iii) is the condition for $(\alpha_1, \alpha_2, a, b)$ to be stable in the sense of geometric invariant theory for the action of the complexification $GL(k)$ of $U(k)$ on V. Donaldson was able to exploit the general principle (cf. §2) that geometric invariant theoretic quotients of complex varieties can also be regarded as symplectic quotients, to show that the moduli space of instantons is also the quotient by the action of $GL(k)$ of the set of $(\alpha_1, \alpha_2, a, b)$ satisfying 8.6(i) and (iii) but not necessarily (ii). This led to its identification with the corresponding moduli space $\mathcal{M}(k,l)$ of holomorphic bundles on $\boldsymbol{P}_2$. In fact it is not hard to show that the action of $U(k)$ on

$$\boldsymbol{C}^{2k(k+l)} = \boldsymbol{H}^{k(k+l)}$$

preserves a natural hyperkähler structure, and that there are associated moment maps μ_1, μ_2, μ_3 for this action such that (after appropriate identifications)

8.7(i) $\mu_1(\alpha_1, \alpha_2, a, b) = [\alpha_1, \alpha_1^*] + [\alpha_2, \alpha_2^*] + bb^* - a^*a$
and
(ii) $(\mu_2 + i\mu_3)(\alpha_1, \alpha_2, a, b) = [\alpha_1, \alpha_2] + ba.$

Thus the moduli space $\mathcal{M}(k,l)$ is the nonsingular part of the hyperkähler quotient $\boldsymbol{H}^{k(k+l)}///K$.

Some significant progress towards understanding the topology of these moduli spaces has been made in the last decade [368], [420], [421], [578], [670], [719], [882]. In particular they are now known [497], [880] to be connected (cf. Appendix 5C).

§ 4. Singular quotients

A symplectic quotient $\mu^{-1}(0)/K$ may be singular for two reasons: firstly 0 may not be a regular value of μ and secondly K may not act freely on $\mu^{-1}(0)$. We have already noted in §2 that the two problems are related because $d\mu(x)$ is surjective if and only if the stabilizer K_x of x in K is finite. We also noted in §2 that, in the situation of geometric invariant theory when X is a nonsingular complex projective variety, the stabilizer K_x is finite for every $x \in \mu^{-1}(0)$ if and only if $X^{ss} = X^s_{(0)}$.

Thus when the stabilizer K_x is finite for every $x \in \mu^{-1}(0)$ the singularities of the symplectic quotient $\mu^{-1}(0)/K$ (or equivalently the geometric invariant theoretic quotient $X//G$) are relatively minor: locally we have the quotient of a manifold by a finite group action (i.e. the symplectic quotient is an orbifold). In general the singularities may be more serious[37], and the appropriate approach to this problem depends on our point of view.

---

[37] Even when X is an arbitrary compact symplectic manifold with a symplectic K-action and moment map μ, the singularities of the symplectic quotient $\mu^{-1}(0)/K$ cannot be pathologically bad. The equivariant DARBOUX theorem ([592] Theorem 22.2) and its generalization due to WEINSTEIN ([592]

In algebraic geometry one can attempt to resolve singularities by a sequence of blow-ups. It turns out that as long as $X^s_{(0)} \neq \emptyset$ there is a fairly straightforward procedure for resolving the more serious singularities of the quotient $X//G$ when X is a nonsingular complex projective variety. One can blow X up along a sequence of G-invariant nonsingular subvarieties to obtain a new variety $\tilde{X}$ with a linearization of the induced G-action such that

$$\tilde{X}^{ss} = \tilde{X}^s_{(0)}$$

and such that the birational morphism $\tilde{X} \to X$ induces a birational morphism to $X//G$ from the quotient $\tilde{X}//G$ which only has finite quotient singularities[38]. This procedure works as follows (for more details see [661]).

Let X be a nonsingular complex projective variety embedded in $\boldsymbol{P}_n$, and let G be a complex reductive group acting linearly on X via a representation $\rho : G \to GL(n+1)$. We assume that $X^s_{(0)}$ is not empty. If $X^s_{(0)} \neq X^{ss}$ then there exists a reductive subgroup R of G of dimension at least one, such that

$$Z^{ss}_R = \{x \in X^{ss} : x \text{ is fixed by } R\}$$

is not empty ([661] 4.4). Let r be the maximal dimension of any such reductive subgroup R and let $\mathcal{R}(r)$ be a set of representatives of the conjugacy classes of the connected reductive subgroups R of dimension r for which Z^{ss}_R is not empty. Then it can be shown ([661] 5.11 and 8.2) that

$$\bigcup_{R \in \mathcal{R}(r)} GZ^{ss}_R$$

is a disjoint union of nonsingular closed subvarieties of X^{ss}. Moreover if Y is the blow-up of X^{ss} along this disjoint union then no reductive subgroup of G of dimension at least r fixes any point of Y which is semistable for a suitable linearization[39] of the induced G-action on Y ([661] 6.2). Equivalently, rather than blowing X^{ss} up

---

Theorem 22.1) can be used to show that there is a K-equivariant diffeomorphism taking $\mu^{-1}(0)$ to the zero set of a homogeneous quadratic function acted on linearly by K [358], and to give a local normal form for the moment map [591], [592] §§39–41, [721], [722].

[38] These cannot in general be resolved by equivariant blow-ups: see [782] p. 31.

[39] The linearization is with respect to the tensor product of $\mathcal{O}(-E)$, where E is the exceptional divisor, with a sufficiently large power of the pullback of the hyperplane line bundle on $X \subseteq \boldsymbol{P}_n$. There is a choice involved here but it does not affect the end result: see [661] §3.

along $\bigcup_{R\in\mathcal{R}(r)} GZ_R^{ss}$ we can resolve the singularities of the closure of $\bigcup_{R\in\mathcal{R}(r)} GZ_R^{ss}$ in X and blow X up along the proper transform of this closure. This makes no difference to the set of semistable points (for an appropriate choice of linearization) since all semistable points in the blow-up lie over semistable points in X.

Repeating this process at most r times gives us a nonsingular complex projective variety $\tilde{X}$ acted on linearly by G, with a G-equivariant birational morphism

$$\pi : \tilde{X} \to X$$

such that

$$\tilde{X}^{ss} = \tilde{X}^{s}_{(0)}$$

and

$$\tilde{X}^{ss} \subseteq \pi^{-1}(X^{ss})$$

and π restricts to an isomorphism from $\pi^{-1}(X^s_{(0)})$ to $X^s_{(0)}$. (Such a morphism $\pi : \tilde{X} \to X$ is sometimes called a stable resolution of X). The induced birational morphism

$$\pi_G : \tilde{X}//G \to X//G$$

is a partial desingularization of $X//G$ in the sense that $\tilde{X}//G$ has only finite quotient singularities[40].

Example 8.8. Let us again consider the action of $G = SL(2)$ on $X = (\boldsymbol{P}_1)^n$ (cf. Example 8.5). When n is odd then $X^{ss} = X^s_{(0)}$, so let us assume that n is even. Then the semistable points of X fixed by reductive subgroups of dimension at least one are those n-tuples $x \in (\boldsymbol{P}_1)^n$ in which just two points of $\boldsymbol{P}_1$ occur, each with multiplicity $n/2$. Thus $r = 1$ and $\mathcal{R}(r)$ has just one element R, which is a maximal torus of G. Its set of semistable fixed points Z_R^{ss} has $\binom{n}{n/2}$ elements. As is in fact always true (see [661]), GZ_R^{ss} is isomorphic to

$$G \times_N Z_R^{ss}$$

where N is the normalizer of R in G; since in this case N/R has order two (it is the Weyl group of G) and acts freely on Z_R^{ss} it follows that GZ_R^{ss} consists of $\frac{1}{2}\binom{n}{n/2}$ G-orbits, which are closed in X^{ss}. After X^{ss} has been blown up along these orbits the other orbits from $X^{ss} \setminus X^s_{(0)}$ are no longer semistable. When these together with the orbits in the

[40] If G acts freely on $X^s_{(0)}$ (which we are still assuming to be nonempty) then the stabilizers in G of points of $\tilde{X}^{ss}$ are finite *abelian* groups [708]. This occurs for example in the construction of the compactified moduli spaces of vector bundles over curves when the rank and degree are not coprime.

exceptional divisors which are contained in their closures have been removed we are left with $\tilde{X}^{ss}$, and after dividing by the action of G we get the partial desingularization $\tilde{X}//G$. Equivalently $\tilde{X}//G$ is obtained from $X//G$ by blowing up its $\frac{1}{2}\binom{n}{n/2}$ singular points.

A different way of constructing stable resolutions is due to REICHSTEIN [819]. He considers the more general situation of a reductive group acting on projective varieties X and Y, linearly with respect to ample line bundles L and K, where there is a G-equivariant morphism $\pi : Y \to X$. He relates stability in X with respect to L to stability in Y with respect to $\pi^* L^d \otimes K$ for sufficiently large d, and derives a relative version of the numerical criterion for stability described in Chapter 2 §1. He uses this to obtain (noncanonical) stable resolutions of X.

From the viewpoint of symplectic geometry, however, these algebro-geometric approaches are not really satisfactory. For one thing, although one can do "symplectic blow-ups" of symplectic manifolds [583], [730], the induced symplectic structures are only defined up to isotopy near the exceptional divisors and do not correspond exactly to the Kähler structures induced on algebro-geometric blow-ups by choosing embeddings in projective spaces. Moreover the singularities of $\tilde{X}//G$ cause problems from the symplectic point of view, and even when $\tilde{X}//G$ is nonsingular as an algebraic variety it may have singularities when regarded as a symplectic quotient. As an example of this phenomenon, consider $G = GL(1)$ acting on $X = \boldsymbol{P}_2$ via the representation $\rho : G \to GL(3)$ such that $\rho(t)$ is the diagonal matrix with entries (t^2, t, t^{-1}). If we take homogeneous cooordinates (x, y, z) on $\boldsymbol{P}_2$ then the ring of invariants is the polynomial ring generated by xz^2 and yz, and $X//G$ is $\boldsymbol{P}_1$ which is nonsingular. On the other hand by Example 8.1(ii) in §1

$$\mu^{-1}(0) = \{(x, y, z) \in \boldsymbol{P}_2 : 2|x|^2 + |y|^2 = |z|^2\}$$

and $K = S^1$ does not act freely on this: the points with homogeneous coordinates in the orbits of $(1, 0, \sqrt{2})$ and $(0, 1, 1)$ are fixed by the cube and square roots of unity respectively. Moreover if we identify $\mu^{-1}(0)/K$ with $X//G = \boldsymbol{P}_1$ then there is no symplectic form on $\boldsymbol{P}_1$ which pulls back under the quotient map $\mu^{-1}(0) \to \mu^{-1}(0)/K$ to the restriction of the Fubini-Study form on $\boldsymbol{P}_2$. An even simpler example which displays the basic problem here is the quotient of $\boldsymbol{C}$ by the multiplicative action of $\{\pm 1\}$. In algebraic geometry the quotient is identified with $\boldsymbol{C}$ via the map $\phi : \boldsymbol{C} \to \boldsymbol{C}$ given by $\phi(z) = z^2$. On the other hand symplectic geometry sees that there is no symplectic form on $\boldsymbol{C}$ which pulls back under ϕ to the standard symplectic form, and that not all invariant *smooth* functions on $\boldsymbol{C}$ are pullbacks via ϕ of smooth functions on $\boldsymbol{C}$.

Because of this, symplectic geometers approach the problem of singularities in $\mu^{-1}(0)/K$ differently [356]. One approach initiated by SNIALYCKI and WEINSTEIN [858] involves working with the Poisson algebra[41] of the symplectic manifold X: that is, the algebra of smooth functions on X equipped with the Poisson bracket $\{\,,\}$ defined by

$$\{f, g\} = \omega(X_f, X_g) = df(X_g) = -dg(X_f)$$

where X_f and X_g are the Hamiltonian vector fields[42] of f and g. Let $\mathscr{I}$ be the ideal[43] in $C^\infty(X)$ generated by the components of the moment map $\mu : X \to \mathfrak{k}^*$. Then K acts on $C^\infty(X)/\mathscr{I}$ and it turns out that the set of invariants

$$(C^\infty(X)/\mathscr{I})^K$$

is naturally a Poisson algebra. It is called the reduced Poisson algebra associated to the symplectic action of K on X, and it coincides with the Poisson algebra of the symplectic quotient $\mu^{-1}(0)/K$ when K acts freely on $\mu^{-1}(0)$. In general, however, it need not be the Poisson algebra of any symplectic manifold.

Another approach due to ARMS, GOTAY and JENNINGS [357] is called by them "geometric reduction". Here there is always an underlying topological space which is $\mu^{-1}(0)/K$ and when K is compact the rôle of "smooth functions" on $\mu^{-1}(0)/K$ is played by the restrictions to $\mu^{-1}(0)$ of smooth functions f on X which are K-invariant on $\mu^{-1}(0)$. The algebra of such functions has a natural Poisson structure, and when K acts freely on $\mu^{-1}(0)$ it coincides with the Poisson algebra of the symplectic quotient $\mu^{-1}(0)/K$. In fact the quotient $\mu^{-1}(0)/K$ has a natural stratification given by orbit type (i.e. by the conjugacy classes of stabilizers in K) such that each stratum is a symplectic manifold ([856] Theorem 3.1.1), and the Poisson bracket can be defined by using the Poisson brackets on the spaces of C^∞ functions on the strata ([855] §3.2).

§ 5. Geometry of the moment map

So far we have mainly considered symplectic quotients of the form $\mu^{-1}(0)/K$, but more generally one can study the symplectic quotients $\mu^{-1}(\zeta)/K_\zeta$ for any $\zeta \in \mathfrak{k}^*$. From some points of view at least, restricting

[41] A Poisson algebra is a Lie algebra whose Lie bracket (usually written $\{\,,\}$) satisfies the Leibniz rule $\{fg, h\} = f\{g, h\} + \{f, h\}g$.

[42] I.e. the vector fields for which f and g are Hamiltonian functions (see §1).

[43] This is using the usual structure of $C^\infty(X)$ as an associative algebra.

to the case $\zeta = 0$ involves no loss of generality. This is because the coadjoint orbit $O_K(\zeta)$ of any $\zeta \in \mathfrak{k}^*$ has a K-invariant symplectic form Ω defined by

$$\Omega_\beta(\xi_\beta, \eta_\beta) = \beta([\eta, \xi])$$

for $\beta \in O_K(\zeta) \subseteq \mathfrak{k}^*$, $\xi, \eta \in \mathfrak{k}$, and the inclusion of $O_K(\zeta)$ in $\mathfrak{k}^*$ is a moment map for this symplectic form[44] ([723] §4.1 or [592] §26). Thus the map

$$\tilde{\mu} : X \times O_K(\zeta) \to \mathfrak{k}^*$$

defined by

$$\tilde{\mu}(x, k{\cdot}\zeta) = \mu(x) - k{\cdot}\zeta$$

is a moment map for the action of K on $X \times O_K(\zeta)$ with respect to the symplectic form $(\omega, -\Omega)$, and the inclusion of $\mu^{-1}(\zeta) \times \{\zeta\}$ in $\tilde{\mu}^{-1}(0)$ induces an identification

$$\mu^{-1}(\zeta)/K_\zeta \to \tilde{\mu}^{-1}(0)/K\,.$$

Note that when X is a nonsingular complex projective variety on which K acts linearly as in Example 8.1(ii), then using Theorem 8.3 the quotient $\tilde{\mu}^{-1}(0)/K$ can be identified with a quotient in the sense of geometric invariant theory, provided that there is an equivariant embedding of the orbit $O_K(\zeta)$ in a projective space with a unitary action of K such that Ω is the restriction of the Fubini-Study form. It follows from [593] that this happens if and only if ζ is a highest weight vector for an irreducible representation of K. In this case,

$$\mu^{-1}(\zeta)/K_\zeta \cong \tilde{\mu}^{-1}(0)/K \cong (X \times G/P)//G$$

for a suitable parabolic subgroup P of the complexification G of K and a suitable G-linearized polarization of G/P.

Sometimes, however, restricting to the case $\zeta = 0$ does involve loss of generality. In particular this is the case if one wants to understand how the topology (and symplectic structure) of the quotient $\mu^{-1}(\zeta)/K_\zeta$ varies as ζ varies in K_ζ. The first obvious question is

"When is $\mu^{-1}(\zeta)/K_\zeta$ nonempty?"

or equivalently

"What is the image $\mu(X)$ of μ?"

In answer to this we have the following convexity theorem.

---

[44]This symplectic structure can in fact be obtained from the canonical symplectic structure on the cotangent bundle T^*K by reduction with respect to the action of K on itself by left multiplication [723]. Moreover if K is connected then any moment map for a K-invariant symplectic form ω on a homogeneous space K/L takes $(K/L, \omega)$ isomorphically onto $(O_K(\zeta), \Omega)$ for some $\zeta \in \mathfrak{k}^*$ such that $K_\zeta = L$.

Theorem 8.9. Suppose that $\mu : X \to \mathfrak{k}^*$ is a moment map for the action of a compact group K on a compact, connected symplectic manifold X. Let $\mathfrak{t}^*_+$ be a positive Weyl chamber in the dual $\mathfrak{t}^*$ of the Lie algebra of a maximal torus T in K.

(i) If $K = T$ is a torus then the image $\mu(X)$ of μ is a convex polytope in $\mathfrak{t}^*$. In fact $\mu(X)$ is the convex hull of the finite set of points $\mu(X^T)$, where X^T is the set of fixed points of the action of T on X.

(ii) In general $\mu(X) \cap \mathfrak{t}^*_+$ is a convex polytope in $\mathfrak{t}^*$.

Proof. (i) was proved independently by ATIYAH [359] (see also [360] and [361]) and by GUILLEMIN and STERNBERG [590]. For the proof of (ii) see [590], [660].

Remarks. (i) Since $\mu(X)$ is K-invariant and every coadjoint K-orbit in $\mathfrak{k}^*$ meets the positive Weyl chamber $\mathfrak{t}^*_+$ in exactly one point, we have

$$\mu(X) = K(\mu(X) \cap \mathfrak{t}^*_+)$$

and we can identify $\mu(X) \cap \mathfrak{t}^*_+$ with the set of coadjoint K-orbits in $\mu(X)$.

(ii) A description of the vertices of $\mu(X) \cap \mathfrak{t}^*_+$ in the general case is possible but is much more complicated than in the case $K = T$ (see [590]).

(iii) Related results include [369], [429], [561], [757], [914].

When $K = T$ is a torus more detailed information is available about the way the symplectic quotient[45] $\mu^{-1}(\zeta)/K_\zeta = \mu^{-1}(\zeta)/T$ varies as ζ varies in $\mathfrak{t}^*$. In this case the convex polytope $\mu(X)$ is the union of subpolytopes of the same dimension, each of which is the convex hull of a subset of the finite set $\mu(X^T)$ and contains no points of $\mu(X^T)$ in its interior. The interior of each such subpolytope consists of regular values of μ. DUISTERMAAT and HECKMAN [519] have shown that as ζ varies in the interior of any of these subpolytopes the diffeomorphism type of $\mu^{-1}(\zeta)/T$ is constant[46] and the cohomology class of the induced symplectic form on $\mu^{-1}(\zeta)/T$ varies linearly in a natural sense. In addition GUILLEMIN and STERNBERG have shown that as ζ crosses the boundary between two subpolytopes the change in diffeomorphism type in $\mu^{-1}(\zeta)/T$ can be described as a composition of symplectic blow-ups

[45] Recall from the first paragraph of this section that if X is a nonsingular complex projective variety on which T acts linearly as in Example 8.1(ii), then provided that ζ is an integral point of $\mathfrak{t}^*$ the symplectic quotient $\mu^{-1}(\zeta)/T$ can be identified with a geometric invariant theoretic quotient $X//T_c$ for a suitable linearization of the action of the complexification T_c of T on X.

[46] Cf. Example 8.2(ii).

and blow-downs, and they explain how the cohomology class of the symplectic form varies (see [593])[47]. In [519] DUISTERMAAT and HECKMAN also prove the beautiful result that when K is a torus the pushforward by the moment map μ of the symplectic measure on X is a piecewise polynomial measure on $\mathfrak{t}^*$. Equivalently if $2n$ is the real dimension of X and μ_a is any component of μ then the integral

$$\int_X e^{-it\mu_a} \frac{\omega^n}{n!}$$

is exactly given by its stationary phase approximation. This turns out to be a consequence of a localization theorem in equivariant cohomology [364], [370], [387], [570].

The convexity theorem of ATIYAH and GUILLEMIN-STERNBERG is related to the theory of (complex) toric varieties[48]. Complex toric varieties are complex varieties X acted on by complex tori with dense open orbits. They are usually constructed from combinatorial objects called fans. When X is projective we can in turn construct[49] the fan $\mathcal{F}$ in $\mathfrak{t}$ from a compact convex polytope $P \subseteq \mathfrak{t}^*$. In fact this fan $\mathcal{F}$ is the set of "dual cones"

$$\check{\sigma}_\Gamma = \{\alpha \in \mathfrak{t} : \alpha.\xi \geq 0 \ \forall \xi \in \sigma_\Gamma\}$$

to the tangent cones σ_Γ of the faces Γ of P defined by

$$\sigma_\Gamma = \boldsymbol{R}_+(P - x_0)$$

for any x_0 belonging to the relative interior of Γ. Since P is compact the fan $\mathcal{F}$ is complete in the sense that the union of its cones is the whole of $\mathfrak{t}$. The toric variety X is defined as a union of affine pieces, one for each Γ, whose rings are the vector spaces of weights of T in σ_Γ.

Given this fan $\mathcal{F}$ we can choose primitive integral vectors $x_1, \ldots, x_N$ in $\mathfrak{t}$ generating its one-dimensional cones. These determine a homomorphism from the compact torus

$$T^{(N)} = \boldsymbol{R}^N / \boldsymbol{Z}^N$$

---

[47] The case when X is a complex projective variety was discussed independently using complex blow-ups and blow-downs by BRION and PROCESI in [434], and the case when the group is not a torus is considered in [490]. [632] contains related ideas. It has been observed by REID [821] (see also [888]) that this phenomenon is related to MORI's theory of flips [748], [676].

[48] For the theory of toric varieties see e.g. [48], [465], [658], [782]. For more details of the relationship with symplectic geometry and the convexity theorem see [370] Chapter VI and the references therein.

[49] The combinatorial relationships between the faces of the convex polytope can be recovered from the fan, but not the sizes of the faces.

to T which maps the standard basis of $Lie(T^{(N)}) = \boldsymbol{R}^N$ to $x_1, \ldots, x_N$. The kernel[50] K of this homomorphism acts on $\boldsymbol{C}^N$ with moment map

$$\mu : \boldsymbol{C}^N \to \mathfrak{k}^*$$

given by the composition with restriction to $\mathfrak{k}^*$ of the map

$$(z_1, \ldots, z_N) \mapsto \frac{1}{2}(|z_1|^2, \ldots, |z_N|^2)$$

to $Lie(T^{(N)})^*$ identified with $\boldsymbol{R}^N$.

The toric variety X associated to the fan $\mathscr{F}$ can now be defined to be the quotient of a certain open subset of $\boldsymbol{C}^n$ by the action of the complexification of this kernel K. This open subset is described in the theory of toric varieties in terms of the fan $\mathscr{F}$. However using a modification of Theorem 8.3 it turns out that if ζ is a regular value of μ then the open subset consists precisely of those points whose orbits under the action of the complexification of K meet $\mu^{-1}(\zeta)$, and its quotient, the toric variety X, can be identified with the symplectic quotient $\mu^{-1}(\zeta)/K$ (see [370] Proposition 3.1.1). Thus the choice of a regular value ζ of μ defines a symplectic structure on X, which is preserved by the action of T induced by identifying T with $T^{(N)}/K$. The image P_ζ of a moment map for this action, with respect to the symplectic structure on X determined by ζ, is a compact convex polytope, and the fan associated to P_ζ is the original fan $\mathscr{F}$.

Thus we have two moment maps for compact torus actions playing crucial rôles here. The first is the standard moment map $\mu : \boldsymbol{C}^N \to \mathfrak{k}^*$ for the action of the compact torus K on $\boldsymbol{C}^N$; when ζ is a regular value of μ the symplectic quotient $\mu^{-1}(\zeta)/K$ can be identified naturally with the toric variety X, and thus X acquires a symplectic structure. The second is the moment map with respect to this symplectic structure for the action of the compact torus T on X. The image of this second moment map is a convex polytope P_ζ from which we can recover the fan $\mathscr{F}$ defining the toric variety X.

§ 6. The cohomology of quotients: the symplectic case

In their important paper [50] Atiyah and Bott applied equivariant Morse theory to the Yang-Mills functional to calculate the Betti numbers of the moduli spaces of semistable bundles of coprime ranks and degrees over Riemann surfaces (see §9). These moduli spaces can be

[50] This kernel is of course a compact torus, but we shall denote it by K (for kernel) since we already have two other compact tori T and $T^{(N)}$.

regarded as symplectic quotients for infinite-dimensional group actions with curvature playing the rôle of the moment map and the Yang-Mills functional playing the rôle of its norm square. It occurred to Atiyah that information about the Betti numbers of symplectic and geometric invariant theoretic quotients[51] for finite-dimensional actions might also be obtained by applying equivariant Morse theory to the norm square of the moment map. This turns out to be the case[52], as we shall now explain[53].

Let X be a compact symplectic manifold on which a compact connected Lie group K acts symplectically, and let $\mu : X \to \mathfrak{k}^*$ be a moment map for this action. As before we shall be concerned mainly with the case when X is a nonsingular complex projective variety embedded in projective space $\boldsymbol{P}_n$ and K acts on X via a representation in $U(n+1)$; this action then extends to a linear action on X of the complex reductive group G which is the complexification of K.

We choose an inner product on the Lie algebra $\mathfrak{k}$ which is invariant under the adjoint action of K (for example the Killing form) and consider the function $f : X \to \boldsymbol{R}$ defined by

$$f(x) = \|\mu(x)\|^2$$

as a Morse function on X. It is not in general a Morse function in the classical sense or even the sense of BOTT [414] – the connected components of the set of critical points of f are not necessarily submanifolds of X. Nonetheless f is sufficiently well behaved[54] to induce Morse inequalities relating the dimensions of the cohomology groups and equivariant cohomology groups[55] of X to those of the connected components of the set of critical points. More precisely, the set of critical points for f on

---

[51] The topology of geometric invariant theoretic quotients was also studied at approximately the same time by NEEMAN [768].

[53] However the argument that ATIYAH and BOTT used to show that the moduli spaces of semistable bundles of coprime ranks and degrees over Riemann surfaces have no torsion in their cohomology does not apply in general.

[53] For further details see [659].

[54] By [659] Proposition 4.15 it is a "minimally degenerate Morse function" in the sense of the appendix to [659].

[55] Recall that the equivariant cohomology $H^*_K(Y;F)$ with coefficients in a field (or ring) F of a space Y acted on by K is defined to be $H^*(Y_K;F)$ where

$$Y_K = Y \times_K EK$$

and $EK \to BK$ is the universal principal K-bundle. If K acts freely on Y then the natural map $Y_K \to Y/K$ induces an isomorphism $H^*(Y/K;F) \to H^*_K(Y;F)$.

X is a disjoint union of finitely many K-invariant closed subsets[56]

$$\{C_{\beta,m} : \beta \in \mathcal{B},\ 0 \leq m \leq \dim_{\mathbf{R}} X\}$$

such that the Poincaré series

$$P_t(X) = \sum_{i \geq 0} t^i \dim H^i(X; \mathbf{Q})$$

and equivariant Poincaré series

$$P_t^K(X) = \sum_{i \geq 0} t^i \dim H_K^i(X; \mathbf{Q})$$

of X can be expressed as

$$P_t(X) = \sum_{\beta,m} t^{d(\beta,m)} P_t(C_{\beta,m}) - (1+t)Q(t)$$

and

$$P_t^K(X) = \sum_{\beta,m} t^{d(\beta,m)} P_t^K(C_{\beta,m}) - (1+t)Q_K(t)$$

where each $d(\beta, m)$ and all the coefficients of $Q(t)$ and $Q_K(t)$ are non-negative integers.

These Morse inequalities can be obtained by considering the subsets

$$\{S_{\beta,m} : \beta \in \mathcal{B},\ 0 \leq m \leq \dim_{\mathbf{R}} X\}$$

of X such that a point $x \in X$ lies in $S_{\beta,m}$ if and only if its path of steepest descent[57] under f has a limit point in $C_{\beta,m}$. Each $S_{\beta,m}$ is a locally closed K-invariant submanifold of X and the inclusion of $C_{\beta,m}$ in $S_{\beta,m}$ induces isomorphisms of cohomology and equivariant cohomology[58]. Moreover X is the disjoint union of the $S_{\beta,m}$ and there is a partial order on the indexing set $\mathcal{B}$ such that

---

[56] The indexing of these subsets will be explained later. It is often sufficient to consider the subsets

$$C_\beta = \bigcup_m C_{\beta,m}$$

instead; in particular when (as is quite often the case) for each $\beta \in \mathcal{B}$ there is at most one value of m such that $C_{\beta,m}$ is not empty.

[57] This is for a suitable choice of K-invariant Riemannian metric on X, which when X is Kähler can be taken to be the Kähler metric.

[58] The gradient flow of f induces retractions of $S_{\beta,m}$ onto arbitrarily small neighbourhoods of $C_{\beta,m}$. Indeed Duistermaat has observed that in fact the gradient flow induces a retraction of $S_{\beta,m}$ onto $C_{\beta,m}$ itself (cf. [659] Remark 10.18); this follows from results of LOJASIEWICZ [711], [712].

$$\overline{S_{\beta,m}} \subseteq S_{\beta,m} \cup \bigcup_{\gamma>\beta} S_\gamma$$

where

$$S_\beta = \bigcup_{m\geq 0} S_{\beta,m}$$

for any $\beta \in \mathcal{B}$. The Morse inequalities and equivariant Morse inequalities can be derived from the Thom-Gysin sequences[59] relating the cohomology of the open subsets of X obtained by removing the $S_{\beta,m}$ one by one, in some order such that before $S_{\beta,m}$ is removed all those $S_{\gamma,\ell}$ with $\gamma > \beta$ have already been removed. The integers $d(\beta, m)$ are just the (real) codimensions of the strata $S_{\beta,m}$ of this "Morse stratification[60]" of X.

In fact the *equivariant* Morse inequalities for the norm square of the moment map are equalities; that is, we have

$$P_t^K(X) = \sum_{\beta\in\mathcal{B},m\geq 0} t^{d(\beta,m)} P_t^K(C_{\beta,m}) \tag{3}$$

with no remainder term $(1+t)Q_K(t)$. To explain why this is true we need the following more detailed description[61] of the critical subsets $C_{\beta,m}$. The indexing set $\mathcal{B}$ can be identified with a finite set of orbits of the adjoint representation of K on $\mathfrak{k}$. When $\mathfrak{k}$ is identified with its dual using the invariant inner product, then for each β in $\mathcal{B}$ and each m such that $C_{\beta,m}$ is not empty, the image under the moment map of the critical subset $C_{\beta,m}$ is just the orbit β which indexes it. If a choice is made of a positive Weyl chamber $\mathfrak{t}_+$ in the Lie algebra $\mathfrak{t}$ of some maximal torus T of K, then each adjoint orbit intersects $\mathfrak{t}_+$ in a unique point. Thus instead of being defined as a finite set of adjoint or co-adjoint orbits, $\mathcal{B}$ can be defined equivalently as a finite set of points in $\mathfrak{t}_+$. When X is a complex projective variety on which K acts linearly via a representation

[59] The Thom-Gysin sequence relating the cohomology of a manifold Y, a closed submanifold Z of Y and its complement $Y-Z$ is a long exact sequence of the form

$$\cdots \to H^{*-d}(Z;\mathbf{Q}) \to H^*(Y;\mathbf{Q}) \to H^*(Y-Z;\mathbf{Q}) \to H^{*+1-d}(Z;\mathbf{Q}) \to \cdots$$

where d is the codimension of Z in Y. The existence of such a sequence implies that

$$t^d P_t(Z) + P_t(Y-Z) = P_t(Y) + (1+t)Q(t)$$

for some $Q(t)$ with nonnegative coefficients.

[60] See [659] §4 and §10 for more details. This stratification was discovered independently by Ness [771].

[61] See [659] §3.

$\rho : K \to U(n+1)$ then these points can be described in terms of the weights $\alpha_0, \ldots, \alpha_n$ of this representation as follows: a point of $\mathfrak{t}_+$ lies in $\mathcal{B}$ if and only if it is the closest point to the origin of the convex hull of a nonempty set of these weights. This is true also in the general symplectic case[62] if "weight" is replaced by "image under the moment map μ of a point fixed by the maximal torus T". Then

$$C_{\beta,m} = K(Z_{\beta,m} \cap \mu^{-1}(\beta))$$

where $Z_{\beta,m}$ is the union of those connected components of the set of fixed points of the subtorus of K generated by β (equivalently the set of critical points for the component μ_β of μ along β defined by $\mu_\beta(x) = \mu(x).\beta$ for all $x \in X$) on which μ_β takes the value $\|\beta\|^2$ and its Hessian[63] has index m. In particular $C_{0,0} = \mu^{-1}(0)$ and $C_{0,m} = \emptyset$ if $m > 0$.

Example 8.10. Consider again the action of $K = SU(2)$ on $X = (\boldsymbol{P}_1)^n$ (cf. Example 8.5). As before we identify $\boldsymbol{P}_1$ with the unit sphere in $\boldsymbol{R}^3$ and then $SU(2)$ acts by rotations. We take T to be the standard maximal torus consisting of the diagonal matrices in $SU(2)$ acting as rotations about the vertical axis in $\boldsymbol{R}^3$, and we take $\mathfrak{t}_+$ to be $[0, \infty)$ after identifying $\mathfrak{t}$ with $\boldsymbol{R}$. We have already noted that when the dual of the Lie algebra of $K = SU(2)$ has been identified equivariantly with $\boldsymbol{R}^3$ then the moment map sends an n-tuple of points on the sphere to a constant scalar multiple of its sum in $\boldsymbol{R}^3$. Thus up to multiplication by a constant (which now depends on n) the function $f : X \to \boldsymbol{R}$ under consideration associates to each n-tuple of points on the sphere the distance squared of its center of gravity from the origin.

Now an n-tuple of points on the sphere is fixed by T if and only if each point in the n-tuple is either the north pole $(0, 0, 1)$ or the south pole $(0, 0, -1)$. Therefore we can identify the indexing set $\mathcal{B}$ with the subset

$$\{0\} \cup \{2r - n : n/2 < r \leq n\}$$

of $[0, \infty)$.

If $\beta = 2r - n \in \mathcal{B}$ is nonzero then the subtorus of K generated by β is T, and the fixed points x of T in X satisfying $\mu(x).\beta = \beta.\beta$ are the $\binom{n}{r}$ n-tuples containing r copies of the north pole and $n - r$ copies of the south pole. At each such x the moment map μ takes the value β

[62] Except for a possible constant scalar factor which depends on the conventions adopted.

[63] μ_β is a nondegenerate Morse function on X in the sense of Bott [414]. The Hessian of μ_β at a critical point x is the quadratic form on its tangent space $T_x X$ given in local coordinates by the matrix of second partial derivatives of μ_β at x, and its index is the number of strictly negative eigenvalues of this matrix.

and the Hessian of μ_β has index $2r$. Therefore $Z_{\beta,m} = Z_{\beta,m} \cap \mu^{-1}(\beta)$ consists of these $\binom{n}{r}$ elements of X if $m = 2r = \beta + n$ and is empty otherwise. So if $m = 2r = \beta + n$ the elements of the critical set $C_{\beta,m}$ are all those n-tuples containing r copies of some point on the sphere and $n - r$ copies of the antipodal point. The elements of the stratum

$$S_\beta = \bigcup_{m \geq 0} S_{\beta,m} = S_{\beta,2r}$$

(i.e. those $x \in X$ whose paths of steepest descent under f with respect to the standard Riemannian metric on the sphere have limit points in $C_{\beta,2r}$) are the n-tuples containing some point on the sphere with multiplicity exactly r.

It follows that if $\beta = 0$ the elements of the stratum S_β are those n-tuples containing no point on the sphere with multiplicity strictly greater than $n/2$. We have already noted (see Example 8.5) that these are the semistable points of X under the associated linear action of the complexification $G = SL(2)$ of $K = SU(2)$, as expected from the discussion in §3.

It seems intuitively reasonable that the gradient flow of f should behave in this way. For by symmetry the gradient flow must preserve the property that an n-tuple contains a point on the sphere with multiplicity r, and given this constraint for $r > n/2$ the closest the center of gravity of the n-tuple can approach the origin is when the remaining $n - r$ components approach the antipodal point on the sphere.

In fact (as is clear in the case of Example 8.10 above and is proved in general in [659]) $C_{\beta,m}$ is always homeomorphic to

$$K \times_{K_\beta} (Z_{\beta,m} \cap \mu^{-1}(\beta))$$

where K_β is the stabilizer of β under the adjoint action of K. Therefore (see [50] §13)

$$H^*_K(C_{\beta,m}; \boldsymbol{Q}) \cong H^*_{K_\beta}(Z_{\beta,m} \cap \mu^{-1}(\beta)); \boldsymbol{Q}) \,. \tag{4}$$

In addition the codimensions $d(\beta, m)$ of the Morse strata $S_{\beta,m}$ are given by

$$d(\beta, m) = m - \dim K + \dim K_\beta \,.$$

In order to prove that the equivariant Morse inequalities are equalities one can now apply the criterion used by Atiyah and Bott in the case of the Yang-Mills functional ([50] §13). It suffices to show that the Thom-Gysin sequences relating the equivariant cohomology of the open subsets of X obtained by removing the $S_{\beta,m}$ one by one in a suitable order, all break up into short exact sequences. Equivalently it is enough to show that each Thom-Gysin map

$$H_K^{*-d(\beta,m)}(S_{\beta,m};\boldsymbol{Q}) \to H_K^*(X - \bigcup_{\gamma>\beta} S_\gamma;\boldsymbol{Q})$$

is injective. This is done by showing that its composition with the restriction map

$$H_K^*(X - \bigcup_{\gamma>\beta} S_\gamma;\boldsymbol{Q}) \to H_K^*(S_{\beta,m};\boldsymbol{Q})$$

is injective. This composition is multiplication by the equivariant Euler class of the normal bundle to $S_{\beta,m}$ in X. Under the isomorphism

$$H_K^*(S_{\beta,m};\boldsymbol{Q}) \cong H_{K_\beta}^*(Z_{\beta,m} \cap \mu^{-1}(\beta));\boldsymbol{Q})$$

coming from (4) this class becomes the equivariant Euler class of the restriction to $Z_{\beta,m} \cap \mu^{-1}(\beta)$ of the normal bundle to $S_{\beta,m}$ in X. There is a torus in K_β (in fact the torus generated by β) which acts trivially on $Z_{\beta,m}$ and whose representation on the normal to $S_{\beta,m}$ in X at any point of $Z_{\beta,m} \cap \mu^{-1}(\beta)$ does not contain the trivial representation[64]. By a remark of ATIYAH and BOTT ([50] 13.4) this implies that the equivariant Euler class is not a zero-divisor, and hence that the equivariant Morse inequalities are equalities.

If $x \in Z_{\beta,m}$ and as usual $\mathfrak{k}$ is identified with its dual, then $\mu(x)$ lies in the Lie algebra of the stabilizer K_β of β under the adjoint action of K (because the moment map is K-equivariant), and the map

$$x \mapsto \mu(x) - \beta$$

can be regarded as a moment map for the action of K_β on the symplectic submanifold $Z_{\beta,m}$ of X (see [659] 5.7). Thus $P_t^{K_\beta}(Z_{\beta,m} \cap \mu^{-1}(\beta))$ is of the same form as $P_t^K(\mu^{-1}(0))$ but with X replaced by a compact symplectic submanifold of X and K replaced by a compact connected subgroup of K. This means that the formula (3), rewritten using (4) as

$$P_t^K(\mu^{-1}(0)) = P_t^K(X) - \sum_{0 \neq \beta \in \mathcal{B},\, m \geq 0} t^{d(\beta,m)} P_t^{K_\beta}(Z_{\beta,m} \cap \mu^{-1}(\beta)), \quad (5)$$

gives us an inductive formula for computing the equivariant Betti numbers of $\mu^{-1}(0)$ in terms of the equivariant Betti numbers of X and

[64] This is a simple consequence of the facts that $Z_{\beta,m}$ is by definition a union of connected components of the fixed point set of the torus generated by β and that the stratum $S_{\beta,m}$ contains an open neighbourhood of $Z_{\beta,m} \cap \mu^{-1}(\beta)$ in $Z_{\beta,m}$.

certain connected components of fixed point sets of subtori of K. This is relevant to the computation of the Betti numbers of the symplectic quotient $\mu^{-1}(0)/K$ when K acts with finite stabilizers on $\mu^{-1}(0)$ (or equivalently by Theorem 8.3 the geometric invariant theoretic quotient $X//G$ when $X^{ss} = X^s_{(0)}$, if X is a nonsingular complex projective variety acted on linearly by the complexification G of K as in Example 8.1(ii)) since then the natural map

$$\mu^{-1}(0) \times_K EK \to \mu^{-1}(0)/K$$

induces an isomorphism[65] of rational cohomology

$$H^*(\mu^{-1}(0)/K; \boldsymbol{Q}) \to H^*_K(\mu^{-1}(0); \boldsymbol{Q}).$$

Moreover when X is a compact symplectic manifold acted on symplectically by a compact connected group K the spectral sequence associated to the fibration

$$\mu^{-1}(0) \times_K EK \to BK$$

degenerates (see [659] Proposition 5.8), so that

$$P_t^K(X) = P_t(X) P_t(BK).$$

Thus the inductive formula (5) gives us a way to compute the Betti numbers of the symplectic quotient $\mu^{-1}(0)/K$ in terms of the Betti numbers of X and certain symplectic submanifolds of X, together with the Betti numbers of the universal classsifying spaces of K and certain compact subgroups of K. (It is also possible in a similar way to compute the Betti numbers of the more general symplectic quotients $\mu^{-1}(\zeta)/K_\zeta$ [667].)

Example 8.11. When $K = SU(2)$ acts on $X = (\boldsymbol{P}_1)^n$ as in Examples 8.5 and 8.10 we do not need induction to apply the formula (5) since each $Z_{\beta,m}$ for nonzero β is either empty or is a finite set of points. We have

$$P_t(X) = (1+t^2)^n$$

and

$$P_t(BK) = (1-t^4)^{-1},$$

so that

$$P_t^K(X) = P_t(X) P_t(BK) = (1+t^2)^n (1-t^4)^{-1}.$$

[65] This follows using spectral sequences from the fact that the universal classifying space of a finite group has trivial rational cohomology.

Moreover since $K_\beta = T$ if $\beta \neq 0$ and

$$P_t(BT) = (1-t^2)^{-1}$$

the sum in (5) reduces to

$$\sum_{n/2<r\leq n} \tbinom{n}{r} t^{2(r-1)}(1-t^2)^{-1}.$$

Thus when n is odd, so that the condition that K acts with finite stabilizers on $\mu^{-1}(0)$ is satisfied, (5) gives us the formula

$$P_t(\mu^{-1}(0)/K) = 1 + nt^2 + \cdots + (1 + (n-1) + \tbinom{n-1}{2} + \cdots \\ + \tbinom{n-1}{\min(j,n-3-j)}))t^{2j} + \cdots + t^{2n-6}$$

for the Betti numbers of $\mu^{-1}(0)/K$.

§ 7. The cohomology of quotients: the algebraic case

The last section described how equivariant Morse theory, applied to the norm square of the moment map, can be used to calculate the Betti numbers of symplectic quotients of the form $\mu^{-1}(0)/K$, where $\mu : X \to \mathfrak{k}^*$ is a moment map for a symplectic action of a compact group K on a compact symplectic manifold X and 0 is a regular value of μ. The results obtained can be translated into algebraic terms to give formulas for the Betti numbers of geometric invariant theoretic quotients $X//G$ when $X^{ss} = X^s_{(0)}$. These formulas can be used, as we shall describe in this section, to calculate the dimensions of the intersection cohomology groups of $X//G$ even when $X^{ss} \neq X^s_{(0)}$, so long as $X^s_{(0)}$ is not empty.

Let us suppose that X is a nonsingular complex variety embedded in $\boldsymbol{P}_n$ and that a maximal compact subgroup K of a complex reductive group G acts on X via a representation $\rho : K \to U(n+1)$. We may also assume that a maximal torus T of K acts diagonally with weights $\alpha_0, \ldots, \alpha_n$. Then the stratification

$$\{S_\beta : \beta \in \mathscr{B}\}$$

and its refinement $\{S_{\beta,m} : \beta \in \mathscr{B}, m \geq 0\}$, defined using symplectic geometry and Morse theory in the last section, can also be defined algebraically as follows. (In fact the stratification can be defined and shown to have similar properties when X is a variety over any algebraically closed field. This can be deduced (see [659] §12 and §13) from results of BIALYNICKI-BIRULA [402], HESSELINK [140] and KEMPF [171]). Let

$$Z_\beta = \{(x_0, \dots, x_n) \in X : x_i = 0 \text{ if } \alpha_i.\beta \neq \|\beta\|^2\}$$

and

$$Y_\beta = \{(x_0, \dots, x_n) \in X : x_i = 0 \quad \text{if } \alpha_i.\beta < \|\beta\|^2 \text{ and } \exists x_i \neq 0 \text{ with } \alpha_i.\beta = \|\beta\|^2\}.$$

These are nonsingular subvarieties of X with $Z_\beta \subseteq Y_\beta$ and with a retraction $p_\beta : Y_\beta \to Z_\beta$ defined by replacing all homogeneous coordinates x_i such that $\alpha_i.\beta \neq \|\beta\|^2$ by 0 (cf. [171] and [402]). For any $m \geq 0$ the $Z_{\beta,m}$ defined using symplectic geometry in §6 is the union of those connected components of Z_β which are contained in connected components of Y_β with real codimension m in X. Moreover (see [659]) the Morse strata S_β and $S_{\beta,m}$ defined as in §6 using the Fubini-Study metric on X are given by

$$S_\beta = GY_\beta^{ss}$$

and

$$S_{\beta,m} = GY_{\beta,m}^{ss}$$

where

$$Y_\beta^{ss} = p_\beta^{-1}(Z_\beta^{ss})$$

and

$$Y_{\beta,m}^{ss} = p_\beta^{-1}(Z_{\beta,m}^{ss})$$

and Z_β^{ss} (respectively $Z_{\beta,m}^{ss}$) is the set of semistable points in Z_β (respectively $Z_{\beta,m}$) with respect to an appropriate linearization of the action of the complexification of K_β (the new linearization corresponds to the moment map $\mu - \beta$). In fact

$$S_\beta \cong G \times_{P_\beta} Y_\beta^{ss}$$

and

$$S_{\beta,m} \cong G \times_{P_\beta} Y_{\beta,m}^{ss}$$

where P_β is the parabolic subgroup of G which is the product of the stabilizer K_β of β (under the coadjoint action of K) and the Borel subgroup B associated to the choice of maximal torus T and positive Weyl chamber $\mathfrak{t}_+$. The inductive formula (5) of the last section can then be rewritten as

$$P_t^K(X^{ss}) = P_t(X)P_t(BK) - \sum_{0 \neq \beta \in \mathcal{B},\, 0 \leq m} t^{d(\beta,m)} P_t^{K_\beta}(Z_{\beta,m}^{ss}) \quad (6)$$

where $d(\beta, m) = m - \dim K + \dim K_\beta$. In the case when $X^{ss} = X^s_{(0)}$ it gives us a formula for

$$P_t(X//G) = P_t^K(X^{ss}).$$

Example 8.12. Consider again the diagonal action of $G = SL(2)$ on $X = (\boldsymbol{P}_1)^n$ as in Examples 8.5 and 8.10. The representation of G which defines the linearization of the action is the nth tensor power of the standard two-dimensional representation of $SL(2)$, and the set of its weights after appropriate identifications is

$$\{2r - n : 0 \leq r \leq n\} = \{-n, -n+2, \ldots, n-2, n\}.$$

Thus as we have already seen (Example 8.11) the indexing set $\mathscr{B}$ can be identified with

$$\{0\} \cup \{2r - n : n/2 < r \leq n\}.$$

Moreover since $\mathfrak{t}$ is one-dimensional the condition $\alpha_i.\beta = \|\beta\|^2$ is equivalent to $\alpha_i = \beta$ or $\beta = 0$. Therefore if $\beta = 2r - n$ is a nonzero element of $\mathscr{B}$ then Z_β consists of all n-tuples of points in $\boldsymbol{P}_1$ containing r copies of 0 and $n-r$ copies of ∞, while Y_β consists of all n-tuples of points in $\boldsymbol{P}_1$ containing exactly r copies of 0. Each component of Y_β has real codimension $2r$ so $Z_{\beta,m} = Z_\beta$ if $m = 2r$ and $Z_{\beta,m}$ is empty otherwise. The action of $K_\beta = T$ on Z_β is trivial and $Z_{\beta,m}^{ss} = Z_{\beta,m}$ so $Y_{\beta,m}^{ss} = Y_{\beta,m}$ for all m, and

$$S_\beta = S_{\beta,2r} = GY_{\beta,2r}^{ss}$$

consists of all n-tuples $x \in (\boldsymbol{P}_1)^n$ containing some point of $\boldsymbol{P}_1$ with multiplicity exactly r. Since $r > n/2$, clearly

$$S_\beta \cong G \times_{P_\beta} Y_\beta^{ss}$$

where P_β is the stabilizer of $0 \in \boldsymbol{P}_1$ under the action of G (i.e. the standard Borel subgroup B of $SL(2)$ consisting of upper triangular matrices). In the case when n is odd so that $X^{ss} = X_{(0)}^s$ we have already calculated $P_t(X//G) = P_t(\mu^{-1}(0)/K)$ using the formula (5) of §6 (see Example 8.11).

In the special case when G acts freely on X^{ss}, formulas equivalent to the formula (6) for $P_t(X//G)$ can be derived via the Weil conjectures by counting points of associated varieties defined over finite fields (see §15 of [659]). This was motivated by the work of HARDER and NARASIMHAN [132] (see also [84]). They gave formulas for the Betti numbers of the moduli spaces of vector bundles of coprime rank and degree over Riemann surfaces, which were subsequently rederived by ATIYAH and BOTT in [50] using equivariant Morse theory.

The calculation of the Betti numbers of $X//G$ when $X^{ss} = X_{(0)}^s$ can also be extended to its Hodge numbers using Deligne's theory [78, 79] of mixed Hodge structures (see §14 of [659]). In particular if the Hodge numbers $h^{p,q}(X)$ of X are zero for $p \neq q$ then the same is true for the Hodge numbers $h^{p,q}(X//G)$ of the quotient $X//G$ when $X^{ss} = X^s_{(0)}$.

Information about the ring structure of the rational cohomology of the quotient $X//G$ when $X^{ss} = X^s_{(0)}$ can also be obtained. First of all it follows directly from the fact that the Morse stratification is equivariantly perfect (i.e. that the equivariant Morse inequalities are actually equalities) that the restriction map

$$H^*_K(X; \boldsymbol{Q}) \to H^*_K(X^{ss}; \boldsymbol{Q})$$

is surjective. Since

$$H^*_K(X^{ss}; \boldsymbol{Q}) \cong H^*(X//G; \boldsymbol{Q})$$

when $X^{ss} = X^s_{(0)}$, this means that generators of the cohomology ring $H^*_K(X; \boldsymbol{Q})$ restrict to give generators of $H^*(X//G; \boldsymbol{Q})$. Moreover we have already noted that the spectral sequence associated to the fibration $X \times_K EK \to BK$ degenerates ([659] 5.8) so that

$$H^*_K(X; \boldsymbol{Q}) \cong H^*(X; \boldsymbol{Q}) \otimes H^*(BK; \boldsymbol{Q}) .$$

Although this isomorphism is not an isomorphism of rings, nonetheless generators of $H^*(X; \boldsymbol{Q})$ and $H^*(BK; \boldsymbol{Q})$ give generators[66] of $H^*_K(X; \boldsymbol{Q})$ and thus of $H^*(X//G; \boldsymbol{Q})$. In order to understand the relations between these generators it is necessary to understand the kernel of the restriction map $H^*_K(X; \boldsymbol{Q}) \to H^*_K(X^{ss}; \boldsymbol{Q})$. This is studied by Brion in [431], [432] and in [669] (see also [506], [887], [918]).

Example 8.13. Consider yet again the diagonal action of $G = SL(2)$ on $X = (\boldsymbol{P}_1)^n$. It turns out (see the final example of [669]) that the kernel of the restriction map

$$H^*_K(X; \boldsymbol{Q}) \to H^*_K(X^{ss}; \boldsymbol{Q})$$

is spanned by all elements of $H^*_K(X; \boldsymbol{Q})$ of the form

$$(q(\xi_1, \dots, \xi_n, \alpha) \prod_{i \in I} (\xi_i + \alpha) - q(\xi_1, \dots, \xi_n, -\alpha) \prod_{i \in I} (\xi_i - \alpha))/2\alpha$$

for some subset I of $\{1, \dots, n\}$ and some polynomial q in $n+1$ variables with rational coefficients, where $\xi_1, \dots, \xi_n$ (coming from $H^2((\boldsymbol{P}_1)^n; \boldsymbol{Q})$) and α^2 (coming from $H^4(BK; \boldsymbol{Q})$) generate $H^*_K(X; \boldsymbol{Q})$ subject to the relations $(\xi_i)^2 = \alpha^2$ for $1 \leq i \leq n$. Equivalent results were obtained by Brion in [431].

---

[63] The generators coming from $H^*(BK; \boldsymbol{Q})$ are algebraic and can be expressed in terms of Chern classes of tautological bundles.

When X is a projective space ELLINGSRUD and STROMME [532] have given a method of computing the Chow ring[67] of $X//G$, showing that it depends only on the following data: the restriction to a maximal torus of G of the representation which defines its linear action, and the action of the Weyl group of G on the maximal torus. They give a complete description of the Chow ring with rational coefficients $A^*(X//G)_Q$ which is valid over any algebraically closed field k. In addition when $k = C$ they prove that the cycle map

$$A^*(X//G)_Q \to H^*(X//G; Q)$$

is an isomorphism. Under strengthened hypotheses this is true for integer coefficients.

The procedure we outlined above only gives us the Betti numbers of the quotient $X//G$ in the very special case when $X^{ss} = X^s_{(0)}$. However, provided that $X^s_{(0)} \neq \emptyset$ it can be applied to the partial desingularization $\tilde{X}//G$ of $X//G$ described in §4. The equivariant Poincaré series $P_t^K(X^{ss})$ of X^{ss} can be calculated using the inductive formula (5). The task is then to relate this to the Poincaré series

$$P_t(\tilde{X}//G) = P_t^K(\tilde{X}^{ss})$$

of the partial desingularization.

To accomplish this[68] let $\mathcal{R}$ be a set of representatives of the finitely many conjugacy classes of reductive subgroups of G which occur as identity components of stabilizers of semistable points $x \in X$ such that the orbit $O_G(x)$ is closed in X^{ss}. We may assume that if $R \in \mathcal{R}$ then $R \cap K$ is a maximal compact subgroup of R. For any $R \in \mathcal{R}$ let N^R denote the normalizer of R in G, let N_0^R be its identity component and let $\pi_0(N^R) = N^R/N_0^R$ be its group of components. Let $\hat{Z}_R^{ss}$ be the proper transform of

$$Z_R^{ss} = \{x \in X^{ss} : R \text{ fixes } x\}$$

at the stage at which it is about to be blown up in the construction of $\tilde{X}^{ss}$ from X^{ss} described in §4 (equivalently by [664] 1.6 $\hat{Z}_R^{ss}$ is the set of semistable points fixed by R at this stage). Let $\mathcal{N}_x^R$ be the normal to $G\hat{Z}_R^{ss}$ at a point $x \in \hat{Z}_R^{ss}$. Let us assume that $\hat{Z}_R^{ss}$ is connected (if it is not we should replace $\hat{Z}_R^{ss}$ by its connected components in what follows). Then the induced representation of R on $\mathcal{N}_x^R$ is independent

[67] VISTOLI [905] has also considered the Chow group of the quotient of any scheme by a group action with finite stabilizers.

[68] For more details see [664].

of x up to isomorphism; let $\boldsymbol{P}(\mathcal{N}_x^R)^{ss}$ denote the set of semistable points for the induced linear action of R on $\boldsymbol{P}(\mathcal{N}_x^R)$. Then it turns out ([664] 1.18) that

$$\begin{aligned} P_t(\tilde{X}//G) = P_t^K(\tilde{X}^{ss}) &= P_t^K(X^{ss}) \\ &+ \sum_{R\in\mathcal{R},\, p,q\geq 0} t^{p+q}(\dim\, [H^p(\hat{Z}_R//N_0^R) \otimes H_R^q(\boldsymbol{P}(\mathcal{N}_x^R)^{ss};\boldsymbol{Q})]^{\pi_0 N^R} \\ &- \dim\, [H^p(\hat{Z}_R//N_0^R;\boldsymbol{Q}) \otimes H^q(BR;\boldsymbol{Q})]^{\pi_0 N^R}) \end{aligned}$$

where $[\,]^S$ indicates the subspace consisting of all elements invariant under the action of a group S. For each $R \in \mathcal{R}$ we also have by [664] Proposition 1.20[69] that

$$\begin{aligned} P_t(\hat{Z}_R//N_0^R) &= P_t^{N_0^R/R}(Z_R^{ss}) \\ &+ \sum_{S\in\mathcal{S},\, p,q\geq 0} t^{p+q}(\dim\, [H^p(\hat{Z}_S//N_0^S;\boldsymbol{Q}) \otimes H_{S/R}^q(\boldsymbol{P}(\hat{\mathcal{N}}_x^S)^{ss};\boldsymbol{Q})]^{\pi_0(N_0^R\cap N^S)} \\ &- \dim\, [H^p(\hat{Z}_S//N_0^S;\boldsymbol{Q}) \otimes H^q(B((N_0^R\cap S)/R);\boldsymbol{Q})]^{\pi_0(N_0^R\cap N^S)}) \end{aligned}$$

where $\mathcal{S}$ is a set of representatives of conjugacy classes relative to N_0^R of connected reductive subgroups S of G which contain R and are the identity components of stabilizers of points of X^{ss} with G-orbits closed in X^{ss}, and if $S \in \mathcal{S}$ then $\hat{\mathcal{N}}_x^S$ is the normal to $N_0^R\hat{Z}_S^{ss}$ in the proper transform of Z_R^{ss} at some $x \in \hat{Z}_S^{ss}$.

This last formula respects the action of the group $\pi_0(N^R)$. Therefore it can be used inductively and combined with the preceding formula to calculate the Betti numbers of the partial desingularization $\tilde{X}//G$. To carry out this calculation one needs knowledge of the rational cohomology of X and certain nonsingular linear sections Z of X, together with the rational cohomology of the classifying spaces of K and certain compact subgroups Γ of K, as well as the action of $\pi_0(\Gamma)$ on $H^*(Z;\boldsymbol{Q})$ for appropriate Γ acting on Z.

Example 8.14. Consider for the penultimate time the diagonal action of $G = SL(2)$ on $X = (\boldsymbol{P}_1)^n$. As in Example 8.8 we assume that n is even, because when n is odd $X^{ss} = X_{(0)}^s$ and so $\tilde{X}//G = X//G$. Then as at 8.8 we find that $\mathcal{R}$ has just one element R which has dimension $r = 1$ and is a maximal torus of G; we choose it to be the standard

[69] Note that in [664] 1.20 $B(S/R)$ should be $B((N_0^R \cap S)/R)$. Moreover in [661] 7.4, 7.13 and 7.14 $1 + t^2 + \cdots$ should be $t^2 + \cdots$, and the last $+$ in 7.13 should be $-$.

maximal torus consisting of diagonal matrices. Then the normalizer N^R of R in G is the subgroup generated by R and the matrix

$$\begin{pmatrix} 0 & 1 \\ -1 & 0 \end{pmatrix};$$

its identity component N_0^R is the maximal torus R and $\pi_0(N^R)$ has order two (it is of course the Weyl group of G). The set Z_R^{ss} of semistable fixed points of R consists of those n-tuples $x \in (\boldsymbol{P}_1)^n$ containing 0 and ∞ each with multiplicity $n/2$, so that $Z_R//N^R$ is finite of order $\frac{1}{2}\binom{n}{n/2}$ while $Z_R//N_0^R$ has order $\binom{n}{n/2}$. The action of R on the normal $\mathcal{N}_x \cong \boldsymbol{C}^{n-2}$ to $GZ_R^{ss} \cong G\times_{N^R} Z_R^{ss}$ at any $x \in Z_R^{ss}$ has weights ± 2 each with multiplicity $(n-2)/2$. Thus

$$\boldsymbol{P}(\mathcal{N}_x)//R \cong \boldsymbol{P}_{(n-4)/2} \times \boldsymbol{P}_{(n-4)/2}$$

and

$$H_R^*(\boldsymbol{P}(\mathcal{N}_x)^{ss};\boldsymbol{Q}) \cong H^*(\boldsymbol{P}_{(n-4)/2} \times \boldsymbol{P}_{(n-4)/2};\boldsymbol{Q}).$$

Let $[H_R^q(\boldsymbol{P}(\mathcal{N}_x)^{ss};\boldsymbol{Q})]^{\pm}$ be the ± 1 eigenspaces for the action of the Weyl group $\pi_0(N^R)$ on $H_R^q(\boldsymbol{P}(\mathcal{N}_x)^{ss};\boldsymbol{Q})$; then

$$[H^p(\hat{Z}_R//N_0^R;\boldsymbol{Q}) \otimes H_R^q(\boldsymbol{P}(\mathcal{N}_x^R)^{ss};\boldsymbol{Q})]^{\pi_0 N^R}$$

is zero when $p \neq 0$ and when $p = 0$ it is isomorphic to the sum of $\frac{1}{2}\binom{n}{n/2}$ copies of $[H_R^q(\boldsymbol{P}(\mathcal{N}_x)^{ss};\boldsymbol{Q})]^+$ and $\frac{1}{2}\binom{n}{n/2}$ copies of $[H_R^q(\boldsymbol{P}(\mathcal{N}_x)^{ss};\boldsymbol{Q})]^-$. It follows that

$$\begin{aligned}\dim[H^0(\hat{Z}_R//N_0^R;\boldsymbol{Q}) \otimes H_R^q(\boldsymbol{P}(\mathcal{N}_x^R)^{ss};\boldsymbol{Q})]^{\pi_0 N^R} \\ = \frac{1}{2}\binom{n}{n/2} \dim H_R^q(\boldsymbol{P}(\mathcal{N}_x^R)^{ss};\boldsymbol{Q}).\end{aligned}$$

Similarly

$$\dim[H^p(\hat{Z}_R//N_0^R;\boldsymbol{Q}) \otimes H^q(BR;\boldsymbol{Q})]^{\pi_0 N^R} = \frac{1}{2}\binom{n}{n/2} \dim H^q(BR;\boldsymbol{Q})$$

when $p = 0$, and the left hand side is zero otherwise. Hence $P_t(\tilde{X}//G)$ is given by the formula

$$P_t^G(X^{ss}) + \frac{1}{2}\binom{n}{n/2}(1 + t^2 + \cdots + t^{n-4})^2 - \frac{1}{2}\binom{n}{n/2}(1 - t^2)^{-1}.$$

The general process described so far still does not tell us directly about the cohomology of $X//G$ itself when $X^{ss} \neq X_{(0)}^s$. However a topological invariant of singular varieties which seems to be at least as interesting as ordinary cohomology is intersection cohomology[70] (introduced by GORESKY and MACPHERSON in [576], [577]). Intersection cohomology coincides with ordinary cohomology for nonsingular varieties, and many of the nice properties of the cohomology of nonsingular complex projective varieties, such as Poincaré duality, the Hodge decomposition and the hard Lefschetz theorem, remain true for singular varieties provided that ordinary cohomology is replaced by intersection cohomology.

Now the intersection cohomology of a blow-up of a complex variety V contains the intersection cohomology of V as a direct summand (this is a special case of the decomposition theorem of BEILINSON, BERNSTEIN, DELIGNE and GABBER [386]). Our partial desingularization $\tilde{X}//G$ is obtained from $X//G$ by a sequence of blow-ups along subvarieties of the form $\hat{Z}_R//N^R$. The decomposition theorem can be used at each stage to relate the intersection cohomology before and after the blow-up. This leads to a formula for the intersection Poincaré series

$$IP_t(X//G) = \sum_{i \geq 0} t^i \dim IH^i(X//G; \boldsymbol{Q})$$

of $X//G$ in terms of

$$IP_t(\tilde{X}//G) = P_t(\tilde{X}//G)$$

together with the rational cohomology of the quotients $\hat{Z}_R//N_0^R$ (already given above) and the intersection cohomology of certain quotients of projective spaces which can be calculated inductively. The formula is (see [664] Theorem 3.1)

$$IP_t(X//G) = P_t(\tilde{X}//G) \\ - \sum_{R \in \mathfrak{R},\, p,q \geq 0} t^{p+q} \dim[H^p(\hat{Z}_R//N_0^R) \otimes IH^{t(q,R)}(\boldsymbol{P}(\mathcal{N}_x^R)//R; \boldsymbol{Q})]^{\pi_0 N^R}$$

where $t(q,R)$ is $q-2$ if $q \leq \dim \boldsymbol{P}(\mathcal{N}_x^R)//R$ and is q otherwise.

Example 8.15. Consider for the last time the diagonal action of $G = SL(2)$ on $X = (\boldsymbol{P}_1)^n$, continuing the discussion of Example 8.14. We assume that n is even. Since

$$IH^*(\boldsymbol{P}(\mathcal{N}_x^R)//R; \boldsymbol{Q}) \cong H^*(\boldsymbol{P}(\mathcal{N}_x^R)//R; \boldsymbol{Q})$$

[70] We shall only consider intersection cohomology with respect to the middle perversity.

we get

$$IP_t(X//G) = P_t(\tilde{X}//G) - \frac{1}{2}\binom{n}{n/2}(t^2 + 2t^4 + 3t^6 + \cdots + (\frac{n}{2} - 2)t^{n-4} + (\frac{n}{2} - 2)t^{n-2} + \cdots + t^{2n-8}).$$

Using the formulas of Examples 8.11 and 8.14 we find that

$$\dim IH^{2j+1}(X//G; \mathbf{Q}) = 0$$

and

$$\dim IH^{2j}(X//G; \mathbf{Q}) = 1 + (n-1) + \binom{n-1}{2} + \cdots + \binom{n-1}{\min(j, n-3-j)}.$$

The general formulas given above for $P_t(X//G)$ when $X^{ss} = X^s_{(0)}$ and for $P_t(\tilde{X}//G)$ and $IP_t(X//G)$ when $X^{ss} \neq X^s_{(0)}$ are very complicated but can be much simplified by judicious use of Poincaré duality, which tells us that we need only work modulo $t^{\dim X//G}$. It is also always the case that the composition

$$H^*_K(X^{ss}; \mathbf{Q}) \to H^*_K(\tilde{X}^{ss}; \mathbf{Q}) \to IH^*(X//G; \mathbf{Q})$$

is surjective ([664] Theorem 2.5), and usually it is an isomorphism in low degrees.

The formula for $IP_t(X//G)$ given above has been applied to calculate the intersection Betti numbers of certain moduli spaces (or compactifications of moduli spaces) which can be represented as geometric invariant theoretic quotients $X//G$ [663], [668], [671–673]. In particular when applying these methods to moduli spaces of vector bundles over Riemann surfaces one is no longer restricted, as Atiyah and Bott were in [50], to the case of coprime rank and degree (cf. §9 below).

An alternative method of calculating the rational intersection cohomology of a quotient $X//G$, in the special case when G is a torus, involves finding a closely related quotient which is a small resolution of $X//G$ (at least if finite quotient singularities are ignored), which means that its rational cohomology is isomorphic to the rational intersection cohomology of $X//G$ [632].

The procedure we have outlined for calculating the intersection Betti numbers of $X//G$ when X is a nonsingular complex projective variety can be extended to the more general case when X is allowed to be singular [665]. Now however intersection cohomology must be used throughout and the formulas become even more complicated.

§ 8. Vector bundles and the Yang-Mills functional

The Yang-Mills equations arose in theoretical physics as generalizations of Maxwell's equations and became important in the study of elementary particles. They began to make an impact on differential geometry in the late 1970s (see for example [415], [418], [510], [705], [760]), and links between Yang-Mills theory and geometric invariant theory were first recognized a few years later[71]. [50], [497]. These links can be understood via symplectic geometry and the moment map. The fundamental theorem of DONALDSON, UHLENBECK and YAU that a holomorphic bundle over a compact Kähler manifold admits an irreducible Hermitian Yang-Mills connection if and only if it is stable (in which case the connection is unique) can be thought of as an infinite-dimensional illustration of the principle that suitable symplectic quotients of complex projective varieties can be identified with geometric invariant theoretic quotients.

The Yang-Mills equations were originally defined by physicists over Minkowski space and were analytically continued to Euclidean 4-space; they were then extended to its conformal compactification, the sphere S^4. They have become important in differential and algebraic geometry formulated over arbitrary compact oriented Riemannian manifolds, and in particular over compact Riemann surfaces and higher dimensional Kähler manifolds. This is the context in which we shall work.

Let M be a compact oriented Riemannian manifold and let E be a fixed complex vector bundle over M with a Hermitian metric. (Equivalently we can work with principal $U(n)$-bundles over M, and more generally with principal G-bundles for any compact connected Lie group G). Recall that a connection A on E (or equivalently its frame bundle) can be defined by a covariant derivative d_A on E, that is a linear map

$$d_A : \Omega^0_M(E) \to \Omega^1_M(E),$$

where $\Omega^p_M(E)$ denotes the space of C^∞-sections of $\bigwedge^p T^*M \otimes E$ (i.e the space of p-forms on M with values in E), satisfying the Leibniz rule

$$d_A(fs) = f\, d_A s + df.s$$

for C^∞ complex valued functions f on M and sections s of E. This covariant derivative extends uniquely to

---

[71] Of course at about the same time Donaldson began applying Yang-Mills theory (in particular using work of UHLENBECK and TAUBES) to the study of 4-manifolds with results that are now famous [496], [500–503], [505], [507], [508]. See also [363], [541–544], [546], [551–554], [916]

$$d_A : \Omega^p_M(E) \to \Omega^{p+1}_M(E)$$

satisfying the extended Leibniz rule

$$d_A(\alpha \wedge \beta) = (d_A\alpha) \wedge \beta + (-1)^p \alpha \wedge d_A\beta$$

for $\alpha \in \Omega^p_M(E)$, $\beta \in \Omega^q_M(E)$. The Leibniz rule implies that the difference of two connections is an $E \otimes E^*$-valued 1-form on M, and hence that the space of all connections on E is an infinite-dimensional affine space $\mathscr{A}$ based on the vector space $\Omega^1_M(E \otimes E^*)$. Similarly the space of all *unitary* connections on E (i.e. connections compatible with the Hermitian metric on E) is an infinite-dimensional affine space based on the space of 1-forms with values in the bundle $\mathfrak{g}_E$ of skew-adjoint endomorphisms of E.

The Leibniz rule also implies that the composition

$$d_A \circ d_A : \Omega^0_M(E) \to \Omega^2_M(E)$$

commutes with multiplication by smooth functions, and thus we have

$$d_A \circ d_A(s) = F_A s$$

for all C^∞ sections s of E, where

$$F_A \in \Omega^2_M(\mathfrak{g}_E).$$

This F_A is the curvature of the unitary connection A. The Yang-Mills functional on the space $\mathscr{A}$ of all unitary connections on E is defined to be the L^2-norm square of the curvature,

$$A \mapsto \int_M |F_A|^2 \,\mathrm{vol}$$

where the volume form on M is defined by the Riemannian metric and the orientation. The Yang-Mills equations are the Euler-Lagrange equations for this functional. They can be expressed in the form

$$d_A * F_A = 0$$

([50] Proposition 4.6) where $*$ is the duality operator characterized by

$$\theta \wedge *\phi = \langle \theta, \phi \rangle \,\mathrm{vol}$$

for all $\theta, \phi \in \Omega^p_M(\mathfrak{g}_E)$, and d_A has been extended in a natural way to $\Omega^*_M(\mathfrak{g}_E)$. The Yang-Mills functional and the Yang-Mills equations are invariant under the action of the gauge group $\mathscr{G}$, that is, the group of unitary automorphisms of E.

Now suppose that M is a complex manifold. Recall that a holomorphic structure $\mathscr{E}$ on the bundle E defines a differential operator

$$\overline{\partial}_{\mathscr{E}} : \Omega_M^{p,q}(E) \to \Omega_M^{p,q+1}(E)$$

such that sections of E are holomorphic if and only if they are annihilated by $\overline{\partial}_{\mathscr{E}}$ ([580] p. 70, [508] §2.1.5). Moreover there is a unique unitary connection

$$d_A : \Omega_M^0(E) \to \Omega_M^1(E)$$

whose $(0,1)$-component is $\overline{\partial}_{\mathscr{E}}$. Conversely the $(0,1)$-component of a unitary connection d_A is of the form $\overline{\partial}_{\mathscr{E}}$ for a (necessarily unique) holomorphic structure $\mathscr{E}$ on E if and only if its curvature $F_A \in \Omega_M^2(\mathfrak{g}_E)$ is of type $(1,1)$ ([508] Proposition 2.1.56). Thus we can identify the space $\mathscr{A}^{(1,1)}$ of unitary connections on E with curvature of type $(1,1)$ with the space of holomorphic structures on E. This space $\mathscr{A}^{(1,1)}$ is an infinite-dimensional complex subvariety of the infinite-dimensional complex affine space $\mathscr{A}$ (the complex structure on $\mathscr{A}$ coming from representing unitary connections by their $(0,1)$-components). It is acted on by the complexified gauge group $\mathscr{G}_c$ (the group of complex C^∞ automorphisms of E) and two holomorphic structures are isomorphic if and only if they lie in the same $\mathscr{G}_c$-orbit.

Next suppose that M is a compact Kähler manifold. For a holomorphic bundle, with corresponding connection $A \in \mathscr{A}^{(1,1)}$, the Kähler identities ([580] Chapter 0§7) mean that one can write

$$d(*F_A) = i(\overline{\partial} - \partial)\Lambda F_A$$

where Λ maps $\Omega^{1,1}(\mathfrak{g}_E)$ to $\Omega^0(\mathfrak{g}_E)$. A connection A such that ΛF_A is a constant multiple of the identity is called a Hermitian Yang-Mills connection. The Hermitian Yang-Mills connections minimize the Yang-Mills functional, and any connection $A \in \mathscr{A}^{(1,1)}$ which satisfies the Yang-Mills equations is a direct sum of Hermitian Yang-Mills connections.

When M is a compact Kähler manifold of complex dimension n with Kähler form ω, there is a Kähler form Ω on the space $\mathscr{A}$ of unitary connections on E defined by

$$\Omega(\alpha,\beta) = \frac{1}{8\pi^2}\int_M tr(\alpha \wedge \beta) \wedge \omega^{n-1}$$

for $\alpha, \beta \in \Omega_M^1(\mathfrak{g}_E)$, which is invariant under the action of the gauge group $\mathscr{G}$. The Lie algebra of $\mathscr{G}$ is the space $\Omega_M^0(\mathfrak{g}_E)$ of sections of $\mathfrak{g}_E$, and it is not difficult to check (see e.g. [508] Proposition 6.5.8) that there is a moment map

$$\mu : \mathscr{A} \to (\Omega^0_M(\mathfrak{g}_E))^*$$

for the action of $\mathscr{G}$ on $\mathscr{A}$ given by the composition of the map

$$A \mapsto \frac{1}{8\pi^2} F_A \wedge \omega^{n-1} \in \Omega^{2n}_M(\mathscr{U}_E)$$

with integration over M which defines a map

$$\Omega^{2n}_M(\mathscr{U}_E) \to (\Omega^0_M(\mathfrak{g}_E))^*.$$

On $\mathscr{A}^{(1,1)}$ the norm square of this moment map agrees up to a constant factor with the Yang-Mills functional. Thus by analogy with the finite-dimensional situation (see §2) one would hope that for a suitable definition of stability the moduli space of stable holomorphic bundles[72] of topological type E over M could be identified with the moduli space of irreducible Hermitian Yang-Mills connections on E (though of course one would expect difficulties to arise because of the infinite-dimensionality and noncompactness). More precisely one would hope that any $\mathscr{G}_c$-orbit in $\mathscr{A}^{(1,1)}$ contains at most one $\mathscr{G}$-orbit of Hermitian Yang-Mills connections, and a $\mathscr{G}_c$-orbit contains an irreducible Hermitian Yang-Mills connection if and only if the orbit is stable. This is indeed the case[73]. For complex dimension one (compact Riemann surfaces) it follows as observed by ATIYAH and BOTT [50] (see §9 below) from Narasimhan and Seshadri's theorem [39] (this was reproved by DONALDSON [495]). For complex projective surfaces it was proved by DONALDSON [499]. The result in general for vector bundles over compact Kähler manifolds was proved by UHLENBECK and YAU [898]. A different proof for nonsingular complex projective varieties was given by DONALDSON [504]. For extensions and analogous results for related moduli spaces see [422], [423], [436], [460], [625], [720], [817], [853].

The theory of symplectic quotients leads us to expect that there is a natural Kähler form induced on the moduli space of stable bundles (see §3) and a natural line bundle with a connection whose curvature gives us the Kähler form. This line bundle can be obtained using Quillen's theory of determinant line bundles ([813], [508] §6.5.4).

---

[72] Equivalently algebraic bundles, if M is a complex projective variety.

[73] An appropriate definition of stability goes as follows. A holomorphic vector bundle $\mathscr{E}$ over M is stable if for every coherent subsheaf $\mathscr{F}$ of $\mathscr{E}$ with rank lower than $\mathscr{E}$ we have

$$\mu(\mathscr{F}) < \mu(\mathscr{E})$$

where the slope $\mu(\mathscr{F}) = \deg_\omega(\mathscr{F})/rank(\mathscr{F})$ is defined in terms of the Kähler form ω on M (see [898] §1 for a definition of $\deg_\omega(\mathscr{F})$). Compare Appendix 5C where the definition when M is algebraic is given in terms of the corresponding line bundle.

For other applications of these sorts of techniques to prove results in algebraic geometry see for example [922] (cf. [744]). Other references include [378], [379], [390], [509], [582], [639], [640], [814], [881], [896], [897].

§ 9. Yang-Mills theory over Riemann surfaces

The relationship between Yang-Mills theory and geometric invariant theory first became apparent in the fundamental paper [50] in which ATIYAH and BOTT considered Yang-Mills theory over compact Riemann surfaces. In the case when M is a compact Riemann surface ATIYAH and BOTT established the identification described in §8 of moduli spaces of irreducible Hermitian Yang-Mills connections and stable holomorphic bundles over M, and then went on to apply equivariant Morse theory to compute cohomology.

In fact when M is a Riemann surface the situation described in the last section simplifies significantly. All connections on the C^∞ bundle E over M have curvature of type $(1,1)$ so $\mathscr{A}^{(1,1)} = \mathscr{A}$ is an infinite-dimensional complex affine space, which can be identified with the space $\mathscr{C}$ of holomorphic structures on E using a fixed Hermitian metric as in §8. A Hermitian Yang-Mills connection is now a connection with constant central curvature[74]. A moment map for the action of the gauge group on $\mathscr{A}$ is given by assigning to a connection $A \in \mathscr{A}$ its curvature $F_A \in \Omega^2_M(\mathfrak{g}_E)$ (cf. §8), but we are free to add a central constant so that the Hermitian Yang-Mills connections become the zeros of the moment map.

A holomorphic bundle $\mathscr{E}$ over M is stable (respectively semistable) if $\mu(\mathscr{F}) < \mu(\mathscr{E})$ (respectively $\mu(\mathscr{F}) \leq \mu(\mathscr{E})$) for every proper subbundle $\mathscr{F}$ of $\mathscr{E}$ where

$$\mu(\mathscr{F}) = deg(\mathscr{F})/rank(\mathscr{F})\,.$$

When the theory of stability of holomorphic vector bundles was first introduced, NARASIMHAN and SESHADRI proved [39] that a holomorphic vector bundle over M is stable if and only if it arises from an irreducible representation of a certain central extension of the fundamental group $\pi_1(M)$. ATIYAH and BOTT translated this in terms of connections to show that a holomorphic vector bundle over M is stable if and only if it admits a unitary connection with constant central curvature. They deduced from this the existence of a homeomorphism between the moduli

[74] That is, when purely imaginary scalars are regarded as skew-adjoint endomorphisms in the obvious way, its curvature is a purely imaginary constant multiple of the Kähler form. The constant is determined by the topological invariants – the rank and the degree – of the bundle E.

space $\mathscr{U}_M(n,d)$ of stable bundles of rank n and degree d over M and the moduli space of irreducible connections with constant central curvature on a fixed C^∞ bundle E of rank n and degree d over M (cf. §8).

ATIYAH and BOTT then used ideas from equivariant Morse theory to compute the Betti numbers[75] of these moduli spaces in the case when n and d are coprime. The methods they used are very closely analogous to those described in §6 for computing the Betti numbers of symplectic quotients (which were in fact motivated by the work of ATIYAH and BOTT). Their inspiration was to consider the Yang-Mills functional as an equivariant Morse function on the space $\mathscr{A}$, and show that it induces Morse inequalities relating the dimensions of the cohomology groups (over any field of coefficients) of the connected components of the set of critical points, which become equalities when equivariant cohomology is used.

Unfortunately analytical problems[76] arise in carrying out this program due to the infinite-dimensionality of the space $\mathscr{A}$ and the fact that the connected components of the set of critical points are not all submanifolds of $\mathscr{A}$. Therefore instead of using Morse theory directly, ATIYAH and BOTT used arguments of HARDER and NARASIMHAN [132] and SHATZ [846] which enabled them to define a stratification of the space $\mathscr{C}$ of holomorphic structures on E playing the rôle of the Morse stratification. In fact any holomorphic vector bundle $\mathscr{E}$ over M has a canonical filtration

$$0 = \mathscr{E}_0 \subset \mathscr{E}_1 \subset \ldots \subset \mathscr{E}_r = \mathscr{E}$$

such that each subquotient $\mathscr{E}_i/\mathscr{E}_{i-1}$ is semistable and

$$\mu(\mathscr{E}_i/\mathscr{E}_{i-1}) > \mu(\mathscr{E}_{i-1}/\mathscr{E}_{i-2})$$

for $1 < i \leq r$. Atiyah and Bott defined the type of such a bundle to be the vector

$$(\mu(\mathscr{E}_1/\mathscr{E}_0), \ldots, \mu(\mathscr{E}_r/\mathscr{E}_{r-1}))$$

in which $\mu(\mathscr{E}_i/\mathscr{E}_{i-1})$ is repeated $rank(\mathscr{E}_i/\mathscr{E}_{i-1})$ times. It follows from results of SHATZ [846], or alternatively from a differential geometric argument of ATIYAH and BOTT ([50] §7), that there is a partial order on the set $\mathscr{M}$ of types such that the holomorphic structures of a given type μ make up a subset $\mathscr{C}_\mu$ of $\mathscr{C}$ satisfying

---

[75] They also showed that the moduli spaces have no torsion in their cohomology.

[76] Progress has since been made towards resolving these sorts of problems [466], [467], [659], [880–882].

$$\overline{\mathscr{C}_\mu} \subseteq \bigcup_{\lambda \geq \mu} \mathscr{C}_\lambda$$

for all $\mu \in \mathscr{M}$. Of course if

$$\mu_0 = (d/n, \dots, d/n)$$

then $\mathscr{C}_{\mu_0}$ is the set of semistable holomorphic structures on E. Moreover whenever

$$\mu = (\mu_1, \dots, \mu_n) \in \mathscr{M}$$

then $\mathscr{C}_\mu$ is a closed submanifold[77] of the open subset

$$U_\mu = \bigcup_{\lambda \leq \mu} \mathscr{C}_\lambda$$

of $\mathscr{C}$ with finite complex codimension

$$d(\mu) = \sum_{i,j \text{ with } \mu_i > \mu_j} (\mu_i - \mu_j + (g-1)) .$$

This means that there are Thom-Gysin sequences of equivariant cohomology[78]

$$\cdots \to H_{\mathscr{G}}^{*-d(\mu)}(\mathscr{C}_\mu; \mathbf{Q}) \to H_{\mathscr{G}}^*(U_\mu; \mathbf{Q}) \to H_{\mathscr{G}}^*(U_\mu - \mathscr{C}_\mu; \mathbf{Q}) \to \cdots .$$

As in the finite-dimensional case (described in §6) one can show that these long exact sequences break up into short exact sequences so that the equivariant Poincaré series of $\mathscr{C}$ is given by

$$P_t^{\mathscr{G}}(\mathscr{C}) = \sum_{\mu \in \mathscr{M}} t^{2d(\mu)} P_t^{\mathscr{G}}(\mathscr{C}_\mu) . \tag{7}$$

For this it is enough to know that the composition of the Thom-Gysin map $H_{\mathscr{G}}^{*-d(\mu)}(\mathscr{C}_\mu; \mathbf{Q}) \to H_{\mathscr{G}}^*(U_\mu; \mathbf{Q})$ with the restriction map $H_{\mathscr{G}}^*(U_\mu; \mathbf{Q}) \to H_{\mathscr{G}}^*(\mathscr{C}_\mu; \mathbf{Q})$ is injective. This composition is multiplication by the equivariant Euler class of the normal bundle to $\mathscr{C}_\mu$. If $\mu = (d_1/n_1, \dots, d_r/n_r)$ we choose a C^∞ unitary decomposition of E as $D_1 \oplus \cdots \oplus D_r$, where D_i has degree d_i and rank n_i, and let $\mathscr{C}(n_i, d_i)$ and $\mathscr{G}(n_i, d_i)$ denote the space of holomorphic structures on D_i and the group of unitary automorphisms of D_i. Then there is an isomorphism

---

[77] Some careful analysis using Sobolev spaces is required to justify the use of infinite-dimensional manifolds (see [50] §14).

[78] See §6 for the definition of equivariant cohomology.

$$H^*_{\mathscr{G}}(\mathscr{C}_\mu; \boldsymbol{Q}) \cong \bigotimes_{1\leq i\leq r} H^*_{\mathscr{G}(n_i,d_i)}(\mathscr{C}(n_i,d_i)^{ss}; \boldsymbol{Q}) \tag{8}$$

induced by the map

$$(\mathscr{D}_1, \dots, \mathscr{D}_r) \mapsto \mathscr{D}_1 \oplus \cdots \oplus \mathscr{D}_r$$

from $\prod_i \mathscr{C}(n_i,d_i)^{ss}$ to $\mathscr{C}_\mu$. The centre of $\prod_i \mathscr{G}(n_i,d_i)$ is an r-dimensional torus acting trivially on $\prod_i \mathscr{C}(n_i,d_i)^{ss}$ and the induced representation on the normal to $\mathscr{C}_\mu$ has characters

$$(t_1, \dots, t_r) \mapsto t_i^{-1} t_j$$

for $i < j$. This means that the representation is primitive, and hence, by the criterion of ATIYAH and BOTT (see [50] §13), that the equivariant Euler class of the normal bundle to $\mathscr{C}_\mu$ is not a zero divisor in $\bigotimes_i H^*_{\mathscr{G}(n_i,d_i)}(\mathscr{C}(n_i,d_i)_\mu; \boldsymbol{Q})$ as required.

Combining (7) and (8) one obtains

$$P_t^{\mathscr{G}}(\mathscr{C}) = \sum_{\mu\in\mathscr{M}} t^{2d(\mu)} \prod_{1\leq i\leq r} P_t^{\mathscr{G}(n_i,d_i)}(\mathscr{C}(n_i,d_i)^{ss}).$$

Since $\mathscr{C}$ is an infinite-dimensional affine space it is contractible so that

$$P_t^{\mathscr{G}}(\mathscr{C}) = P_t(B\mathscr{G}) = \prod_{1\leq l\leq n} (1 - t^{2l})^{-1}.$$

Therefore one has an inductive formula

$$P_t^{\mathscr{G}}(\mathscr{C}^{ss}) = P_t(B\mathscr{G}) - \sum_{\mu\neq\mu_0} t^{2d(\mu)} \prod_{1\leq i\leq r} P_t^{\mathscr{G}(n_i,d_i)}(\mathscr{C}(n_i,d_i)^{ss}) \tag{9}$$

for the equivariant Betti numbers of $\mathscr{C}^{ss}$.

In the case when n and d are coprime every semistable bundle is stable, and the quotient $\overline{\mathscr{G}}$ of $\mathscr{G}$ by its constant central subgroup $U(1)$ acts freely on $\mathscr{C}^{ss}$ with quotient the moduli space $\mathscr{U}_M(n,d)$ of semistable bundles over M of rank n and degree d. From this ATIYAH and BOTT deduced that

$$P_t^{\mathscr{G}}(\mathscr{C}^{ss}) = P_t^{\overline{\mathscr{G}}}(\mathscr{C}^{ss}) P_t(BU(1)) = P_t(\mathscr{U}_M(n,d))(1-t^2)^{-1},$$

and hence that (9) gives a way to calculate the Betti numbers of the moduli space[79]. In fact the same formulas hold for cohomology with

[79] Equivalent inductive formulas for the Betti numbers of the moduli spaces had earlier been obtained [83], [132] via the Weil conjectures in number theory. The formulas of ATIYAH and BOTT were also rederived in [662] using the methods of §6 and §7.

coefficients in the field Z_p for any prime p, so the cohomology of the moduli space has no torsion.

ATIYAH and BOTT also studied the ring structure of the cohomology of this moduli space. Since the restriction map

$$H^*(B\overline{\mathscr{G}}; Z) \cong H^*_{\overline{\mathscr{G}}}(\mathscr{C}; Z) \to H^*_{\overline{\mathscr{G}}}(\mathscr{C}^{ss}; Z) \cong H^*(\mathscr{U}_M(n,d); Z)$$

is surjective, generators for the ring $H^*(B\overline{\mathscr{G}}; Z)$ (which is the tensor product of a polynomial algebra and an exterior algebra) induce generators for $H^*(\mathscr{U}_M(n,d); Z)$ which are all Chern classes or Künneth components of Chern classes of universal bundles. Conjectures about the relations between these generators were made by MUMFORD[80] (that a certain set of relations, arising from suitable applications of the Grothendieck-Riemann-Roch theorem to the universal bundles, should be a complete set) and by NEWSTEAD and RAMANAN (that when $n = 2$ the Chern classes and Pontryagin ring of the tangent bundle should vanish in degrees strictly greater than $4(g-1)$). These conjectures have now been proved in the case when $n = 2$ by various different methods, some of which have exciting new links with theoretical physics and conformal field theories [416], [422], [468], [506], [566], [641], [669], [776], [887], [902], [917], [918]. New ideas here come from the 'Verlinde formulas' concerning the sections of line bundles over the moduli spaces, and there has been substantial activity recently in giving geometric proofs of these formulas [390], [417], [641], [761], [874], [888].

Other related references include [626], [628], [642], [781], [794], [910]. For recent results on extending classical Brill-Noether theory on Jacobians (described in [352]) to moduli spaces of bundles of higher rank, see for example [383], [389], [391–393], [400], [514], [777], [872], [884–886].

When n and d are not coprime then not all semistable bundles are stable and the moduli space $\mathscr{U}_M(n,d)$ is singular (except in the special case when n and the genus of M are both 2). Then the method of ATIYAH and BOTT for calculating the Betti numbers of the moduli space breaks down. However the methods described in §7 can be used to calculate the Betti numbers of a partial desingularization $\tilde{\mathscr{U}}_M(n,d)$ of $\mathscr{U}_M(n,d)$ and the intersection Betti numbers of $\mathscr{U}_M(n,d)$ itself [663]. This is relevant (see [556], also [555]) to the study of CASSON's invariant for 3-manifolds [346].

---

[80] MUMFORD has also conjectured that a similar approach gives generators and relations for the cohomology rings of moduli spaces of curves (cf. [737] and [595–601], [919]).

Appendix to Chapter 1

A. Geometric reductivity [3]

We start by clarifying our use of the word 'reductive' and its cognates. In the original text, an algebraic group G was called reductive if all its rational representations are completely reducible. This has been changed in the new edition to conform to the current standard terminology. A linear algebraic group G is called

i) *reductive* if the radical of G (= maximal connected solvable normal subgroup) is a torus,

ii) *linearly reductive* if every rational representation of G is completely reducible,

iii) *geometrically reductive* if, whenever $G \to GL(V)$ is a rational representation and $0 \neq v \in V^G$, then for some $r > 0$, there is an $F \in (\mathrm{Symm}^r\, V^*)^G$ with $F(v) \neq 0$.

It has been known for a long time that (i) and (ii) are equivalent in char. 0 (see [343], p. 178). However NAGATA [28] showed that in char. $p > 0$, the only linearly reductive groups are extensions of a torus by a finite group of order prime to p. This motivated the search for a weakening of (ii), equivalent to (i) in all characteristics: property (iii), also called semi-reductive by NAGATA [26]. In characteristic zero, all three properties are equivalent (see Th. A. 1. 0 below), and it is evident that, in any characteristic, ii) $\Rightarrow$ iii). It was conjectured in the preface of the original edition that i) $\Rightarrow$ iii). This has been proved by HABOUSH [128], and means that almost all of the theory in the 1st edition of this book is valid for the action of a reductive group in any characteristic.

The history of proofs that i) $\Rightarrow$ iii) can be summarized as follows:

ODA [253]	for $G = SL(2)$, char. 2
SESHADRI [300]	for $G = SL(2)$, any char.
FORMANEK and PROCESI [105]	for $G = GL(n)$, any char., or over Z
HABOUSH [128]	for G reductive, any char.
HUMPHREYS [155]	for G semisimple, any char.
HUMPHREYS and JANTZEN [156]	
SESHADRI [305]	for G reductive groupscheme over a scheme S

(See sect. G for the meaning of iii) for groupschemes over general bases.) Combining these with recent results of POPOV [269], one now can assert

[3] More recent references related to this section include [586], [587], [655], [710], [773], [798–809], [820], [836], [878], [889], [908]. Others related to Chapter 1 in general include [347], [380], [428], [481], [522], [537], [588], [646–648], [717].

Theorem A.1.0. Let G be an affine algebraic group over any field k. The following are equivalent:

a) G is reductive
b) G is geometrically reductive
c) for all finitely generated k-algebras R on which G acts rationally by k-algebra automorphisms, R^G is finitely generated.

We will outline the proofs a) $\Rightarrow$ b) and c) $\Rightarrow$ a). The assertion b) $\Rightarrow$ c) follows from Theorem A.1.1 in the next section. Details for a) $\Rightarrow$ b) can be found in [128] and [305]. c) $\Rightarrow$ a) is POPOV's result. a) $\Rightarrow$ b): the basic idea of the proof is: given V and v as in iii) above construct an 'explicit' representation V_0 such that there exists

1) a G-linear map $\phi: V \to V_0$ with $\phi(V) \neq 0$,
2) an invariant form f_0 on V_0, of positive degree, with $f_0(\phi(v)) \neq 0$.

$f_0 \circ \phi$ will be the F in iii). One reduces the general case easily to the case where G is semisimple and simply connected and where V is contained in $H^0(G, O_G)$, G acting on itself by left multiplication. Fix a maximal torus T in G and $B \supset T$ a Borel subgroup. Let ϱ stand for half the sum of the positive roots. If m is a positive integer, let $W_{m\varrho} = H^0(G/B, \mathscr{L}(m\varrho))$, where $\mathscr{L}(m\varrho)$ is the invertible sheaf on G/B corresponding to the character $m\varrho$ of B. According to results of STEINBERG [318], in characteristic p, $W_{m\varrho}$ is irreducible of dimension a power of p, and self-dual, for $m = p^r - 1$. Then $W_{m\varrho} \otimes W_{m\varrho}$ carries the G-invariant form given by $W_{m\varrho} \otimes W_{m\varrho} \xrightarrow{\sim} \mathrm{Hom}\,(W_{m\varrho}, W_{m\varrho}) \xrightarrow{\det} k$. Because $W_{m\varrho}$ is irreducible, the only G-invariant elements of Hom $(W_{m\varrho}, W_{m\varrho})$ are scalar multiples of the identity, on which det is non-zero! Next it is shown that, with T acting on the right, $H^0(G, 0_G)^T$ is an increasing union of copies of $W_{m\varrho} \otimes W_{m\varrho}$ (m increasing). Via the action of G on the left, one gets a G-linear projection $H^0(G, \underline{o}_G) \to H^0(G, \underline{o}_G)^T$, hence a G-linear map $\phi: V \to W_{m\varrho} \otimes W_{m\varrho}$ for some m. Then $\phi(v) = a \cdot (id)$, $a \neq 0$, hence det $(\phi(v)) \neq 0$. An important remark is that this proves the strong form of geometric reductivity, viz., the r which is the degree of $F \in (\mathrm{Symm}^r V^*)^G$ is actually a power of the characteristic p.

c) $\Rightarrow$ a). Suppose the unipotent radical H of G is not trivial. By the counterexample of NAGATA [224], there is an action of $\mathbf{G}_a^r$ on $\mathbf{A}^n$ with $H^0(\mathbf{A}^n, \underline{o}_{\mathbf{A}^n})^{\mathbf{G}_a^r}$ *not* finitely generated. Let m be the largest integer such that $H^0(\mathbf{A}^n, \underline{o}_{\mathbf{A}^n})^{\mathbf{G}_a^m}$ is finitely generated and let X be the spectrum of this ring. Then $H^0(X, \underline{o}_X)^{\mathbf{G}_a}$ is not finitely generated. Choosing a homomorphism from H onto $\mathbf{G}_a$, the same holds for H acting on X. G acts on the 'twisted product' $Y = G \times_H X$ (= quotient of $G \times X$ by the action $(g, x) \to (gh^{-1}, hx)$ of H) via left multiplication in the first factor. Y is in fact a locally trivial fibre space over G/H with fibre X. This follows from a result of ROSENLICHT [283] that the map $G \to G/H$ has a local section, which implies that $Y \to G/H$ has a local section. Thus Y is an affine variety and $H^0(Y, \underline{o}_Y)^G$ is isomorphic to $H^0(X, \underline{o}_X)^H$, which is not finitely generated.

The problem of finite generation of rings of invariants — Hilbert's fourteenth problem — is discussed from a historical viewpoint in [221]. Prior to NAGATA's famous counterexample, [224], it seems to have been generally believed that if G is any algebraic group acting rationally on a finitely generated k-algebra R, the ring R^G is finitely generated. Beside the cases covered by Theorems 1.1 and A.1.1, it has been shown by WEITZENBOCK that if $G = G_a$ acts homogenoeusly on a polynomial ring R, then R^G is finitely generated. For a modern proof of this result by SESHADRI, see [298].

The assertion in Theorem 1.1 that if a reductive group G acts on a noetherian Q-algebra R, then R^G is noetherian, is false in characteristic p. NAGATA [225], gives a counter-example where G is the finite group Z/pZ. It has, however, been shown [101], that if a finite group G acts on a noetherian k-algebra R and if the module $\Omega_{R/k}$ of Kähler differentials of R over k is a finite R-module, then R^G is noetherian and $\Omega_{R^G/k}$ is a finite R^G-module.

Nagata's counter-example to Hilbert's 14th problem has been analyzed by GROSSHANS [125], [126] using the following concept:

Definition-Theorem: A subgroup H of a connected algebraic group G is said to satisfy the *codimension 2 condition* if one of the following equivalent conditions holds:

i) $G/H \cong X - Y$, where X is an affine variety and Y is a closed subset of codimension at least 2,
ii) G has a representation V with a point x such that $H = \text{Stab}\,(x)$ and $\overline{O(x)} - O(x)$ has codimension at least 2 in $\overline{O(x)}$.
iii) G/H is quasi-affine and $\Gamma(\underline{o}_G)^H$ is finitely generated.

He then proves that if $H \subset G$ satisfies the codimension 2 condition and G is reductive, then whenever G acts on a normal affine variety X, $\Gamma(\underline{o}_X)^H$ is finitely generated. In particular, if $G_a^r \subset GL(V)$ is the example of Nagata, where $k[V]^{G_a^r}$ is not finitely generated, it follows that G_a^r in $GL(V)$ does not satisfy the codimension 2 condition.

B. Stability

We wish to clarify the use of the words "stable", "semi-stable" and "unstable" which have come to be used somewhat differently from their use in the text (which has not been changed in this edition). In the text, we start with a reductive group G over k acting on a scheme X of finite type over k and on an invertible sheaf L over k. Then open subsets

$$X^s_{(0)}(L) \subset X^{ss}(L) \subset X$$

are defined by the existence of G-invariant sections $s \in \Gamma(X, L^n)$, $n > 0$, such that X_s is affine. Namely

$$X^{ss}(L) = \bigcup_{\substack{s \in \Gamma(X, L^n)^G \\ X_s \text{ affine}}} X_s, \qquad X^s_{(0)}(L) = \bigcup_{\substack{\text{as above and all} \\ \text{stabilizers finite}}} X_s$$

However, by far the most important case of this is when $X \subset \mathbf{P}^n$, $L = \underline{o}_X(1)$ and G acts on $\mathbf{P}^n$, $\underline{o}_{\mathbf{P}^n}(1)$ as well. In this case, we consider the affine cone $\hat{X}$ over X:

$$\begin{array}{ccc} \hat{X} & \subset & \mathbf{A}^{n+1} \\ \cup & & \cup \\ \hat{X} - (0) & \subset & \mathbf{A}^{n+1} - (0) \\ \downarrow & & \downarrow \\ X & \subset & \mathbf{P}^n \end{array}$$

and G acts on all these spaces equivariantly. Then Prop. 2.2 gives a much more elementary definition of stability and semistability. Namely

$$x \in X^s_{(0)}(\underline{o}_X(1)) \Leftrightarrow \text{for one (and hence all) } \hat{x} \in \hat{X} \text{ over } x,\ 0^G(\hat{x}) \text{ is closed in } \mathbf{A}^{n+1} \text{ and } \dim 0^G(\hat{x}) = \dim G$$

$$x \in X^{ss}(\underline{o}_X(1)) \Leftrightarrow \text{for one (and hence all) } \hat{x} \in \hat{X} \text{ over } x,\ 0 \notin \overline{0^G(\hat{x})}$$

The properties on the right are interesting topological properties to study in any linear representation. Hence the terminology has been extended as follows.

Definition: Let a reductive group G act linearly on a vector space V. Then an element $x \in V$ is called (e.g., [218], p. 41):

unstable if $0 \in \overline{0^G(x)}$

semi-stable if $0 \notin \overline{0^G(x)}$

stable if $0^G(x)$ *is closed* and $\dim 0^G(x) = \dim G$.

Thus in any such V, there is a Zariski closed cone $\mathcal{N}$ of unstable points and an open (possibly empty) set $U \subset V$ of stable points, closed under multiplication by k^*, which are basic topological invariants of the representation (G, V).

C. Extension of the results in the text to characteristic p [4]

Theorem A.1.1. Let X be an affine algebraic scheme over a field k of characteristic $p > 0$, and let $\sigma: G \times X \to X$ be an action of a reductive algebraic group G on X. Then a uniform categorical quotient

[4] Recent references related to this section include [400], [589].

(Y, ϕ) of X by G exists, Y is an affine algebraic scheme over k and ϕ is submersive. Moreover, if $\mathcal{F}$ is a coherent G-linearized $\underline{o}_X$-module, then $\mathcal{F}^G$ is a coherent $\underline{o}_Y$-module.

In place of the Reynolds operator, we have

Lemma A.1.2. Let G be a geometrically reductive group over the field k of characteristic $p > 0$. Let G act on the k-algebra R rationally by k-algebra automorphisms. If I is a G-stable ideal in R and if $f \in (R/I)^G$, then for some n, $f^{p^n} \in R^G/I^G$.

Proof. Choose an f_0 in R whose I residue is f. Let E be a finite-dimensional G-stable k-subspace of R with $f_0 \in E$. Let $V = kf_0 + (E \cap I)$. Let $\lambda \in V^*$ be the linear form which maps $E \cap I$ to 0 and f_0 to 1. Then $\lambda \in (V^*)^G$ and for some n, there is an $f_1 \in (\mathrm{Symm}^{p^n} V)^G$ with $f_1(\lambda) = 1$. Then the image of f_1 in R lies in R^G and maps to f^{p^n} in R/I.

To prove A.1.1, take $R = H^0(X, \underline{o}_X)$ and $Y = \mathrm{Spec}\, R^G$. Let ϕ be the morphism induced by the inclusion $R^G \to R$. Then

1) if S is a flat R^G-algebra, then

$$S = \left(R \underset{R^G}{\otimes} S\right)^G$$

(Without the flatness assumption, it is true that if $f \in \left(R \underset{R^G}{\otimes} S\right)^G$, then, for some n, $f^{p^n} \in S$).

2) if (I_α) is a set of G-stable ideals in R and if $f \in (\sum I_\alpha)^G$, then for some n, $f^{p^n} \in \sum I_\alpha^G$.

1) follows from remark 7 on p. 9 of the text. To prove 2), observe that since we deal with one element at a time, we may assume that there is only a finite number of the I_α, and then by induction, we need only prove: if $f \in (I_1 + I_2)^G$, then for some n, $f^{p^n} \in I_1^G + I_2^G$. If $f = F_1 + F_2$ where $F_1 \in I_1$ and $F_2 \in I_2$, let f_1 and f_2 be the $(I_1 \cap I_2)$-residues of F_1 and F_2, resp. Applying Lemma A.1.2, we see that there is an n such that $f_i^{p^n} \in I_i^G/(I_1 \cap I_2)^G$, $i = 1, 2$, i.e., $(F_1 + F_2)^{p^n} = H + L + K$, with $H \in I_1^G$, $L \in I_2^G$, $K \in I_1^G \cap I_2^G$. Thus $f^{p^n} \in I_1^G + I_2^G$.

The proof that (Y, ϕ) is a categorical quotient and ϕ is submersive is now as in the text, except to show that (Y, ϕ) is a uniform quotient, we need only consider flat base extensions, $Y' \to Y$.

For the rest of the theorem, we show that R^G is a finitely generated k-algebra and if M is a noetherian R—G-module, then M^G is a noetherian R^G-module. If G is a finite group, both assertions are well known (see [212] § 7), so we shall assume that G is connected.

Step 1. Assume that R is graded, $R_0 = k$, and G acts homogeneously. First we reduce to the case where R^G is a domain and R is torsion-free over R^G. If f is a homogeneous element of R^G which is a zero-divisor in R, then by noetherian induction, we may assume that $(R/fR)^G$ and

$(R/(0:fR))^G$ are finitely generated. By Lemma A.1.2, there is an n with $((R/fR)^G)^{p^n} \subset R^G/(fR)^G \subset (R/fR)^G$, from which it follows directly that $R^G/(fR)^G$ is finitely generated and *a fortiori* is a noetherian R^G-module. Via multiplication by f, $(fR)^G$ is isomorphic to $(R/(0:fR))^G$ so that $(fR)^G$ is also a noetherian R^G-module. But then R^G itself is a noetherian R^G-module, i.e., R^G is a noetherian ring. Since R^G is also graded and $(R^G)_0 = k$, it follows that R^G is a finitely generated k-algebra.

If $0 \neq g \in R^G$ is a homogeneous non-unit (such must exist unless $R^G = k$), we may assume that $(R/gR)^G$ is finitely generated and, as above, that $R^G/(gR)^G$ is also finitely generated. Since g is a non-zero-divisor in R, $(gR)^G = gR^G$. Let $I = R^G_+ = \sum_{n=0} R^G_n$. Then I/gR^G is a finitely generated ideal in R^G/gR^G so I is a finitely generated ideal in R^G. Thus R^G is a finitely generated k-algebra.

Step 2. Let R be as in step 1. We show that if M is a noetherian R—G-module (not necessarily graded) then M^G is a noetherian R^G-module. Let $\boldsymbol{M}$ be the category of noetherian $R-G$-modules and let $\boldsymbol{M}_0$ be the full subcategory of those M in $\boldsymbol{M}$ for which M^G is a noetherian R^G-module. By left exactness of invariants, if $0 \to M' \to M \to M'' \to 0$ is an exact sequence in $\boldsymbol{M}$, then $M', M'' \in \boldsymbol{M}_0 \Rightarrow M \in \boldsymbol{M}_0$. If $M \in \boldsymbol{M}$, then in order to show that $M \in \boldsymbol{M}_0$, by noetherian induction, we can assume that all N in $\boldsymbol{M}$ with support properly contained in the support of M are in $\boldsymbol{M}_0$. If I is the annihilator of M, set $R_1 = R/I$. If R_1 is a domain, let $T(M)$ be the torsion submodule of M. The exact sequence $0 \to T(M) \to M \to M/T(M) \to 0$ shows that we may assume M is torsion-free over R_1. If $M^G = (0)$, then clearly $M \in \boldsymbol{M}_0$. If $M^G \neq (0)$, we can find a G-linear injection $R_1 \to M$. Then rank $(M/R_1) <$ rank(M) and, by induction on the rank, we are reduced to the case $M = R_1$.

If R_1 is reduced, write $(0) = P_1 \cap \cdots \cap P_r$, with P_i a prime ideal of R_1. As G is connected, each P_i is G-stable and there is a G-linear map $j: M \to \bigoplus M/P_iM$. Both ker j and coker j have smaller support than M, and by the case where R_1 is a domain, we can assume that each M/P_iM is in $\boldsymbol{M}_0$. Thus $M \in \boldsymbol{M}_0$. If R_1 is not reduced, let J be its nilradical. Then $J^sM = (0)$ for some s, and by the reduced case, $J^rM/J^{r+1}M \in \boldsymbol{M}_0$. From this it follows readily that $M \in \boldsymbol{M}_0$.

Thus we are reduced to proving that if P is a G-stable prime ideal of R, then $R/P \in \boldsymbol{M}_0$. We show, in fact, that $(R/P)^G$ is finitely generated over k. By Lemma A.1.2, $(R/P)^G$ is integral over R^G/P^G and the latter is finitely generated by step 1. Let K be the fraction field of R/P, L the fraction field of $(R/P)^G$ and K_0 the fraction field of R^G/P^G. Since K is finitely generated over K_0 and L is algebraic over K_0, L is a finite extension of K_0. Since $(R/P)^G$ is contained in the integral closure of

R^G/P^G in L, $(R^G/P)^G$ is a finite R^G/P^G-module and hence a finitely generated k-algebra.

Step. 3. R arbitrary. Choose a finite-dimensional G-stable k-subspace E of R, generating R as k-algebra. Let S be the symmetric algebra of E. The canonical surjection $S \to R$ is G-linear and maps S^G onto a finitely generated subring R_0 of R^G. By step 2, R^G is a noetherian R_0-module, and hence a finitely generated k-algebra. This completes the proof of Theorem A.1.1.

There is a recent proof of Theorem A.1.1. which avoids the graded case (see [102]).

Corollary A.1.3. If W_1 and W_2 are two disjoint closed G-stable subsets of X, then, there is an $f \in H^0(X, \underline{o}_X)^G$ such that $f|W_1 \equiv 0$ and $f|W_2 \equiv 1$.

Proof. Let J_i be the ideal in R defining the reduced structure on W_i, $i = 1, 2$. Since $W_1 \cap W_2 = \emptyset$, $J_1 + J_2 = R$. By 2) on p. **149**, if $h \in R^G$, then for some n, $h^{p^n} \in J_1^G + J_2^G$. Taking $h = 1$, we see that $R^G = J_1^G + J_2^G$. Thus $1 = f + g$ with $f \in J_1^G$, $g \in J_2^G$ and f is the required invariant.

Several proofs in the text need amplification to cover the characteristic p case.

i) **Proposition 1.16.** Using the notation in the text, let $\alpha: \bigoplus_n H^0(X, L^n) \to H^0(X_{\text{red}}, f^*L^n)$ be the homomorphism of graded rings induced by f. Let R be the image of α. By EGA 2, 4.5.13.1, given $s \in H^0(X_{\text{red}}, f^*L^n)^G$ there is a k with $s^k \in (R_{nk})^G$. Furthermore, for some ν, s^{kp^ν} is the image of a section $t \in H^0(X, L^{nkp^\nu})^G$ and X_t is affine.

ii) **Proposition 1.18.** The last part of the proof should read: there is a ν such that $s'^{p^\nu} \in H^0\big(X_s, (f^*M)^{nkp^\nu}\big)^G$ and s'^p maps to $(s^kf)^{p^\nu} \in H^0\big(X_s, (f^*M)^{nkp^\nu}\big)$. Then $t = ss'^p$ is zero outside X_s and at points of X_s where $f = 0$. Thus X_t is an affinite neighborhood of x contained in X_s. Since t is invariant and every stabilizer is 0-dimensional, x is properly stable.

iii) **Theorem 1.19.** Beginning with "Since S is a finite R-module", the proof should read, "S^G is a finite R^G-module. Thus there is an equation of integral dependence $s^m + b_i s^{m-1} + \cdots + b_m = 0$, where the b_i are homogenous elements of R^G. Since $s(x) \neq 0$, we have $b_i\big(f(x)\big) \neq 0$ for some i, i.e., $f(x) \in (P_n)_{b_i}$." The rest is the same as in the text.

D. Luna's results on the orbit structure for the action of a reductive group[5]

In this section we describe some of LUNA'S results in the important paper [188]. Throughout, G is a reductive group.

(I) If $\phi: X \to Y$ is a quasifinite G-morphism of affine G-varieties such that the image of a closed orbit is closed and $\phi/G: X/G \to Y/G$ is finite, then ϕ is finite.

[5] More recent references related to this section include [643–645], [715].

This yields Prop. 0.8 and Cor. 2.5 in the text as Corollaries.

(II) We define a G-morphism $\phi\colon X \to Y$ of affine G-varieties to be *strongly étale* if $\phi/G\colon X/G \to Y/G$ is étale and $(\varphi, \pi_X)\colon X \to Y \times_{Y/G} X/G$ is an isomorphism, $\pi_X\colon X \to X/G$ being the quotient morphism. (The definition implies that ϕ is étale.)

Fundamental lemma. Let $\phi\colon X \to Y$ be a G-morphism of affine G-varieties. Let T be a closed orbit of G in X such that

i) ϕ is étale at some point of T,
ii) $\phi(T)$ is closed in Y,
iii) ϕ is injective on T,
iv) X is normal along T.

Then there are G-stable open sets U in X and V in Y, with $T \subset U$, such that ϕ/U is a strongly étale G-morphism of U onto V.

A close examination of LUNA's proof shows that it uses the characteristic 0 hypothesis in [188] only via MATSUSHIMA's result, (see [198]), that $0(x)$ is affine $\Leftrightarrow G_x$ is reductive. HABOUSH [129] and RICHARDSON [281], have extended this to characteristic p, so the fundamental lemma holds in any characteristic.

(III) **Étale slice theorem.** (characteristic 0) Let X be an affine G-variety and let T be a closed orbit of G in X along which X is normal. If $x \in T$, then there is a locally closed G_x-stable affine subvariety W of X with $x \in W$, such that $U = G \cdot W$ is open in X and the G-morphism $\phi\colon G \times_{G_x} W \to U$ is strongly étale. Moreover, if X is smooth at x, then W can be taken to be smooth and there is a strongly étale G_x-morphism η from W onto a neighborhood of 0 in $N_x(T, X)$, the normal vector space $T_x(X)/T_x(T)$ to T at x.

Thus if X is smooth at x, we may say that 'locally in the étale topology' the action of G at x is like the action of G on $G \times_{G_x} N_x$. More precisely, if $V = W/G_x$, then V is étale over both X/G and N_x/G_x and we have a pair of linked cartesian product diagrams.

$$\begin{array}{ccccccc}
G \times_{G_x} N_x & \xleftarrow{\text{étale}} & G \times_{G_x} W & \xrightarrow{\text{étale}} & U & \subset & X \\
\downarrow & \square & \downarrow & \square & \downarrow & & \downarrow \\
N_x/G_x & \xleftarrow{\text{étale}} & V & \xrightarrow{\text{étale}} & U/G & \subset & X/G.
\end{array}$$

If X is smooth at x, the proof of the étale slice theorem runs (in outline) as follows. Since T is closed, G_x is reductive. The canonical map $d\colon \boldsymbol{m}_x \to \boldsymbol{m}_x/\boldsymbol{m}_x^2$ is G_x-linear and since the characteristic is 0, d has a G_x-linear section, which induces a G_x-homomorphism $\operatorname{Symm}^{\cdot}(\boldsymbol{m}_x/\boldsymbol{m}_x^2) \to H^0(X, \underline{o}_X)$, i.e., a G_x-morphism $\zeta\colon X \to \boldsymbol{T}_x(X)$ which is étale at x. Let N be a G_x-complement to $\boldsymbol{T}_x(T)$ in $\boldsymbol{T}_x(X)$. Set $W = \zeta^{-1}(N)$. The action of G on X induces a G-morphism $G \times_{G_x} W \to X$ which is étale at the image of (e, x) in $G \times_{G_x} W$. The result now follows from the fundamental lemma.

As one can see, the characteristic zero assumption is essential. The theorem has many consequences (as does its analogue in the differentiable category), among which is:

Corollary (char. 0). Let X be an affine G-scheme of finite type. Then X is a principal étale fibre space over X/G if and only if $G_x = (e)$ for all $x \in X$.

(IV) LUNA uses the étale slice theorem in the following series of results:

Let H be a reductive subgroup of G, let X be a normal affine G-variety, let $X^H = \{x \in X \mid H \text{ fixes } x\}$ and let $N_G(H)$ be the normalizer of H in G. Then:

a) If $x \in X^H$, then $0_G(x)$ is closed $\Leftrightarrow$ $0_{N_G(H)}(x)$ is closed. (See [187])

b) The map β: $X^H/N_G(H) \to X/G$ is finite. (See [188])

c) (LUNA- RICHARDSON [190]) If H is an isotropy group of a generic closed orbit and X^H is irreducible (e.g., if $X = A^n$ and G acts linearly) then β is an isomorphism.

E. The Hochster-Roberts theorem and its extensions [6]

As the culmination of a series of papers ([142], [143], [144]), HOCHSTER and ROBERTS prove that if a linearly reductive group over the field k acts on a regular noetherian k-algebra R, then R^G is Cohen-Macauley. This has corollaries such as vanishing theorems for the cohomology of line bundles on various homogeneous spaces in char. 0. The proof is based on their result that if B is a regular noetherian domain of characteristic $p > 0$ and A is a noetherian subring of B such that B is a pure A-module i.e., for all A-modules M, $M \to M \otimes_A B$ is injective, then A is Cohen-Macauley. In [172], KEMPF proves this directly for finitely generated algebras in any characteristic, giving a short proof of the Hochster-Roberts theorem.

In the graded case, the Hochster-Roberts theorem takes the following very explicit form (since a regular graded ring is a polynomial ring):

Let V be a rational representation of the linearly reductive group G. Then there exist algebraically independent homogeneous elements $I_1, \ldots, I_d$ in $S = (\text{Symm } V)^G$ such that S is a finite free graded $k[I_1, \ldots, I_d]$-module.

In [172], KEMPF, using the same techniques which prove the Hochster-Roberts theorem, shows the degrees of the generators of S over $k[I_1, \ldots, I_d]$ are bounded by

$$\deg I_1 + \cdots + \deg I_d.$$

[6] More recent references related to this section include [473–475], [674], [685], [715], [810].

In [170], KEMPF shows that if G is semisimple, the characteristic is 0 and G acts on a smooth affine variety X with $\Gamma(\underline{o}_x)$ a UFD with units k^* (e.g., $X = A^n$), then X/G has only rational singularities. BOUTOT [419] has proven in fact that if G is reductive, the characteristic is 0 and G acts on an affine variety X with only rational singularities, then X/G has only rational singularities. Finally, in [173], KEMPF shows that if $X = A^n$ and $\dim X/G \leqq 2$, then $X/G \approx A^m$.

F. A.criterion for generic stability

An action of an algebraic group G on an algebraic variety X is said to be generically stable if there is a non-empty G-invariant open set U in X such that for all $x \in U$, $0(x)$ is closed in X. We say that the generic stabilizer is reductive if there is a non-empty G-invariant open set W in X such that for all $x \in W$, the stabilizer G_x of x is reductive. The following has been proven by NAGATA [27], (in a weaker form), POPOV [267], and LUNA-VUST [191]:

Theorem. Let the semisimple group G act on the affine variety X over the algebraically closed field k.If $\Gamma(\underline{o}_x)$ is a UFD, then the action is generically stable if and only if the generic stabilizer is reductive.

We give here an outline of the proof in [267], observing that the restriction of characteristic 0 is not necessary. That generic stability implies the generic stabilizer reductive follows from the results of MATSUSHIMA and RICHARDSON cited in sect. C (see [129]). For this, the factoriality is not needed.

For the converse, let V be the open set of points $x \in X$ such that $0(x)$ has maximum dimension m. Let W be a non-empty open set which is G-invariant and such that G_x is reductive if $x \in W$. Let $U = V \cap W$ and let $E = X - U$. If E is empty, there is nothing to prove, as all orbits have dimension m, and so all orbits are closed.

Let $\phi: X \to Y$ be the quotient by the action of G. According to NAGATA [27], since X is factorial, $k(X)^G = k(Y)$. This is seen as follows: if $f \in k(X)^G$, write $f = a/b$, $a, b \in \Gamma(o_X)$, a, b relatively prime. Then for all $\sigma \in G$, $a^\sigma/b^\sigma = a/b$, hence $a^\sigma = e(\sigma) \cdot a$, $b^\sigma = e(\sigma) \cdot b$, where e is a 1-cocyde on G with values in $\Gamma(\underline{o}_X^*)$. By result of ROSENLICHT (see [32] or [182]), $\Gamma(\underline{o}_X^*)/k^*$ is a finitely generated group. Therefore the action of G G on $\Gamma(\underline{o}_X^*)$ is trivial, and e is just a homomorphism $G \to \Gamma(\underline{o}_X^*)$ which is itself trivial as G is semisimple. Therefore a and b are G-invariant, hence $f \in k(Y)$. From this it follows easily (e. g. see ROSENLICHT [284]) that there is a non-empty G-invariant open set U_0 in X and a non-empty open set V_0 in Y such that, for $y \in V_0$, $\phi^{-1}(y) \cap U_0$ consists of a single orbit. Thus it is clear that by restricting to a suitable affine open subset of Y, we may assume that each (closed) fibre of φ is the closure of an orbit of dimension m whose points have reductive stabilizers. Since these orbits are affine, it follows that:

(*) if $y \in Y$, then $E \cap \phi^{-1}(y)$ is either empty or is of pure codimension 1 in $\phi^{-1}(y)$.

Now either $\overline{\phi E} = Y$ or the theorem is obvious. We show that $\overline{\phi E} = Y$ is absurd. If x is a point of E lying on only one irreducible component, say E_1, of E, then for some $z \in U$, there is an irreducible component of $\overline{O(z)} - O(z)$ containing x and lying in E_1. Hence $\dim\left(\phi^{-1}(\phi(x)) \cap E_1\right) = m - 1$. It follows then, e.g., using generic flatness, that E_1 has codimension 1 in X. Since $\Gamma(\underline{o}_x)$ is a UFD, E is the locus of zeroes of some non-zero $f \in H^0(X, \underline{o}_X)$. Since E is invariant, the ideal (f) is invariant. Thus, for all $\sigma \in G$, $f^\sigma = e(\sigma)\, f$, where e is a 1-cocycle on G with coefficients in $\Gamma(\underline{o}_x^*)$. As before, f is invariant, which means that $\overline{\phi E} \neq Y$

G. Geometric reductivity over more general rings

In [305], SESHADRI extends the basic theorems for the action of a reductive group from the case in the text (and appendices) where the base is a field to the case where the base is $S = \operatorname{Spec}(R)$ with R a noetherian ring (or, more generally, where S is a locally noetherian scheme). A reductive group G over S is an affine groupscheme, smooth over S, with connected fibres which are reductive groups. Geometric reductivity is formulated as follows:

given a linear action of the affine groupscheme G over S on affine space $\boldsymbol{A}^n_S$ over S, and a G-invariant geometric point x of $\boldsymbol{A}^n_S$ with $x \neq 0$, there is an invariant form $F \in R[t_1, \ldots, t_n]$ of positive degree, with $F(x) \neq 0$.

Then SESHADRI proves the following:

i) reductivity of $G \Rightarrow$ geometric reductivity of G.

Semi-stability and stability of geometric points are defined for linear actions of G on $\boldsymbol{P}^{n-1}_S$ as in Prop. 2.2 or Appendix 1B. If a reductive groupscheme G over S acts linearly on $\boldsymbol{P}^{n-1}_S$ and X is a G-stable closed subscheme of $\boldsymbol{P}^{n-1}_S$, then he proves:

ii) the set of semi-stable (resp. stable) geometric points of X is the set of geometric points of a G-stable open subscheme X^{ss} (resp. $X^s_{(0)}$) of X, which is the union of the open sets $(X_a)^{ss}$, $(X_a)^s_{(0)}$ of the fibres X_a over $\operatorname{Spec} \boldsymbol{k}(a)$, $a \in S$, defined in the text.

iii) a uniform categorical quotient (Y, ϕ) of X^{ss} by the action of G exists and is universally closed over S. Moreover, if

$$X = \operatorname{Proj}\left(R[t_1, \ldots, t_n]/I\right),$$

I a homogeneous ideal, then

$$Y = \text{Proj}\left((R[t_1, \ldots, t_n]/I)^G\right),$$

ϕ is surjective, and for k algebraically closed, ϕ induces a bijection

$$Y(k) \xrightarrow{\sim} X^{ss}(k) \Big/ \left\{ \begin{array}{l} x_1 \sim x_2 \Leftrightarrow \\ O(x_1) \cap O(x_2) \cap X^{ss} \neq 0 \end{array} \right.$$

iv) if S is universally japanese*, then $(R[t_1, \ldots, t_n]/I)^G$ is a finitely generated R-algebra, hence Y is proper over S.

v) There is an open set $Y^s \subset Y$ such that

$$X^s = \phi^{-1}(Y^s) \text{ and } \phi \text{ induces a bijection}$$

$$Y^s(k) \xrightarrow{\sim} X^s(k)/G(k).$$

Note that by (ii), Prop. 1.18, 1.19, 1.20 extend automatically to the case of a general base.

Appendix to Chapter 2 [7]

A. The proof of Theorem 2.1 in characteristic p

This is given by SESHADRI in [303]. The only point in the text that requires modification is the proof of IWAHORI's theorem on p. 52. We replace it by the following. Let $R_n = k[[t^{p^{-n}}]]$, $K_n = k((t^{p^{-n}}))$. Let G be a reductive group over k and let θ be a K_0-valued point of G. Then:

For a suitable n, there exist $U, V \in G(R_n)$ and a 1-*PS* λ of G with

$$U\langle\lambda\rangle\, V = \theta \circ j,$$

where j: Spec $(K_n) \to$ Spec (K_0) is the canonical map.

Using the notations in the text, the difficulty arises because $\pi\colon G \to G'$ may not be etale, so that $\pi(R)\colon G(R) \to G'(R)$ may not be surjective. However, factoring the isogeny π into separable and purely inseparable parts, the argument in the text applies to the separable part, so we need only consider the case where π is purely inseparable. Now given $\phi \in G'(R)$, then for some n, there is a $\psi \in G(R_n)$ mapping to the image of ϕ in $G'(R_n)$. This means that, given θ in $G(K_0)$, then for some n, we can find $U, V \in G(R_n)$ and a 1-*PS* λ' of G' so that $V^{-1}\theta U^{-1}$ is a K_n-valued point of $\pi^{-1}(\lambda'(G_m))$. But $V^{-1}\theta U^{-1}$ must factor through the subgroup $(\pi^{-1}(\lambda'(G_m)))_{\text{red}}$ which is the image of a 1-*PS* λ of G. It is then an easy matter of taking roots to verify that $U\langle\lambda\rangle\, V = \theta \circ j$.

In the proof on p. 53 using the valuative criterion of properness, we can replace R with R_n and K with K_n and proceed as in the text. The same remark applies to the proof of proposition 2.4.

---

[7] More recent references related to Chapter 2 include [348], [405–407], [409], [410], [440–442], [584], [677], [678].

B. Construction of canonical destabilizing flags [8]

In [285], ROUSSEAU proves TITS' center conjecture in a special case, which is enough to get the corollary conjectured on p. 64 of the text. In [171], KEMPF independently proved this corollary and further developed the ideas of Ch. 2 (without explicitly using the TITS complex, $\Delta(G)$). Their main ideas are essentially the same, but KEMPF's viewpoint is closer than ROUSSEAU's to that in the text. Before describing KEMPF's results, we give his set-up and main definitions in the following table where they are compared to the set-up and definitions of the text. As usual, G is a reductive group.

KEMPF [171]	MUMFORD, SESHADRI [303]
k alg. cl. X = affine G-variety S = closed G-subscheme of X $x \in X - S$	k alg. cl. $\mathbf{A}^{n+1} \supset X$ = cone over a G-stable $Y \subset \mathbf{P}^n$, G acting linearly on $\mathbf{P}^n$. $S = (0)$ = vertex of X $x \neq 0$
$\overline{O(x)} \cap S \neq \emptyset$ $\Updownarrow$ $\exists$ 1-PS λ with $\lim_{t \to 0} \lambda(t) \cdot x \in S$	$0 \in \overline{O(x)}$ $\Updownarrow$ $\exists$ 1-PS λ with $\lim_{t \to 0} \lambda(t) \cdot x = 0$
$\lvert X, x \rvert = \left\{ \begin{array}{c} \text{1-}PS \\ \lambda \end{array} \middle\vert \begin{array}{l} \psi_x \circ \lambda \text{ has an extension } \eta_x: \\ \mathbf{G}_m \xrightarrow{\psi_x \circ \lambda} X \\ \cap \quad \nearrow \eta_x \\ \mathbf{A}^1 \end{array} \right\}$	If λ is given by $\lambda(t) = \begin{pmatrix} t^{r_0} & 0 \\ 0 & t^{r_n} \end{pmatrix}$,
$a_{S,x}(\lambda) = \deg_0 \left(\eta_x^{-1}(S)\right)$	$-\mu(x, \lambda) = \inf \{r_i : x^{(i)} \neq 0\}$
$b_{S,x}(\lambda) = \dfrac{a_{S,x}(\lambda)}{\lVert\lambda\rVert}$	$\nu(x, \delta) = \dfrac{\mu(x, \lambda)}{\lVert\lambda\rVert}$, $\lambda \in \delta \in \Delta(G)$

Here, $\deg_0\left(\eta_x^{-1}(S)\right)$ = [degree at 0 of the closed subscheme $\eta_x^{-1}(S)$ of $\mathbf{A}^1$]. As $x \notin S$, this subscheme will be supported at 0. The link between these functions is that $a_{\{0\},x}(\lambda) = -\mu(x, \lambda)$ in the set-up of the text, and $\lvert X, x \rvert$ maps to the semi-convex subset of $\Delta(G)$ where $\nu(x, \delta) \leq 0$, (see prop. 2.3 and cor. 2.16 of the text).

[8] More recent references related to this section include [402–404], [474], [585], [618–620], [699], [771], [815], [823], [824], [826].

Kempf's main theorem is that *if* $\overline{O(x)} \cap S \neq \emptyset$, $b_{S,x}$ defines a function on the image of $|X, x|$ in $\Delta(G)$ with a unique maximum on that set.

A direct consequence of this uniqueness is that if $k_0 = k$ is perfect, and $x \in X(k_0)$ and $\overline{O(x)} \cap S \neq \emptyset$, then there is a 1-*PS* $\lambda \in |X, x|$, defined over k_0, giving this maximum. Another result of KEMPF's work is:

> If k has characteristic 0 and $H \subset G_x$ is a reductive subgroup, then if $\overline{O(x)} \cap S \neq \emptyset$, there is a 1-*PS* λ of $Z_G(H)$ defined over k such that $a_{S,x}(\lambda) > 0$. Hence, if any 1-*PS* λ of $Z_G(H)$, defined over k, lies in H, then $O_G(x)$ is closed. (Compare LUNA [187].)

Another corollary is: If V is a representation of the semisimple group G, and $x \in V$, then $O(x)$ is closed whenever G_x is not contained in any proper parabolic subgroup of G. This yields the stability of the Hilbert and Chow points of an abelian variety embedded by a complete linear system of dimension prime to the characteristic, or their semi-stability in the case of a "flag manifold" G/P, G semi-simple, P parabolic, embedded by a complete linear system.

KEMPF'S ideas have been further developed by HESSELINK [140] and RAMANAN-RAMANATHAN [340]. Another analysis of the unstable cone is due to BOGOMOLOV [334].

C. Stable vectors in complex representations [9]

KEMPF and NESS [175] characterize stable vectors in a complex representation space V of a reductive group G as follows. Let K be a maximal compact subgroup of G. Fix a K-invariant Hermitian norm $\| \ \|$ on V. For $v \in V$, $g \in G$, set $p_v(g) = \|g \cdot v\|^2$. If G_v is the stabilizer of v, then p_v defines a function on the double coset space $K \backslash G / G_v$ which is also denoted by p_v. Then all critical points of p_v are minima and if p_v attains a minimum, it does so on exactly one double coset. Finally, p_v attains a minimum $\Leftrightarrow$ $O_G(v)$ closed. Thus every stable orbit $O_G(v)$ in V contains a *canonical* sub-K-orbit $O_K(v_1)$ characterized by requiring that $\|v_1\| \leqq \|g \cdot v_1\|$ for all $g \in G$. It would be interesting to determine these canonical K-orbits in various cases.

This theory has been linked by GUILLEMIN-STERNBERG [335], [336] to the concept of the moment map arising in symplectic geometry: for the action of K on $\boldsymbol{P}(V)$ with symplectic structure defined by its Kahler 2-form, the moment map becomes

$$m : \boldsymbol{P}(V) \to \mathfrak{p} \subset \mathfrak{g} = \text{Lie}\,(G)$$

$$m(v) = \frac{1}{\|v\|^2} \cdot dp_v(1)$$

---

[9] More recent references related to this section include [356–361], [364], [369], [408], [429], [590], [591], [659], [723], [757], [771]. See also Chapter 8, and the references given there.

(because of K-invariance, the differential of $p_v(g)$ lies in the perpendicular complement $\mathfrak{p}$ to Lie K in Lie G). Kempf-Ness's results may be stated as: $\forall x \in \boldsymbol{P}(V)$, x is semi-stable $\Leftrightarrow$ $0 \in m\left(\overline{O_G(x)}\right)$. This theory suggests many interesting questions.

Appendix to Chapter 3

We give an alternative proof of proposition 3.6, using the following result of EDMUNDS [94] on partitions into independent sets. (This approach was suggested to the author by G.-C. ROTA.)

Let $r = k(n + l)$. A sequence $(x_1, \ldots, x_r)$ of geometric points of $\boldsymbol{P}_n$ can be partitioned into k subsequences of $n + 1$ independent points if and only if, for all subsequences, $(x_{i_1}, \ldots, x_{i_j})$, we have

$$j \leqq k \cdot l(i_1, \ldots, i_j),$$

where $l(i_1, \ldots, i_j) = 1 +$ (dimension of linear subspace spanned by $x_{i_1}, \ldots, x_{i_j}$).

Let $x = (x^{(0)}, \ldots, x^{(m)})$ be a geometric point of $(\boldsymbol{P}_n)^{m+1}$. Let $N_0 =$ g.c.d $(n + 1, m + 1)$. Write $N_0 = N(m + 1) - M(n + 1)$ with $N \geqq n + 1$. Let y_i be the geometric point of $(\boldsymbol{P}_n)^{M(n+1)}$ gotten by repeating $x^{(j)}$ N times, $j \neq i$, and repeating $x^{(i)}$ $N - N_0$ times.

Lemma. y_i is stable, $0 \leqq i \leqq m$, $\Leftrightarrow$ x is stable.

Proof. Let L be an l-dimensional linear subspace of $\boldsymbol{P}_n$. If x is stable, one sees easily that

$$(\text{no. of } x^{(j)} \text{ in } L) \leqq \frac{m + 1}{n + 1}(l + 1) - \frac{N_0}{n + 1}.$$

Therefore

$$\begin{aligned}(\text{no. of } y_i^{(k)} \text{ in } L) &\leqq N(\text{no. of } x^{(j)} \text{ in } L)\\ &\leqq N\left(\frac{m + 1}{n + 1}(l + 1) - \frac{N_0}{n + 1}\right)\\ &\leqq \frac{M(n + 1) + N_0}{n + 1}(l + 1) - N_0\\ &\leqq M(l + 1),\end{aligned}$$

and y_i is stable. Conversely, if for some L_0, all $x^{(j)}$ are in L_0, then for all i,

$$(\text{no. of } y_i^{(k)} \text{ in } L_0) = M(n + 1) \geqq M(l_0 + 1)$$

and y_i is not stable. Hence, for each L, some $x^{(j)} \notin L$.

Then if $x^{(i)} \notin L$ and y_i is stable,

$$(\text{no. of } y_i^{(k)} \text{ in } L) = N(\text{no. of } x^{(j)} \text{ in } L),$$

i.e., $M(l+1) > N(\text{no. of } x^{(j)} \text{ in } L)$. Therefore

$$\frac{m+1}{n+1}(l+1) > (\text{no. of } x^{(j)} \text{ in } L),$$

and x is stable.

Proposition 3.6 now follows from EDMUND'S result, because if y_i is stable, then it can be partitioned into M sequences of independent points. But this is the same as the existence of M determinants $D^{(i)}_{\alpha_0\ldots\alpha_n}$ with

$$P_i = \Pi D^{(i)}_{\alpha_0\ldots\alpha_n} \in \Gamma(L_0^N \otimes\cdots\otimes L_i^{N-N_0} \otimes\cdots\otimes L_m^N),$$

and $P_i(x) \neq 0$.

Appendix to Chapter 4

A. Representations with a polynomial ring of invariants or other simplifying properties

A great deal of work has been done finding complete lists of representations of reductive groups for which various especially nice things happen, such as the ring of invariants being a polynomial ring. This line of research has been carried out chiefly by KAC, POPOV and VINBERG. Recent advances are due to G. SCHWARZ, ADAMOVIC and GOLOVINA. One of the key ideas behind this classification is to make use of points with non-trivial stabilizers, applying LUNA'S étale slice theorem near them (see appendix D to Ch. 1). The main result of this work is that there is a "relatively short" list of representations in which the ring of invariants is a polynomial ring and that outside this list various pathologies make a complete description of the orbit structure quite difficult. To be concrete, let a reductive group G act linearly on a vector space V, let $k[V]$ be the algebra of polynomial functions on V, let $k[V]^G$ be the ring of invariants, let $V/G = \operatorname{Spec} k[V]^G$ and let $\pi: V \to V/G$ be the canonical map. We assume the characteristic is 0. Then, following KAC, we may compare the following conditions on G and V:

i) V is "visible", i.e., for $x \in V/G$, $\pi^{-1}(x)$ is a finite union of G-orbits,
ii) all fibres $\pi^{-1}(x)$ of π have the same dimension,
iii) $k[V]$ is a free $k[V]^G$-module,
iv) $k[V]^G$ is a polynomial ring,
v) every point $x \in V$ has a non-trivial stabilizer $G_x \subset G$.

Between these conditions, we have the following easy implications:

(i) $\Rightarrow$ (ii)
(iii) $\Rightarrow$ (ii) + (iv).

The striking result, due to KAC, POPOV, VINBERG ([163], [166], [265], [268]) is the following:

Theorem. When G is simple and V is irreducible, all 5 conditions on (G, V) are equivalent. Moreover, a complete list of (G, V) satisfying these consists in the infinite families*

a) $SL(n)$ on k^n, Hom (k^n, k), Symm^2 (k^n), $\Lambda^2(k^n)$, Hom $(k^n, k^n)'$,
b) $SO(n)$ on k^n, Symm^2 $(k^n)'$, $\Lambda^2(k^n)$,
c) Sp$(2n)$ on k^{2n}, Symm^2 (k^{2n}), $\Lambda^2(k^{2n})'$,

and the special cases:

d) $SL(n)$ on Symm^m (k^n), $(n, m) = (2, 3)$, $(2, 4)$ and $(3, 3)$,
e) $SL(n)$ on $\Lambda^m(k^n)$, $(n, m) = (6, 3)$, $(7, 3)$, $(8, 3)$, $(9, 3)$, $(8, 4)$,
f) $SO(n)$ in its spin representation, $n \leq 14$, $n = 16$,
g) Sp$(2n)$ on $\Lambda^m(k^{2n})'$, $(n, m) = (3, 3)$, $(4, 4)$,
h) each of the exceptional groups in its lowest dimensional and its adjoint representation.

This theorem has been proven by a separate classification of the representations satisfying each condition. Thus in their joint paper [166], KAC, POPOV and VINBERG outline the classification of cases where $k[V]^G$ is a polynomial ring. The details of their argument were published by SCHWARZ [292], where the much larger list of cases of simple groups G, and possibly reducible representations V with $k[V]^G$ a polynomial ring is found. His results were discovered independently by ADAMOVIC and GOLOVINA [41]. To give a sample of his results, he finds that for $SL(2)$, the only such V are the 8 representations:

$$\begin{aligned} &\text{Symm}^n\,(k^2),\ n = 1, 2, 3, \text{ and } 4,\\ &k^2 + k^2,\ k^2 + k^2 + k^2\\ &\text{Symm}^2\,(k^2) + k^2\\ &\text{Symm}^2\,(k^2) + \text{Symm}^2(k^2). \end{aligned}$$

For $SL(3)$, he finds in addition to the 5 irreducible ones in the Theorem, 24 reducible ones. To give an idea of the method, to show that $k[V]^G$ is *not* a polynomial ring, one may use induction** and the following Corollary of Luna's slice theorem:

Lemma. If $x \in V$ has a closed G-orbit $0(x)$ and stabilizer G_x, let $N_x = T_{x,V}/T_{x,0(x)}$ be the normal space to the orbit as a representation space for G_x. Then:

$$k[V]^G \text{ a polynomial ring} \Rightarrow k[N_x]^{G_x} \text{ a polynomial ring.}$$

*′ indicates the irreducible piece of highest weight.

** V. Kac has pointed that out the most difficult point in [166] is that G_x may not be connected so you cannot use induction unless you show that passing to G_x^0 does not spoil the picture.

Also, because the Kac-Popov-Vinberg-Schwarz-Adamovic-Golovina list contains many representations that, although fairly low-dimensional, were not investigated by classical invariant theory, they need some quite new methods of determining $k[V]^G$. The following method is due to LUNA and RICHARDSON [190]:

Proposition. Let $\bar{x} \in V/G$ be a sufficiently general* point and let x over $\bar{x}$ have a closed G-orbit. Let G_x = stabilizer of x, $N_x = \mathrm{Norm}_G(G_x)/G_x$. Then the natural map Φ:

$$\begin{array}{ccc} V^{G_x} & \hookrightarrow & V \\ \downarrow & & \downarrow \\ V^{G_x}/N_x & \xrightarrow{\Phi} & G/V \end{array}$$

is an isomorphism.

SCHWARZ also makes extensive use of an analog of this Proposition, where $\bar{x}$, instead of being "generic", is a special "1-subprincipal" point, which means that the image of V^{G_x} in V/G has codimension 1, and $k[V]^G$ mod a principal ideal is a subring of $k[V^{G_x}]^{N_x}$.

Looking at the other conditions that may be imposed on (V, G), KAC [163] classified all reductive groups G and irreducible representations V which are *visible* in the sense described above.

Moreover, he shows that all these representations can be obtained by a general construction — the so called θ-groups-except for Spin_{11} and Spin_{13}.

To give a sample of his methods, the following is one of the key steps: let $H \subset G$ be a semi-simple subgroup corresponding to a proper subset of the Dynkin diagram. Let U be the linear span of the H-orbit of the highest weight vector $x \in V$. Then U belongs to the cone $\mathcal{N}$ of unstable points of A with respect to G, and if V is G-visible, then U can have only a finite number of orbits under $H \cdot k^*$ (k^* acting by scalars).

POPOV [268] classifies the irreducible V and simple G such that (ii) or (iii) are satisfied. The above lemma for sub-Dynkin diagrams is again used. Various interesting inequalities on the dimension of the cone $\mathcal{N}$ of unstable points are also given, e.g.,

$$\dim \mathcal{N} \geqq \frac{\dim V - \dim V^T}{2} + 1 \geqq \frac{\dim V}{3} + 1$$

(T = maximal torus in G), which enable him to show rapidly that almost all representations cannot satisfy (ii).

---

* More precisely, there is an open set $(V/G)_0$ such that all $x \in V$ over $(V/G)_0$ with $0(x)$ closed have conjugate stabilizer groups. We want $\bar{x} \in (V/G)_0$. The closed orbits $0(x)$ over such $\bar{x}$ are called the *principal orbits*, cf. [188].

ÉLASHVILI [95] and POPOV [265], [266] classified all representations V of simple groups G with non-trivial generic stabilizer subgroup (complete proofs are not given and there is at least one gap, cf. [292], p. 170). SATO and KIMURA [288] classified all pairs (G, V), where G is reductive and V is irreducible with G having an open orbit in V^*. Also IGUSA [159], [161] has given an analytic characterization of a class of relatively simple representations that he calls "admissible".

So far, all these lists indicate that for any semi-simple G and any representation V, (ii) $\Rightarrow$ (iv), hence that representations with any of the properties i), ii), iii) or iv) have $k[V]^G$ a polynomial ring. But this has not been proven in general. In any case, these lists are not far from each other, and this research suggests that there is a small list of "tame" representations for which orbits can easily be classified, beyond which lie "wild" representations for which classification, esp. of unstable orbits, is substantially harder.

B. Representations associated to sets of linear maps and linear subspaces [10]

One of the most natural situations in which to apply invariant theory is when one is given some data of the type: 1) a set of vector spaces V_i, 2) a set of linear maps $f_j: V_{i_1} \to V_{i_2}$ or $V_{i_2}^*$, and/or 3) a set of vectors $x_j \in V_i$ or subspaces $W_j \subset V_i$, and one seeks to classify this data, up to isomorphisms between the V_i's. Thus, for instance, such problems arise:

a) in "classical invariant theory" where one seeks the invariants of $x_1, \ldots, x_n \in V$, $\xi_1, \ldots, \xi_m \in V^*$ mod $GL(V)$, or, if $f: V \to V^*$ defines a fixed quadratic or alternating form on V, mod $0(V)$ or Sp (V)

b) in seeking a normal form for 2 (non-commuting) matrices A, B mod $(A, B) \sim (SAS^{-1}, SBS^{-1})$, $S \in GL(n, \mathbf{C})$**

c) in control theory, where one seeks to describe the situation

$$k^m \xrightarrow[B]{} V \xrightarrow[C]{} k^p \qquad (A \circlearrowleft V)$$

where B is the control, A describes the system to be controlled, and C is the set of observations that you can make, and one wants to classify triples (A, B, C) modulo $GL(V)$.

---

* One of their main discoveries is a procedure for generating examples of this type that they call "castling". Kac has pointed out to me that this is essentially the same as the reflection of BERNSTEIN-GEL'FAND-PONOMAREV [58] (see [165]).

** In applications, this usually occurs in the closely related real form: find a normal form for a single matrix A mod unitary equivalence, $A \sim SAS^{-1}$, S unitary. If we associate to A the pair $(A, {}^tA)$, we can reduce this problem to the one above.

[10] More recent references related to this section include [427], [482–484], [487], [613], [614], [616], [646], [647], [704], [879], [893], [907].

d) in applications of invariant theory to moduli of bundles, and in many other cases, one is led to ask for the invariants of a set of subspaces $W_i \subset V$.

On all these and on many related topics there is an immense literature and we will only touch on a few results that have come to our attention.

Classical invariant theory in char. 0 can be considered to be essentially complete, as it is described in H. WEYL's famous book [331]. The so-called "First fundamental theorem" states that in the situations described in (a), the inner products $\langle x_i, \xi_j\rangle$, resp. $\langle x_i, f(x_j)\rangle$ generate the rings of invariants, and if $GL(V)$ is replaced by $SL(V)$, then we must add $\det(x_{i_1}, \ldots, x_{i_k})$, $\det(\xi_{j_1}, \ldots, \xi_{j_k})$, $k = \dim V$. An elementary proof is given in GARDNER [113]. The "Second fundamental theorem" gives generators for the ideal of relations between these invariants. However the proof of these results in char. p has been carried out only very recently by DE CONCINI and PROCESI [77], based on a key lemma of P. DOUBILET, G. C. ROTA, J. STEIN [91]. Their work overlaps with a parallel investigation by V. LAKSHMIBAI, C. MUSILI and C. S. SESHADRI (see [180] and earlier work referred to there). A summary of these ideas is in PPROCESI'S Helsinki Congress talk [275].

In this connection, there is one general result which follows directly from Capelli's identity of classical invariant theory, but which is applicable for invariants of arbitrary groups. This is "Pascal's Theorem" (see WEYL [331], p. 251). If $G \subset GL(V)$ is any subgroup, then generators for the ring of invariants of G acting on $V^{(n)} = V \oplus \cdots \oplus V (n \times)$ can be obtained by (i) permutations of factors and (ii) polarization, from the invariants of G acting on $V^{(\dim V)}$. An analysis of this result and an extension of it to include the *relations* on these generators is given by VUST [328].

The invariants associated to situation (b) were determined by C. PROCESI [274] in char. O. More generally, let $A_1, \ldots, A_n : V \to V$ be linear maps. Then he proved that the ring of polynomial functions in the coefficients of the A_i invariant under $GL(V)$ acting by conjugation is generated by the traces

$$\operatorname{tr}(A_{i_1}, \ldots, A_{i_N}), \text{ all words } x_{i_1}, \ldots, x_{i_N} \text{ in n letters.}$$

He also determines the ideal of relations between these traces and gives a bound on the length of the words that are needed. Many of his results were discovered independently by RASMYSLEV [279]. That the traces generate the ring of invariants had been conjectured by M. ARTIN [44], who proved the qualitative results:

a) $(0) \in$ Closure ($GL(V)$-orbit of $A_1, \ldots, A_n$)

iff V has a flag with respect to which all A_i are nilpotent

b) If $(A.)^S =$ "semi-simplification" of $(A.)$ (i.e., let $V_0 \subset V_1 \subset \cdots \subset V_n = V$ be a maximal flag invariant under all A_i, then $(A.)^s$ is the induced action of $\{A_i\}$ on $\bigoplus V_i/V_{i-1}$), then

(Closure of orbit of $(A.)$) $\cap$ (Closure of orbit of $(B.)$) $\neq \emptyset$ iff

$(A.)^S \cong (B.)^S$ for some isomorphism of the spaces on which they act.

The case of most interest for applications is the classification of *one* complex matrix A mod unitary equivalence:

$$A \sim B \Leftrightarrow \exists\, U \in U(n, \boldsymbol{C}), \quad UAU^{-1} = B.$$

Here there are no "unstable" points and W. SPECHT [315] proved earlier that

$$A \sim B \Leftrightarrow \left\{\begin{matrix}\text{for all words } w(x, y) \\ \operatorname{tr} w(A, {}^t\bar{A}) = \operatorname{tr} w(B, {}^t\bar{B})\end{matrix}\right\}$$

More explicit results for 3×3 matrices can be found in PEARCY [262], LEW [184] and SMITH [314]. These results were motivated by their applications to mechanics: see RIVLIN [282], SPENCER [316].

PROCESI had studied the ring of invariants in n linear maps $A_1, \ldots, A_n$ earlier [273], proving that it was isomorphic to the center of the non-commutative ring obtained by adjoining to the field k n independent generic $r \times r$ matrices $A_i (r = \dim V)$. Developing his ideas, FORMANEK [103], [104] has proven that the function field of this ring of invariants is a purely transcendental extension of k if $r \leqq 4$. This suggests the possibility of a reasonable normal form for an arbitrary set of k $r \times r$ matrices $\{A_i\}$ mod conjugation if $r \leqq 4$ and the A_i do not leave fixed any common subspace of k^r.

The application of invariant theory to control theory, in situation c. above, was the idea of R. KALMAN [168], which has been developed by BYRNES [66], BYRNES-HURT [68], FALB [97], HAZEWINKEL [136], HAZEWINKEL-KALMAN [137], and HERMANN-MARTIN [139] among others. A survey can be found in BYRNES-FALB [67]. They prove, for instange, that in situation c., the $GL(V)$-stable points are the triples (A, B, C) such that

$$\sum_n \operatorname{Im}(A^n B) = V, \qquad \bigcap_n \operatorname{Ker}(CA^n) = (0),$$

and the quotient of the set of these by $GL(V)$ is isomorphic to a smooth rational quasi-affine variety which may be embedded by the functions $(CA^n B)_{ij}$.

Situation d. is known as the N-subspace problem. Invariant theoretically, one may set the problem up in 2 essentially equivalent ways:

d1) Given $r_i < n$, consider the action of

$$G = SL(r_1) \times \cdots \times SL(r_N) \times SL(n)$$

$$\text{on } \operatorname{Hom}(k^{r_1}, k^n) \oplus \cdots \oplus \operatorname{Hom}(k^{r_N}, k^n)$$

d2) Let $G = SL(n)$ and consider its action on the subvariety of

$$\Lambda^{r_1}(k^n) \oplus \cdots \oplus \Lambda^{r_N}(k^n)$$

consisting of N-tuples of decomposable forms $(\omega_1, \ldots, \omega_N)$.
If all r_i's are equal, we have determined in the text the stable and unstable points for these representions. Others have addressed the question of giving a complete description of all orbits, esp. the unstable ones which lead to proliferating special cases and combinatorial difficulties. L. A. NAZAROVA [239] and I. M. GELFAND-V. A. PONOMAREV [115] solved this for 4 subspaces. BRENNER [63] showed that the 5 subspace problem can be reduced to the classification problem for non-commuting linear transformations $A_1, A_2 : V \to V$. The link between the 2 classification problems is given by associating to the pair $A_1, A_2 : V \to V$ the 5 subspaces of $V \oplus V$: $(0) + V$, $V + (0)$, diagonal, graph of A_1, graph of A_2. An interesting remark, due to L. A. NAZAROVA, is that the problem of classifying N-subspaces W_i of a fixed V is identical to the problem of classifying torsion-free modules over the one-dimensional complete local ring

$$R_N = \{(f_1, \ldots, f_N) \mid f_i \in k[[t]],\ f_1(0) = \cdots = f_N(0)\}.$$

A quite general analysis of the classification problem for arbitrary sets of linear maps was initiated by P. GABRIEL [112]. He starts with an oriented finite graph Γ with integers $n(v)$ attached to each vertex. A *quiver* based on Γ is a set of vector spaces $W(v)$, one for each vertex v, with dim $W(v) = n(v)$, and a set of linear maps

$$\lambda(e) : W(v_1) \to W(v_2),$$

one for each edge starting at v_1 and ending at v_2. In this language, the N subspace problem is the classification of quivers for the graph

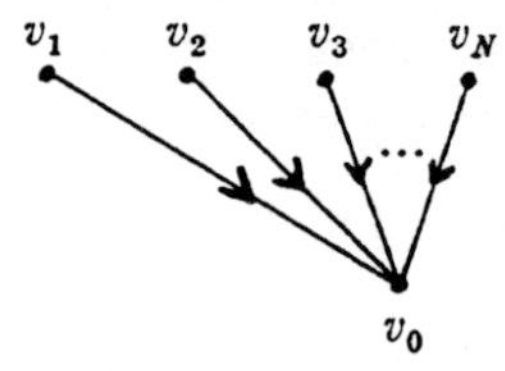

and the 2 non-commuting linear maps problem are quivers for

GABRIEL proved that for all graphs Γ, there are only a finite set of quivers up to isomorphism for each choice of weights $n(v)$ iff the graph Γ is the Dynkin diagram A_n, D_n, or E_6, E_7, E_8. This theory has been developed by I. M. GEL'FAED-V. A. PONOMAREV [115], I. BERNSTEIN-GEL'FAND-PONOMAREV [58], L. A. NAZAROVA [240], V. KAC [165], V. DLAB-C. M. RINGEL [86], and others (see bibliographies of the above papers). Their results echo the theme of the previous appendix: there is a small list of "tame" graphs for which a complete explicit classification of quivers can be made. Beyond this are "wild" graphs for which classification is as hard as for the 2 non-commuting maps problem*.

In yet another direction, D. GIESEKER ([117], see § 2) has given a partial description of stability for non-decomposable tensors in

$$\Lambda^r(k^n) \oplus \cdots \oplus \Lambda^r(k^n) \qquad (N \text{ copies}).$$

If $(\omega_1, \ldots, \omega_N)$ is in this space, and ω_i are considered as alternating forms in r vectors, he proves:

$$\left[\exists\, W \subset k^n \text{ s.t. for all } k,\ \omega_k(a_1, \ldots, a_r) = 0 \text{ if } a_1, \ldots, a_d \in W,\ d \geqq \frac{r}{n} \dim W\right]$$

$$\Downarrow$$

$$(0) \in \text{closure of } SL(n)\text{-orbit of } (\omega_1, \ldots, \omega_N)$$

$$\Downarrow$$

$$\left[\begin{array}{c} \exists\, W_1 \subset W \subset k^n \text{ s.t. } \dim W_1 < \frac{r}{n} \dim W; \\ \text{for all } k,\ \omega_k(a_1, \ldots, a_r) = 0 \quad \text{if } a_1, \ldots, a_{d_1} \in W_1,\ a_{d_1+1} \in W; \\ \text{for some } k,\ \exists\, a_1, \ldots, a_r, \text{ such that } a_1, \ldots, a_{d_1} \in W_1 \\ \text{and } \omega_k(a_1, \ldots, a_r) \neq 0 \quad (\text{here } d_1 = \dim W_1). \end{array}\right]$$

When the ω_i are decomposable, the 1st implication becomes $\Leftrightarrow$, as proved in Ch. 4, § 4. This, as well as the results in Ch. 4, § 4, are motivated by their applications to the construction of moduli spaces of vector bundles. We will discuss this in Appendix 5B.

The orbit space structure and explicit invariants for $GL(n)$ acting on $\Lambda^3(k^n)$ has been determined for $n \leqq 9$ only ([96], [127], [291]). Already in this case the problem looks hard for general n.

---

* The classification of 2 non-commuting linear maps $f, g\colon V \to V$ mod $GL(V)$ is sometimes referred to as an "impossible" problem. It is not clear to me why this is said. If $\dim V = n$, then for each n, there might be a finite number of explicitly describable algebraic varieties $W_i^{(n)}$ whose points are in natural $1-1$ correspondence with suitable "strata" $S_i^{(n)}$ in the full set S_n of pairs $(f, g)\colon k^n \to k^n$ mod $GL(n)$.

C. Stability of hypersurfaces and Chow forms/Hilbert points of other projective varieties[11]

The invariant theory problems with the most direct application to moduli problems are concerned with forming a quotient of a variety $\mathscr{H}$ which parametrizes all subvarieties V of $\boldsymbol{P}^n$ of some type by the canonical action of $PGL(n+1)$. The simplest case is when V is a hypersurface of some degree d. Then the parameter space $\mathscr{H}$ is the projective space $|dH|_{\boldsymbol{P}^n}$, known as the complete linear system of divisors D on $\boldsymbol{P}^n$ linearly equivalent to d times a hyperplane.

Beginning with $\boldsymbol{P}^1$, T. Shioda [313] succeeded in completely describing the ring of invariants as well as their relations and syzygies for the case of "binary octics", i. e., $|8H|_{\boldsymbol{P}^1}$. His method is a very determined application of all the old techniques of 19th century invariant theory. Shioda (op. cit.) also worked out this generating function for the dimension of the invariants in each degree for $PGL(3)$ acting on $|4\mathrm{H}|_{\boldsymbol{P}^2}$. W. Seiler [295] gives a Fortran program for computing the unstable plane curves, i.e., points of $|nH|_{\boldsymbol{P}^2}$ with group $PGL(3)$, for any n and describes the result for $n \leq 7$.

The 2 cases of $PGL(3)$ acting on sextic curves in $\boldsymbol{P}^2$, i.e., $|6H|_{\boldsymbol{P}^2}$ mod projectivities, and of $PGL(4)$ acting on quartic surfaces in $\boldsymbol{P}^3$, i.e., $|4H|_{\boldsymbol{P}^3}$ mod projective transformations, have attracted a lot of attention because of their connection to the moduli problem for $K3$ surfaces. A $K3$ *surface* is a smooth surface X with $\Omega^2_X \cong \underline{o}_X$ and $H^1(\underline{o}_X) = (0)$. There are an infinite number of algebraic families of projective $K3$ surfaces, parametrized by the degree $2d = (D^2)$ of the ample divisors D generating a polarization of X. The simplest cases are $d = 1$ and 2. In these cases, almost all such surfaces are obtained:

a) if $d = 1$, as double covers of $\boldsymbol{P}^2$, branched in a smooth sextic curve, mapped to $\boldsymbol{P}^2$ by $|D|$, and

b) if $d = 2$, as smooth quartic surfaces in $\boldsymbol{P}^3$, embedded by $|D|$. The problem is to understand the moduli spaces $\mathscr{K}_d$ of these $K3$ surfaces for $d = 1, 2$ by explicitly compactifying them $\mathscr{K}_d \subset \overline{\mathscr{K}}_d$ in such a way that the points of $\overline{\mathscr{K}}_d - \mathscr{K}_d$ correspond to singular limits of $K3$-surfaces. $\overline{\mathscr{K}}_d$ is obtained by starting from the full orbit space

$$|6H|^{ss}_{\boldsymbol{P}^2}/PGL(3) \qquad \text{if d} = 1$$

$$|4H|^{ss}_{\boldsymbol{P}^3}/PGL(4) \qquad \text{if d} = 2$$

and blowing this space both up and down to obtain $\overline{\mathscr{K}}_d$. One must start by describing $|6H|^{ss}_{\boldsymbol{P}^2}$ and $|4H|^{ss}_{\boldsymbol{P}^3}$, then one must reconstruct what happens to the sextic branch curve, or quartic image in $\boldsymbol{P}^3$ if various bad things happen to the $K3$ surface (e.g., $|D|$ gets base points, X acquires

[11] More recent references related to this section include [375], [445], [481], [485], [574], [617], [638], [650–652], [903].

various singularities, etc.) and then one must translate this into a birational modification of the orbit space. This work has been carried on mainly by J. SHAH in a difficult series of papers [308], [311], [312]. Some of his results were obtained independently by E. HORIKAWA [145].

Another example of geometric interest is the invariant theory of pencils of binary and ternary cubics: i.e., $PGL(2)$ and $PGL(3)$ acting on $Grass_1(|3H|_{P^1})$ and $Grass_1(|3H|_{P^2})$.

P. NEWSTEAD [248] described the full situation for P^1 using classical results for $SL(2)$, and R. MIRANDA [201] described the set of stable points in the case of P^2 and applied his results to the construction of a moduli space for rational elliptic surfaces, i.e., smooth rational surfaces X with a morphism $\pi: X \to P^1$ with elliptic fibres.

For projective varieties of higher codimension, there are 2 specific actions to study:

a) Given $V^r \subset P^n$ a projective variety. Let $u^{(1)}, \ldots, u^{(n-r)}$ be $(n-r)$ sets of dual variables $u^{(i)} = (u_0^{(i)}, \ldots, u_n^{(i)})$. Then the *Chow form* Φ_V of V is the multi-homogeneous form $\Phi_V(U^{(1)}, \ldots, U^{(n-r)})$ defined by:

$$\left(\Phi_V(U^{(1)}, \ldots, U^{(n-r)}) = 0\right) \Leftrightarrow \left(\exists x \in V^r \text{ such that } \sum_{i=0}^{n} x_i u_i^{(k)} = 0, l \leqq k \leqq n-r\right)$$

Φ_V lies in a vector space $V_{d,r}$ depending on the degree d of V as well as the dimension, and we wish to know when Φ_V is stable with respect to $SL(n+1)$.

b) Given $V^r \subset P^n$ as above, let $I(V) \subset k[X_0, \ldots, X_n]$ be its homogeneous ideal and let

$$I(V)_N \subset k[V]_N$$

be the degree N pieces of the ideal and the graded ring.

If

$$P(N) = \dim I(V)_N,$$

then we obtain a one-dimensional subspace:

$$\Lambda^{(P)N}\left(I(V)_N\right) \subset \Lambda^{P(N)}(k[X]_N).$$

An element of this subspace is called the Nth *Hilbert point* $h_N(V)$ of V. The problem is whether it is a stable point of $\Lambda^{P(N)}(k[X]_M)$ with respect to $SL(n+1)$.

The 2 problems are related:

Φ_V stable (resp. unstable) implies $h_N(V)$ stable (resp. unstable) for $N \gg 0$

This follows from the results of J. FOGARTY [100], applying Prop. 2.18 to the morphism from the Hilbert scheme to the Chow scheme. After the publication of the 1st edition of this book, no progress at all was made in these difficult questions until D. GIESEKER [116] overcame truly huge difficulties and succeeding in proving that if V is a surface of general type (i. e., $|m \cdot K_V|$ defines a birational morphism from V to a surface $\Phi_{mK}(V)$ in $\mathbf{P}^n$ for m sufficiently large), then for m sufficiently large and N sufficiently large compared to m, $h_N(\Phi_{mK}(V))$ is stable. Furthermore, m and N depend only on the numerical characters (K_V^2) and $\chi(\underline{o}_V)$ of V. This implies immediately that a quasi-projective moduli space for such surfaces exists. His proof is based on verifying the following numerical criterion which is obtained by specializing the general numerical criterion of Ch. 2 to Hilbert points:

$$h_N(V) \text{ is stable} \Leftrightarrow \left\{\begin{array}{l} \forall\, 1-PS\ \lambda\colon \mathbf{G}_m \twoheadrightarrow SL(k^{n+1}), \text{ change coordinates to} \\ \text{write } \lambda(t) = \begin{pmatrix} t^{r_0} & 0 \\ 0 & t^{r_n} \end{pmatrix}. \text{ Then there exist monomials} \\ M_1, \ldots, M_l \text{ of degree } N \text{ which form a basis of} \\ k[X]_N/I(X)_N \text{ such that } \sum_i \text{weight}_\lambda(M_i) < 0 \end{array}\right\}$$

(Here, the *weight* of $\prod x_{x_i}^{d_i}$ is $\sum r_i d_i$.) Inspired by Gieseker's success, the author analyzed the stability of Chow forms again [218], obtaining a numerical criterion in terms of a number measuring the degree of "contact" between a variety $V \subset \mathbf{P}^n$ and a weighted flag $L_0 \subset L_1 \subset \cdots \subset L_{n-1} \subset \mathbf{P}^n$. From this, it is proved that if V is a reduced connected curve with at most ordinary double points and no smooth rational components E meeting $\overline{V-E}$ in only one or two points, then the Chow form $\Phi_{mK}(V)$ is stable if $m \geqq 5$ and K is the canonical divisor class of V. More recently I. MORRISON [205] has proven the stability of the Nth Hilbert point of a "stable ruled surface". Here a stable ruled surface is a smooth surface V with a morphism to a curve

$$\pi\colon V \to C$$

with all fibres isomorphic to $\mathbf{P}^1$ such that every curve $C' \subset V$ which is a section of π satisfies $(C'^2) > 0$. In his theorem, he must embed V in $\mathbf{P}^n$ by a very ample divisor D such that

$$\big(D.\, \pi^{-1}(x)\big) = 1$$

(D^2) suficiently large (depending only on the genus of C).

Furthermore, it follows as a Corollary of the results of either LUNA [189] KEMPF [171] or ROUSSEAU [285] that if $V \subset \mathbf{P}^n$ is an abelian variety embedded by any complete very ample linear system, then its Chow form and all its Hilbert points are stable. This is because the group $K \subset PGL(n+1)$ of projectivities mapping V to itself is finite and its inverse image $H \subset SL(n+1)$ acts irreducibly on k^{n+1}. Hence, following

LUNA, (using the fact that in this case the normalizer N of K in $SL(n+1)$ is finite) or following KEMPF (using the fact that $H \not\subset$ any parabolic of $SL(n+1)$) its Chow forms and Hilbert points are stable.

As far as surfaces go, it remains to analyze the stability of Chow forms and Hilbert points for rational, elliptic and $K3$-surfaces, and, above all, for singular surfaces.

Appendix to Chapter 5 [12]

A. Moduli spaces in the category of algebraic spaces

ARTIN and MOIŠEZON'S introduction of objects more general than algebraic varieties, called *algebraic spaces* by ARTIN, (see [46], [177], [203]) provides a very convenient setting in which to begin the study of moduli spaces. We use the terminology of Ch. 5, § 1. A very general class of moduli problems can be set up as follows:

i) Fix a polynomial $P(x)$ of degree r and an algebraically closed field k.

ii) Consider the set $\mathscr{V}_p(k)$ of pairs consisting of a smooth projective r-dimensional variety X defined over k such that

a) no smooth deformation of X is birational to $Y \times \boldsymbol{P}^1$, any Y,

b) $H^0(T_X) = 0$, $T_X =$ tangent bundle to X

and of an inhomogeneous polarization $U \subset Pic(X)$ with Hilbert polynomial $P(X)$, modulo isomorphism.

One seeks a coarse moduli space V_p whose k-valued points are in $1-1$ correspondence with $\mathscr{V}_p(k)$. (One makes this precise via functors as in Def. 5.4, 5.6). It is only in special cases that V_p is known to exist as a variety, but in complete generality when char $k = 0$, V_p exists as an algebraic space, separated and of finite type over k. Surprisingly, this does not seem to appear in the literature (see POPP [270], [271], NARASIMHAN-SIMHA [238] for special cases), hence we give a sketch of the construction of V_p here. In particular, if $k = \boldsymbol{C}$, V_p exists as a Hausdorff analytic space. We sketch the construction of V_p as an analytic space first as it is easier to follow.

By a fundamental theorem of MATSUSAKA ([196], [186]), there is an integer n such that for all $(X, U) \in \mathscr{V}_p$ and for all $L \in U$, L^n is very ample and $H^i(L^n) = (0)$, $i > 0$. Let $N = P(n) - 1$ and consider the smooth subvarieties $X \subset \boldsymbol{P}^N$ such that $\left(X, \underline{o}_X(1)\right) \cong (Y, L^n)$ for some $(Y, U) \in \mathscr{V}_p(k)$, and $L \in U$. This is the same as the set of smooth $X \subset \boldsymbol{P}^n$ such that

$$\chi\left(\underline{o}_X(k)\right) = P(nk), \quad \text{all } k$$

$$\underline{o}_X(1) \text{ is divisible by } n \text{ in } Pic(X).$$

[12] Recent references related to Chapter 5 include [443], [444], [538], [547], [615], [635–637], [650–652], [729], [738], [765], [833–835], [860].

Such X's will be classified by a Zariski open subset H of the Hilbert scheme

$$Hilb_{P^N}^{P(nx)}.$$

We now consider the related scheme:

$$R = \left\{ \begin{array}{l} \text{scheme representing triples } X_1, X_2, \Phi \text{ where } X_1 \in H, X_2 \in H \\ \text{and } \Phi\colon X_1 \to X_2 \text{ is an isomorphism such that} \\ \Phi^*\big(\underline{o}_{X_2}(1)\big) \text{ is num. eq. to } \underline{o}_{X_1}(1) \end{array} \right\}$$

The image of R in $X \times X$ is clearly an equivalence relation and the analytic, or algebraic, space that we seek is the quotient H/R. We first note a few properties of R:

i) the projection $R \to H$, taking (X_1, X_2, Φ) to X_1, is smooth. This is seen as follows. Fixing X_1, every pair (X_2, Φ) defines a sheaf $M = \Phi^*\big(\underline{o}_{X_2}(1)\big)$ on X_1 numerically equivalent to $\underline{o}_{X_1}(1)$ and a basis $s_0 = \Phi^*(X_0), \ldots, s_N = \Phi^*(X_N)$ of $\Gamma(X, M)$, modulo scalars. Conversely, M and s_i define an immersion ψ of X_1 into P^N, hence define $X_2 = \mathrm{Im}\, \psi$, $\Phi = \mathrm{res}\, \psi$. But the set of M's is parametrized by $Pic^\tau(X_1)$ and the set of s_i's by an open subset of $\Gamma(X_1, M)^N$ mod k^*. These form a smooth variety varying in a smooth family with X_1.

ii) the projection $R \to H \times H$ is finite and unramified. For all $(X_1, X_2) \in H \times H$, the set of all $(X_1, X_2, \Phi) \in R$ over (X_1, X_2) is either empty or a principal homogeneous space over the group of automorphisms of the polarized variety (X_1, U). Since $H^0(T_X) = (0)$, the Lie algebra of this group is trivial, hence the group is finite and reduced. Thus $R \to H \times H$ is quasi-finite and unramified. It is proper by the theorem of MATSUSAKA and MUMFORD [197], since X is not birationally ruled.

Now work specifically over $k = \mathbf{C}$. Using the fact that $R \to H \times H$ is finite, we first construct H/R as a Hausdorff topological space, in the complex topology. We need local charts to make H/R analytic. For each $x \in H$, let $0^R(x) = \{y \in H \mid \exists\, \Phi \text{ such that } (x, y, \Phi) \in R\}$. Then $0^R(x)$ is a smooth subvariety of H. Let W be a local section of H through x transverse to $0^R(x)$, (i.e., if $\dim 0^R(x) = k$, and $f_1, \ldots, f_k \in m_{x,H}$ generate $m_{x,0^R(x)}$, let W be the analytic subspace of H defined by $f_1 = \cdots = f_k = 0$ in a neighborhood of x). Restricting R to $W \times W$, we get an equivalence relation R_1 over $W \times W$ such that the projection $R_1 \to W$ is étale. If W is small enough, we may assume R_1 is isomorphic to k copies of W. This gives us k analytic maps $f_i = pr_2 \circ \sigma_i$ from W to W:

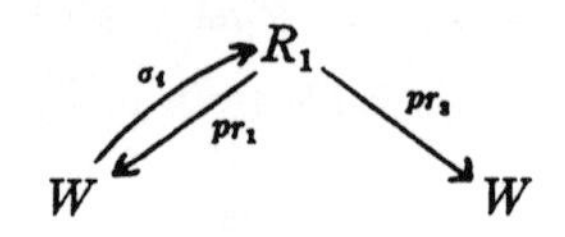

Consider the map

$$W \to \mathrm{Symm}^k W$$

$$x \mapsto \left(\text{0-cycle } \sum_{i=1}^{k} f_i(x)\right).$$

The image W_1 of this map is a quotient W/R_1 and gives a local analytic structure on H/R near x.

Only the last part of this argument needs to be modified to work for algebraic spaces. Recall that an algebraic space is essentially a gadget whose local charts are formal quotients Y/S where Y is a scheme and $S \subset Y \times Y$ is an equivalence relation such that $S \to Y$ is étale. In our setting, we construct local charts for "H/R" of this type as follows: start with any $x \in H$ and let W be a closed subscheme of an affine neighborhood of $x \in H$ such that W is transverse to $0^R(x)$ at x. Replace W by an open piece which is everywhere transverse to R. Let R_1 be the restriction of R to $W \times W$. Then there is an étale map

$$p: W_1 \to W$$

with $x = p(x_1)$, $x_1 \in W_1$ such that

$$R \times_W W_1 = (\text{disjoint union of } k \text{ copies of } W_1) \quad \amalg \; R^*$$

where $x_1 \notin \mathrm{Im}\,(R^* \to W_1)$. As in the analytic case, the first part of $R \times_W W_1$ defines a morphism

$$W_1 \to \mathrm{Symm}^k W_1.$$

If W_2 is the image, one sees easily that W_2 is étale over the formal quotient H/R in a neighborhood of x.

In characteristic p, various things make the problem harder. First of all, it is not known whether Matsusaka's finiteness theorem is true, so we must be content with coarsely representing $\mathscr{V}_p$ by a space *locally* of finite type over k. Secondly, the family of Picard schemes $Pic(X)$ is not in general smooth over the base space S of a family of smooth projective varieties X_s, $s \in S$. Thus $R \to H$ will not in general be smooth. This will complicate the construction of local charts, and should be investigated.

B. General discussion of moduli spaces[13]

What types of questions can be asked about the moduli spaces V_p, and what techniques are available to answer them? This section is meant to be a brief sketch of the situation at present.

The first question we ask is to describe the *local* structure of V_p. In particular, one would like to know when V_p is smooth or is a

[13] More recent references related to this section include [579], [697] and the many references therein, [853].

V-manifold, i.e., locally a smooth space modulo a finite group. Following Grothendieck, this can be studied by the formal deformation theory of the pair (X, U): i.e., the functor Def (R) which assigns to an Artin local k-algebra R, with residue field k, the set of isomorphism classes of quadruples $(\mathscr{X}, \mathscr{U}, \Phi, \pi)$ where

a) $\pi: \mathscr{X} \to \operatorname{Spec} R$ smooth

b) $\mathscr{U} \subset Pic_{\mathscr{X}/R}$, a coset of $Pic^{\tau}_{\mathscr{X}/R}$

c) $\Phi: \mathscr{X} \times_R \operatorname{Spec}(k) \xrightarrow{\approx} X$ an isomorphism over k such that $\phi(\mathscr{U} \times_R \operatorname{Spec}(k)) = U$ (both being subsets of Pic^{τ}_X).

For background, see GROTHENDIECK [13], exp. 195, SCHLESSINGER-LICHTENBAUM [289], [185], ARTIN [45]. One says that the complete local k-algebra $\mathcal{O}$ *pro-represents* the functor Def of formal deformations if the functor $R \mapsto$ Def (R) is isomorphic to the functor:

$$R \mapsto \operatorname{Hom}_k(\mathcal{O}, R).$$

If $\mathcal{O}$ is isomorphic to $k[[t_1, \ldots, t_n]]$, the functor is said to be *unobstructed*. In general, $\mathcal{O}$ is related to the structure of V_p at the point x corresponding to (X, U) by the isomorphism

$$\mathcal{O}^G \cong \hat{\underline{o}}_{x, V_p}$$

where G is the finite group of automorphisms of the pair (X, U). There is an extensive theory on these questions and we will only cite a few results. For curves (any char.) and for abelian varieties (in char. 0), the deformations are unobstructed ([259]), but for abelian varieties in char. p, they are usually obstructed ([249], [250], [251]). For surfaces with $H^2(T_X) = H^2(\underline{o}_X) = (0)$, deformations are unobstructed ([13], exp. 195). For surfaces in general, deformations *may* be obstructed ([65], [169], [329]).

Turning to global questions about V_p, one asks next whether or not V_p is quasi-projective. This is the question that geometric invariant theory can sometimes answer. In particular, in Appendix 4C we have seen that V_p is quasi-projective if.

a) $\dim X = 1$, or

b) $\dim X = 2$, K_X ample and $U =$ polarization containing K_X, or

c) X an abelian variety of any dimension.

A third technique is available sometimes to study the global structure of V_p. This is the method of periods. To explain this, we take the ground field k to be $\mathbf{C}$ and note that on each connected component V_p^0 of V_p the topological type of the corresponding varieties X_t, $t \in V_p^0$, is the same for each $t \in V_p^0$. Fix a base point $t_0 \in V_p^0$ and let R^* be the graded cohomology ring $H^*(X_{t_0}, \mathbf{Z})$, $h \in H^2(X_{t_0}, \mathbf{Z})$ the first chern class of a line bundle $L \in U(c_1(U)$ for short). For any $t \in V_p^0$, there is a countable, non-empty set of graded ring isomorphisms

$$\alpha: H^*(X_t, \mathbf{Z}) \xrightarrow{\sim} R^*$$

such that

$$\alpha\big(c_1(U)\big) = h.$$

The set of pairs (t, α) forms a topological covering space of V_p^0:

$$\pi: \tilde{V}_p^0 \to V_p^0.$$

For each, t however,

$$H^k(X_t, \boldsymbol{C}) \cong H^k(X_t, \boldsymbol{Z}) \otimes \boldsymbol{C}$$

comes with a filtration induced by its representation by differential forms:

$$F^p\big(H^k(X_t, \boldsymbol{C})\big) = \left\{\begin{array}{l}\text{the cohomology classes represented by diff.} \\ \text{forms } \omega \text{ with at least } p \text{ holomorphic forms } dz_i \\ \text{in each term.}\end{array}\right\}$$

Although

$$d(p, k) = \dim F^p\big(H^k(X_t, \boldsymbol{C})\big)$$

is independent of t, the actual subspaces

$$\alpha\big(F^p(H^k(X_t, \boldsymbol{C}))\big)$$

of $R^k \otimes \boldsymbol{C}$ vary with t. This gives us a map

$$\text{per}: \tilde{V}_p^0 \to \prod_{k=0}^{2\dim X} \left(\prod_{p=0}^{k} \mathit{Grass}\big(d(p, k), b(k)\big)\right)$$

where $b(k)$ is the kth Betti number of the X_t's and $\mathit{Grass}(d, b)$ is the Grassmannian of d-planes in the $b(k)$-dimensional space R^k. This map turns out to be holomorphic and is called the period map.

We will not pursue this theory, but refer the reader to GRIFFITH'S articles [121], [122], [124], and DELIGNE'S articles [80], [82]. This technique applies directly to the study of the moduli spaces of abelian varieties and of $K3$-surfaces (over $\boldsymbol{C}$). In fact, if $P(X) = d \cdot X^g$, then V_p has various components $V_p^{(\delta)}$ classifying pairs (X, U) where X is an abelian variety. The various $V_p^{(\delta)}$ are distinguished by their polarized cohomology ring (R, h): on $V_p^{(\delta)}$, there is an isomorphism

$$\beta: R^* \xrightarrow{\approx} \Lambda^*(\boldsymbol{Z}^{2g})$$

with

$$\beta(h) = \sum_{i=1}^{g} \delta_i(e_i \wedge e_{i+g})$$

where $\delta_1 | \delta_2 | \cdots | \delta_g$ are integers and $d = \prod_1^g \delta_i$. For these $V_p^{(\delta)}$, $\tilde{V}_p^{(\delta)}$ is the universal covering space of $V_p^{(\delta)}$ and the period map for $p = k = 1$:

$$\text{per}: \tilde{V}_p^{(\delta)} \to \mathit{Grass}\,(g, 2g)$$

is an *isomorphism* of $\tilde{V}_p^{(\delta)}$ with a subset $\mathfrak{H}_g$ of $\mathit{Grass}(g, 2g)$ known as "Siegel's upper half-space of dimension g". Concretely, $\mathfrak{H}_g$ is the set of g-planes W_Ω:

$$W_\Omega = \left\{\text{span of } e_i + \sum_{j=1}^{g} \Omega_{ij} e_{j+g}\right\}$$

for arbitrary symmetric complex $g \times g$ matrices Ω with $\operatorname{Im}\Omega$ positive definite. The final result is that

$$V_p^{(\delta)} \cong \mathfrak{H}_g/\Gamma_\delta$$

where Γ_δ is the group of integral $2g \times 2g$ matrices which preserve the 2-form $\sum \delta_i(e_i \wedge e_{i+g})$. Thus the theory of periods leads to an explicit analytic description of all the moduli spaces of abelian varieties over $\mathbf{C}$. These results go back to Riemann and are easily proven due to the fact that the abelian variety X with periods Ω can be explicitly reconstructed from Ω: i. e.,

$$X \cong \mathbf{C}^g/L_\Omega$$

L_Ω = lattice generated by columns of the $g \times 2g$ matrix:

$$\left(\begin{array}{ccc|c} \delta_1 & & 0 & \\ & \ddots & & \Omega \\ 0 & & \delta_g & \end{array}\right)$$

See [123] or [220]).

Much subtler is the 2nd case mentioned above: that of $K3$-surfaces. For every integer $d \geqq 1$, let $P(X) = 2dX^2$. Then V_p has a unique component $\mathscr{K}_d$ representing $K3$-surfaces X (possibly with rational double points) and polarizations U such that if $U = \{L\}$, then $c_1(L)^2 = 2d$, $c_1(L)$ indivisible in $H^2(X, \mathbf{Z})$. Then the period map with $p = k = 2$ induces an isomorphism of the universal cover $\tilde{\mathscr{K}}_d$ of $\mathscr{K}_d$ with a bounded symmetric domain D:

$$\text{per}: \tilde{\mathscr{K}}_d \xrightarrow{\approx} D \subset \mathit{Grass}(1, 22),$$

D is a homogeneous space under the orthogonal group S0(2,19) and $\mathscr{K}_d$ itself is given by a quotient:

$$\mathscr{K}_d \xrightarrow{\approx} D/\Gamma_d$$

where Γ_d is a discrete subgroup of $S0(2,19)$. This theorem is harder because there is no known way to explicitly reconstruct the $K3$-surface from its periods, hence an elaborate combination of algebraic and analytic subterfuges is needed to prove this. The injectivity of per is due to Šafarevich-Pjatetskij-Shapiro [272] (see also [64]), the surjectivity of per to Kulikov [179].

Indirectly, the period map leads to descriptions of many other moduli spaces. Thus the period map for curves is essentially the map which associates a curve to its Jacobian variety; this gives a map from the moduli space of curves to the moduli space of abelian varieties. It is injective by Torelli's theorem (see Appendix 7D). Similarly we may associate to Fano 3-folds their "intermediate jacobians" by the period map with $p = 3$, $k = 2$ and in some cases this gives an injective map of moduli spaces (see TJURIN [324]). Although Enriques surfaces have no periods, they are doubly covered by a $K3$-surface, and this leads to a period description of their moduli space (HORIKAWA [147]). For surfaces of general type, it is still very unclear whether or not the period map (in the broad sense: looking at periods not only of X but of finite covers of X), usually is injective. Recent counter-examples make this thesis hard to defend ([71], [72], [326]).

With these techniques in hand, one would like to study the following questions:

a) what are the components of V_p?
b) give an explicit compactification $\overline{V}_p$ of V_p,
c) evaluate, or bound, invariants of V_p or $\overline{V}_p$ or a resolution of these (such as its Hodge numbers, or Chern numbers).

A few comments on these: the extensive work of HORIKAWA on particular surfaces of general type [146] makes it clear that (a) in general does not have an easy answer. (b) is important because without it, it is almost impossible to work on (c). We shall discuss (b) in appendices 5D and 4E. (c) is particularly promising for the cases directly described by period maps, due to the recent progress in computing the cohomology of discrete groups (see, for instance, [62], [200], [261], [108] but there is a very rapidly growing literature on this).

C. Moduli of vector bundles [14]

One of the main applications of geometric invariant theory has been to the construction of moduli spaces of vector bundles on a fixed projective variety X. The idea that such moduli spaces exist (and have arithmetic applications) goes back to the fundamental paper [330] of A. WEIL. The theory of these spaces when X is a curve has been intensively developed in the last 20 years by the Tata Institute group of M. S. NARASIMHAN, S. RAMANAN and C. S. SESHADRI, as well as by P.

[14] p-adic versions of some of the results described are due to BURNOL [437]. Other recent references related to this section include [375], [396], [417], [425], [426], [514], [515], [564], [569], [608–610], [622], [630], [631], [657], [699], [704], [724], [744], [775], [777], [781], [816], [827], [845], [846], [872], [886], [899], [901]. On bundles over projective spaces we have [376], [388], [435], [450], [451], [471], [512], [525], [531], [612], [634], [706], [725], [734], [739–741], [746], [762–764], [766], [772], [774], [784–789], [830], [831], [854], [863–865], [871]. On bundles over other surfaces and higher dimensional varieties we have [412], [567], [568], [621], [683], [716], [726], [728], [732], [733], [745], [750], [752], [753], [790], [861], [894], [926]. References on the cohomology of moduli spaces of bundles over curves include [371–373], [511], [566], [662], [663], [669], [778–780], and on rationality questions [374], [413], [489], [847]. On Yang-Mills theory and instantons we have [365], [366], [378], [379], [415], [418], [436], [460], [466], [467], [611], [705], [727], [794], [853], [866], [900], and on stable pairs, parabolic bundles and theta divisors, [385], [394], [397], [398], [401], [422], [423], [468], [469], [524], [702], [709], [883]. See also Chapter 8 for many more references.

NEWSTEAD, G. HARDER, A. TJURIN, and (over higher dimensional X) by D. GIESEKER, M. MARUYAMA and others. Over P^n, much work has been done on these spaces by W. BARTH, G. HORROCKS and others. These spaces have been applied to arithmetic by V. DRINFELD and G. HARDER, and to non-linear partial differential equations by M. ATIYAH and R. BOTT, and by V. DRINFELD and Y. MANIN.

We describe first the existence theorem for these spaces. First, say X is a smooth curve, and E is a vector bundle on X.

Definition: E is stable (resp. *semi-stable*) if for all proper subbundles $F \subsetneqq E$,

$$c_1(F) < c_1(E) \cdot \frac{\operatorname{rk} F}{\operatorname{rk} E}$$

$$(\text{resp.} \quad ,, \quad \leqq \quad ,, \qquad).$$

If E is semi-stable but not stable, we consider those subbundles $F \subset E$ for which equality holds above. Let $(0) \subset F_1 \subset \cdots \subset F_n = E$ be a maximal chain of such bundles.
Define $\operatorname{gr} E$ to be the bundle

$$\bigoplus_{k=1}^{n} F_k/F_{k-1}.$$

This is independent of the choice of the F_i, and SESHADRI [299] proved:

Theorem: Fix X, r and d. Then the set of semi-stable vector bundles E on X of rank r with $c_1(E) = d$, modulo the equivalence relation $E \sim F$ if $\operatorname{gr} E \cong \operatorname{gr} F$, is in a natural way the set of points of a normal projective variety $\mathscr{U}_X(r, d)$*. $\mathscr{U}_X(r, d)$ has a smooth open set corresponding bijectively to the isomorphism classes of stable bundles.

Secondly, say X is a smooth n-dimensional projective variety with hyperplane section H. Then following GIESEKER, if E is a torsionfree sheaf on X, define:

Definition: E is stable (resp. semi-stable) if for all proper subsheaves $F \subsetneqq E$,

$$\chi\big(F(nH)\big) < \frac{\operatorname{rk} F}{\operatorname{rk} E} \cdot \chi\big(E(nH)\big) \text{ for } n \gg 0$$

$$(\text{resp.} \quad ,, \quad \leqq \qquad ,, \qquad).$$

* In the interest of easy communication, it would be good if one notation for this space were generally agreed upon. The most common notations in use are $U(r, d)$ and $M(r, d)$. But M is merely short for "moduli" and might refer to almost any moduli space. In general, I like to use script capital letters for moduli spaces.

If E is semi-stable but not stable, define gr E as before using subsheaves F such that $=$ holds (for large n, hence for all n).
Then GIESEKER [117] and MARUYAMA [193], [194] proved:

Theorem: Fix X, H, r and algebraic cycles $a_1, \ldots, a_k$, $k = \min(r, L)$ up to numerical equivalence (codim $a_i = i$). Then the set of semistable torsion-free sheaves E with $c_i(E)$ numerically equivalent to a_i, modulo the equivalence relation $E \sim F$ if $\operatorname{gr} E \cong \operatorname{gr} F$, is in a natural way the set of points of a scheme $\mathcal{U}_X(r; a_1, \ldots, a_k)$ locally of finite type. If $n = 2$, $r = 2$ or char $(k) = 0$, $\mathcal{U}_X$ is projective.

Here one conjectures that $\mathcal{U}_X$ is always projective. The only missing step is to prove that $\mathcal{U}_X$ is of finite type in all cases. Whenever it is of finite type, properness follows by the methods of LANGTON [183] as generalized by MARUYAMA [194]. $\mathcal{U}_X$ of finite type is proven in char. 0 by FORSTER-HIRSHOWITZ-SCHNEIDER [106], and for dim 2, any char. by MARUYAMA [192] and for rk 2, any char. by MARUYAMA [195]. There are two quite different methods that have been used to reduce these existence theorems to invariant theory. The first consists in choosing a large number N of points $P_i \in X$ and associating to every vector bundle E of rank r on X the collection of maps

$$\begin{array}{ccc} \Gamma(X, E(n)) & \to E(n)\,(P_i) \to 0 & \qquad (n \gg 0) \\ s & \mapsto s(P_i) & \end{array}$$

or equivalently, N subspaces of codimension r:

$$W_i = \{s \in \Gamma(X, E(n)) \mid s(P_i) = 0\}.$$

In the second, we assume a line bundle L on X is given and consider vector bundles E of rank r plus isomorphisms $\Lambda^r E \xrightarrow{\sim} L$. To such data we can associate the canonical map

$$(*) \qquad \Lambda^r \Gamma(X, E(n)) \to \Gamma(X, L(nr)).$$

In both cases, let $G = SL(\Gamma(X, E(n)))$. Then one proves that if E is stable and the n and the P_i are chosen suitably, the collection of subspaces $\mathbf{P}(W_i) \subset \mathbf{P}(\Gamma(X, E(n)))$ is G-stable (as in Ch. 4, § 4); or that if a basis of $\Gamma(X, L(nr))$ defines $\Gamma(X, L(nr) \xrightarrow{\sim} k^M$, hence $(*)$ defines M elements $\omega_i \in \Lambda^r \Gamma(X, E(n))^*$, then $(\omega_1, \ldots, \omega_M)$ is G-stable, i.e., has a closed G-orbit and finite G-stabilizer (compare Appendix 4B).

There are two cases in which the varieties $\mathcal{U}_X$ have been intensively studied: when X is a curve and when $X = \mathbf{P}^n$. To outline the results in these cases, if L is a fixed line bundle of appropriate degree, let $S\mathcal{U}_X$ be the subscheme of $\mathcal{U}_X$ corresponding to vector bundles E with $\Lambda^r E \xrightarrow{\sim} L$.

Let X be a smooth curve of genus g and consider the space $S\mathcal{U}_X(r, d)$ (recall $r = \operatorname{rank} E$, $d = c_1E$). Then if $(r, d) = 1$, the following are known:

a) $S\mathcal{U}_X$ is a smooth unirational variety, hence $h^{0,p} = 0$, all p, and if $d \equiv \pm 1 \pmod r$ then $S\mathcal{U}_X$ is even rational (NEWSTEAD [246]).

b) $Pic(S\mathcal{U}_X) \cong \mathbf{Z}$ and the canonical divisor K corresponds to -2 (ample generator) (RAMANAN [276], TJURIN [324]).

c) $H^3(S\mathcal{U}_X, \mathbf{Z}) \cong \mathbf{Z}^{2g}$, the associated intermediate jacobian of $S\mathcal{U}_X$ is Jac (X). In fact, $S\mathcal{U}_{X_1} \cong S\mathcal{U}_{X_2}$ implies $X_1 \cong X_2$ and all small deformations of $S\mathcal{U}_{X_1}$ are varieties of the form $S\mathcal{U}_{X_2}$ (NARASIMHAN-RAMANAN [233], TJURIN [322], MUMFORD-NEWSTEAD [223]

d) The Betti numbers of $S\mathcal{U}_X$ are known and if X is defined over a finite field, the zeta function $\zeta_{S\mathcal{U}_X}$ can be calculated from ζ_{X_1} (NEWSTEAD [244], HARDER [130], HARDER-NARASIMHAN [132], DESALE-RAMANAN [84]).

e) $\chi(S\mathcal{U}_X, \Omega^p)$ has been calculated (NARASIMHAN-RAMANAN [234]) and a method which, in principle, calculates the Chern numbers is known (GIESEKER [566]).

f) if $r = 2$, then H^2, H^3 and H^4 generate $H^*(S\mathcal{U}_X, \mathbf{Q})$ as a *ring* (NEWSTEAD [244]).

g) if $r = 2$ and X is hyperelliptic, $S\mathcal{U}_X$ can be interpreted as the space of $\mathbf{P}^{g-2}$'s in the intersection of 2 quadrics

$$\sum_{i=0}^{2g+1} x_i^2 = \sum_{i=0}^{2g+1} \lambda_i x_i^2 = 0$$

in $\mathbf{P}^{2g+1}(\{\lambda_i\} = \text{branch points})$ (DESALE-RAMANAN [83]).
This is analogous to M. REID'S description of Jac (X) as the space of $\mathbf{P}^{g-1}$,s in the same intersection (unpublished). Further generalizations of this are developed in RAMANAN [277].
When $(r, d) \neq 1$, semi-stable but not stable bundles exist and the spaces $S\mathcal{U}_X$ are more complicated.

a) Unless $g = r = 2$, d even, $S\mathcal{U}_X$ is always singular on the points corresponding to bundles which are not stable. (NARASIMHAN-RAMANAN [232]).

b) The varieties $S\mathcal{U}_X$ are only known to be unirational.

c) If $r = 2$, d even, then an explicit desingularization of $S\mathcal{U}_X$ is known (NARASIMHAN-RAMANAN [235], SESHADRI [307]).

We have only mentioned some of the main results of this theory: the reader is referred to other papers of the above named authors in the bibliography for more information. These spaces are also used in the amazing solution of Langlands' conjecture for $GL(2)$ over function fields by V. DRINFELD [92], (see [131]), and in ATIYAH-BOTT'S study [50] of the

YANG-MILLS equation in dimension 2. At the present time, the analytic theory of these spaces lags behind as no good theory of theta functions attached to higher dimensional representations of $\pi_1(X)$ has been found.

The other case where $\mathcal{U}_X$ has attracted interest is when $X \cong \boldsymbol{P}^n$. An extraordinary idea goes back to SCHWARZENBERGER who asked, around 1960, whether if E is a vector bundle of rank r on $\boldsymbol{P}^n$ and $n \gg r$, then must E be a direct sum of line bundles? In this direction TJURIN [325] proved that if you linked $\boldsymbol{P}^1 \subset \boldsymbol{P}^2 \subset \boldsymbol{P}^3 \subset \cdots$ as successive hyperplane sections and if for all n, E_n is a vector bundle on $\boldsymbol{P}^n$ such that

$$E_n|_{\boldsymbol{P}^{n-1}} \cong E_{n-1},$$

then each E_n is a direct sum of line bundles. Another reason for this conjecture is that there are no known rank 2 vector bundles on $\boldsymbol{P}^5$ except for direct sums $E = L_1 \oplus L_2$, and only one other (up to twists) on $\boldsymbol{P}^4$ (G. HORROCKS and D. MUMFORD [150]). Bundles on $\boldsymbol{P}^2$ were studied by SCHWARZENBERGER [294], see also MARUYAMA [194], § 7. Stable rank 2 bundles with c_1 even on $\boldsymbol{P}^2$ were classified by W. BARTH [55]: he gave a quite beautiful explicit description of the space

$$\mathcal{U}_{\boldsymbol{P}^2}(2;\, c_1,\, c_2)$$

when c_1 is even. In particular, he proves that $\mathcal{U}_{\boldsymbol{P}^2}(2;\, 2k,\, c_2)$ is irreducible and rational. His theory and this result have been extended to the case c_1 odd by K. HULEK [153], who also proved that $\mathcal{U}_{\boldsymbol{P}^2}(r;\, 0,\, c_2)$ is irreducible [154]. Many examples of rank 2 bundles on $\boldsymbol{P}^3$ were discovered by G. HORROCKS [148]. A particular class of stable rank 2 bundles with $c_1 = 0$ on $\boldsymbol{P}^3$, called instanton bundles, turn out very remarkably to be in $1-1$ correspondence with so-called "self-dual" solutions of the Yang-Mills equation: a non-linear *PDE* on $\boldsymbol{R}^4$: see ATIYAH [49] and a fairly explicit description of these bundles has been given (although it is not explicit enough to enable one to check whether for given c_2, the space of these bundles is connected[15]!) A general procedure for analyzing vector bundles von $\boldsymbol{P}^n$ was given by G. HORROCKS [149]: this has motivated much of the above work. Two recent surveys of the theory of vector bundles on $\boldsymbol{P}^n$, and their applications are HARTSHORNE [133] and OKONEK-SCHNEIDER-SPINDLER [257]. The question of the existence of non-trivial rank 2 vector bundles on $\boldsymbol{P}^n$, $n \geqq 5$, is the most interesting unsolved problem in projective geometry that I know of.

To complete this survey, I should also mention the recent result of F. BOGOMOLOV ([60], see REID [280]): if X is a smooth algebraic surface and E is a rank 2 vector bundle on X such that $c_1(E)^2 > 4c_2(E)$, then E is unstable for any polarization on X.

---

[15] See [880] and also [368], [420], [421], [497], [508], [578], [670], [817] for more recent related advances.

D. Stable curves and the compactification of $\mathcal{M}_g$ [16]

The moduli space $\mathcal{M}_g$ classifying curves of genus g, constructed in Theorem 5.11, is not compact, i. e., is a quasi-projective but not projective variety. The reason for this is quite plainly that a family of curves of genus g may become singular for special values of the parameters and these singular curves ought to correspond to certain "points at infinity" on a compactification of $\mathcal{M}_g$. The concept of stability in invariant theory, however, makes it clear that one cannot just take *every* singular curve of arithmetic genus g, and associate to its isomorphism class a boundary point. This would be like taking all the G-orbits in a vector space V on which G acts and trying to make them into a variety. The resulting object is highly non-separated. It is necessary to pick very carefully and add in just the right amount of such boundary points! I discussed this problem in various expository talks intended to explain the "philosophy" behind the ideas of stable limit points [208] 1964, [213] 1970, [218] 1976. The appropriate choice of boundary points for $\mathcal{M}_g$ was discovered in an unpublished joint work with A. MAYER in 1963. We defined a *stable curve* X to be a connected, reduced curve, possibly with more than one component, but with only ordinary double points as singularities and with one more condition to exclude "unnecessary" rational components: if a component E of X is smooth and isomorphic to $\boldsymbol{P}^1$, then E meets the other components of X in at least 3 points. The main result is that the set of isomorphism classes of stable curves is a projective variety $\bar{\mathcal{M}}_g$. This was proven by F. KNUDSEN and myself. The proof has been greatly simplified by incorporating ideas of C. S. SESHADRI and D. GIESEKER and appears in [218] and [176]. This holds over any field, and the varieties $\bar{\mathcal{M}}_g$ in each characteristic are part of one scheme $\bar{\mathcal{M}}_g$ projective over $\boldsymbol{Z}$.

The first step in verifying this theorem is to check that the functor of stable curves satisfies the valuative criterion for separation and properness. More precisely, let R be a discrete valuation ring with fraction field K, residue field k. For separation one wants to know:

i) if C_1, C_2 are 2 proper flat schemes over Spec R whose fibres are stable curves, and if $\Phi: (C_1)_\eta \to (C_2)_\eta$ is an isomorphism of their generic fibres, then Φ extends to an isomorphism $\Phi: C_1 \to C_2$.

This is an immediate consequence of the theory of minimal models for surfaces: see lemma 1.12 [82]. For properness, one wants to know:

---

[16] More recent references related to this section include [210], [337], [350], [452], [453], [456], [457], [461–463], [477], [478], [520], [526], [527], [529], [530], [558], [559], [563], [571–573], [575], [602–607], [652], [654], [701], [707], [713], [747], [749], [792], [818], [832], [839–844], [847], [849], [851], [876], [892], [893], [920], [921]. In particular there has been a breakthrough in working out facts about the homology of the moduli spaces $\mathcal{M}_g$ and $\bar{\mathcal{M}}_g$; see [455], [523], [534], [595–601], [679], [684], [735], [737], [756], [919], and also [736], [795], [796].

ii) if C_η is a stable curve over Spec K, then there exists a finite algebraic extension field $K_1 \subset K$, R_1 = integral closure of R in K_1 and a proper flat scheme C_1 over Spec R_1 with stable curves as fibres and $(C_1)_\eta = C \times_K K_1$.

This was proven by A. MEYER and the author when char $k = 0$, by P. DELIGNE and the author in general [82] using an analogous theory of stable reduction of abelian varieties, and again by M. ARTIN and G. WINTERS [47] by direct arguments using the theory of curves on surfaces. Result (ii) had an immediate application, possible even without knowing that a scheme $\bar{\mathcal{M}}_g$ projective over $\boldsymbol{Z}$ exists, to proving the *irreducibility* of $\mathcal{M}_g$ in char. p (see [82]). The method is easily explained using the later result that $\bar{\mathcal{M}}_g$ exists: by local deformation theory, one checks that the fibres of $\bar{\mathcal{M}}_g$ over $\boldsymbol{Z}$ are geometrically unibranch. By the classical analysis of curves over $\boldsymbol{C}$ as branched covers of $\boldsymbol{P}^1$ with prescribed glueing, one knows that the geometric fibre of $\mathcal{M}_g$ over Spec $\boldsymbol{C}$ is irreducible. Therefore by Zariski's connectedness theorem, *all* the geometric fibres are connected, hence irreducible. In [82], this argument is modified in 2 ways to work without knowing that $\bar{\mathcal{M}}_g$ exists.

There are at present 2 proofs that $\bar{\mathcal{M}}_g$ exists. The first relies on the Torelli map $t\colon \bar{\mathcal{M}}_g \to \bar{\mathcal{A}}_g$ as in Ch. 7. Here we take for $\bar{\mathcal{A}}_g$ the Satake compactification of the moduli space $\mathcal{A}_g$ of abelian varieties (constructed in char. p in [209], see Appendix 7B). The idea is to construct 2 line bundles μ and δ on $\bar{\mathcal{M}}_g$, where $\mu = t^*(\mu')$, μ' ample on $\bar{\mathcal{A}}_g$ and

$$\delta = \underline{o}_{\bar{\mathcal{M}}_g}(\Delta)$$

where Δ = the divisor $\bar{\mathcal{M}}_g - \mathcal{M}_g$. The key point is to show that δ^{-1} is relatively ample for t. The proof uses a detailed analysis of the fibres of t and was carried out by F. KNUDSEN in his thesis. To make the correct induction, he uses a generalization of the theorem to ampleness of an appropriate line bundle on the important scheme $\bar{\mathcal{M}}_{g,n}$ of "n-pointed stable curves of genus g". The proof has been published after some delay in [176]. The second proof is based on the verification that stable curves, embedded by n-canonical maps, $n \geqq 5$, have stable Chow points (see MUMFORD [218]), or stable Hilbert points (see GIESEKER [119]).

Basic to all these papers is a preliminary study of the line bundles on $\bar{\mathcal{M}}_g$ (or — if one doesn't know this scheme exists — on the moduli functor of stable curves: see [207]). Thus one can canonically associate to a stable curve X the one-dimensional vector spaces:

$$\lambda_1 = \Lambda^g \Gamma(X, \omega_X)$$

$$\lambda_n = \Lambda^{(2n-1)(g-1)} \Gamma(X, \omega_X^{\otimes n}), \ n \geqq 2$$

where ω_X = the dualizing sheaf on X (Ω_X^1 if X is smooth).

* N is the l.c.m. of the orders of the automorphisms of the spaces λ_n induced by automorphisms of X's.

As X varies, these λ_k "patch together" to form a line bundle on the moduli functor. More concretely, an appropriate power* $\lambda_n^{\otimes N}$ patches together to form a line bundle λ_n^* on $\bar{\mathscr{M}}_g$. In fact, λ_1^* is the μ mentioned above (see ARAKELOV [43]). This is a special case of the following general situation:

$f: X \to S$ proper flat
L a relatively ample invertible sheaf on X
$\lambda_n = \Lambda^{r_n} f_*(L^{\otimes n})$, where $r_n = rk(f_* L^{\otimes n})$ and n is large enough so that $R^i f_* L^{\otimes n} = (0)$, $i > 0$.

The essential result is that these λ_n's are like values of a Hilbert polynomial with values in $Pic(S)$, i.e., there are invertible sheaves μ_0, $\mu_1, \ldots, \mu_{r+1}$ ($r = \dim X/S$) on S such that in $Pic(S)$

$$(*) \qquad \lambda_n \cong \bigotimes_{i=0}^{r+1} \mu_i^{\otimes \binom{n}{i}}.$$

This has been analyzed in FOGARTY [100] and KNUDSEN [176] in connection with the map from the Hilbert scheme to the Chow variety. Both of these papers develop further the theory of § 5.3 and § 5.4, generalizing the construction of Div. In [100], these ideas are applied to factor the morphism *Hilb* $\to$ *Chow* through schemes which classify mixed objects (Z, ζ), $Z \subset \mathbf{P}^n$ a subscheme all of whose components have dimension $> k$, and ζ a cycle of dimension k.

In the case of stable curves, one has the beautiful result

$$\lambda_n \cong (\lambda_1^{12} \otimes \delta^{-1})^{\binom{n}{2}} \otimes \lambda_1$$

as a consequence of GROTHENDIECK'S RIEMANN-ROCH theorem ([218], theorem 5.10). The final result is that $\lambda^a \otimes \delta^{-b}$ is ample on $\bar{\mathscr{M}}_g$ if $a > 0$, $b > (11.2)$ a and not ample if $a \leqq 0$ or $b \leqq 11a$. ([218], Corollary 5.18).

Now that the existence of the projective variety $\bar{\mathscr{M}}_g$ is so well understood the time seems ripe to try to compute some of its birational invariants. See the references in the footnote at the beginning of this section.

E. Compactification of other moduli spaces [17]

Encouraged by success with $\mathscr{M}_g$, it is natural to ask whether there is an appropriate generalization of the concept of stable but singular curves to surfaces and other higher dimensional varieties and/or a corresponding compactification of some of the moduli spaces V_p. The results in this direction are still fragmentary, but they do indicate a considerably more complicated picture. We will discuss this only over $\mathbf{C}$ or other char. 0 field k.

---

[17] More recent references related to this section include [439], [446], [447], [549], [671], [672], [697] and the many references therein, [829], [869], [924].

There are 2 methods of attacking this problem: in the cases where V_p is described via the period map as a locally symmetric variety D/Γ (D a bounded symmetric domain, $\Gamma \subset \mathrm{Aut}(D)$ a discrete group: see appendix 5B) one can use the analytic theory of such domains to construct a compatification of $\overline{D/\Gamma}$, and then go on to interpret this compactification in terms of the moduli problem $\mathscr{V}_p$. The other approach is to study algebro-geometrically all possible degenerations of polarized varieties (X, U) in $\mathscr{V}_p$ and single out the "best possible" ones which lead to a functor of families of this type satisfying the valuative criteria for separation and properness (compare the previous appendix in the case of curves).

The first method has been applied chiefly in the case of abelian varieties and $K3$-surfaces, but there is a general theory of compactifications of all varieties D/Γ. Every variety D/Γ has a canonical "minimal" compactification, called its Baily-Borel compactification after their fundamental paper [52]. It generalizes Satake's compatification [287] in the case $\mathscr{A}_g = \mathfrak{H}_g/\mathrm{Sp}\,(2g, \mathbf{Z})$ ($\mathfrak{H}_g$ = Siegel's upper half space), which, set-theoretically, puts together all the $\mathscr{A}_h$'s, $0 \leqq h \leqq g$:

$$\bar{\mathscr{A}}_g = \mathscr{A}_g \amalg \mathscr{A}_{g-1} \amalg \cdots \amalg \mathscr{A}_0.$$

It is economical, but at the expense of often introducing rather bad singularites. In another direction, the author, with the collaboration of Ash, Rapaport and Tai (see [217], [48], [229]), has constructed a whole class of compactifications of D/Γ with mild singularities, including, when Γ is "neat", non-singular compactifications: we call these the toroidal compactifications of D/Γ. This construction generalized earlier work of Igusa [158] and Hirzebruch [141] in special low-dimensional cases. In the case of $\mathscr{A}_g$, these toroidal compactifications can be related back either to degenerations of abelian varieties or to degenerations of the Hodge structure:

$$F^1(H^1(X, \mathbf{C})) \subset H^1(X, \mathbf{C}) \supset H^1(X, \mathbf{Z})$$

known as "mixed Hodge structures". The degenerate limits of the abelian varieties have been studied by the author [214], Namikawa [227], [228] Oda-Seshadri [254] and Nakamura [226]. The Hodge-theoretic interpretation of the boundary $\bar{\mathscr{A}}_g - \mathscr{A}_g$ of toroidal compactifications is due to Carlson-Cattani-Kaplan [70]. In all of this, because the abelian variety can be reconstructed so simply from its periods over $\mathbf{C}$, one finds that on the boundary, too, one can really "write down" their singular limits. A major problem, however, is to extend this theory to char. p. Satake's compactification of $\mathscr{A}_g$ can be carried out in char. p (or over $\mathbf{Z}$) using the theory of algebraic theta functions (appendix 7B). But the toroidal compactifications have not yet been constructed in char. p. If they can be constructed, one expects to prove as a Corollary the irreducibility of $\mathscr{A}_g$ in char. p as well as the finite generation of the ring of Siegel modular forms with integral Fourier coefficients (compare appendix 5D and IGUSA [162]). The compactification of the $K3$-moduli spaces $\mathscr{K}_d$ has been studied by a combination of analysis and the algebro-geometric method, to which we turn next.

In any dimension, F. KNUDSEN and the author [174] proved that if R is a discrete valuation ring with char. 0 residue field k and X is a smooth projective variety over the fraction field K of R, then there is a finite algebraic extension $K_1 \supset K$ and a regular scheme X_1, projective over Spec R_1 (R_1 = integral closure of R in K_1), such that

$$(X_1)_\eta \cong X \times_K K_1$$

$(X_1)_s$ is reduced with smooth components crossing transversely.

A similar result holds when R is replaced by the unit disc Δ with coordinate z, K by the punctured disc Δ^* and X by a smooth projective family of varieties over Δ^*. Then R_1 becomes the new disc Δ_1 with coordinate $z_1 = z^{1/n}$ and X_1 is a complex manifold, projective over Δ_1 on which the divisor $z_1 = 0$ has normal crossings, multiplicity one. In this situation, the classical cohomology of one of the smooth fibres of X over Δ^* and the monodromy map associated to the family X can be related to the cohomology of the limit $(X_1)_s$. Moreover, mixed Hodge Structures can be put on these groups. This situation has been much studied and gives a very important description of the cohomological degeneration encountered when a variety becomes singular. Among many papers related to this, we mention [73], [290], [79], [263], [69], [124], [317], [342].

The problem with this class of singular limits is that there is no analog of the prohibition in the dimension one case of rational components with only one or two double points on them. Thus they have a lot of garbage caused by excessive blowings up. Only in one set of cases have more minimal limits been described. These are the cases where the fibres of X over Δ^* have dim = 2, Kodaira dim = 0, i.e., they are either abelian or $K3$-surfaces or a quotient of one of these by a finite group acting freely, a hyperelliptic or Enriques' surface respectively. This work was started by PERSSON [263], but the main breakthrough is due to KULIKOV [179]. His work has been clarified and extended by PERSSON-PINKHAM [264], and DAVID MORRISON [204]. In all these cases, if you start with a family X/Δ^* of surfaces of one of the above types, it is shown that after an nth root, there exists an X_1/Δ whose special fibre $(X_1)_s$ belongs to a relatively small list: *for example*, it can be a finite union of rational surfaces glued* according to a triangulation Γ of

* One component for each vertex of Γ, one rational curve for each edge of Γ, one triple point for each 2-simplex of Γ.

S^2 (if X_η is a $K3$ surface)
$S^1 \times S^1$ (if X_η is an abelian surface)
$\boldsymbol{R}P^2$ (if X_η is an Enriques surface)
Kleinbottle (if X_η is hyperelliptic)

For details, see MORRISON, op. cit. The existence of these limits was applied by KULIKOV to prove that the period map for $K3$-surfaces is surjective.

However from other points of view these limits still seem to be excessively blown-up: roughly, the objects we have are like those in the dim. one case after prohibiting rational components with only one double point, but while still allowing them if they contain two double points. The beginnings of a theory of even more contracted limiting surfaces has been made by the author [218] and J. SHAH [309], [310]. This work was motivated by the idea of letting the stability of the Chow form of the whole surface dictate what singularities can be included. This leads to a natural definition of "stable" singular points. For surfaces, SHAH has essentially classified these and has illustrated the theory by examples of compactifications of moduli spaces of $K3$-surfaces of low degree for which many of these stable singular points occur on the singular $K3$-surfaces associated to "boundary" points. However, for this approach to be successful in general, the following *conjecture* is absolutely essential:

Existence of asymptotically stable limits: Let R, K, k be as above. Let (X_η, L_η) be a smooth projective surface over K and an ample line bundle. Then there exists R_1, K_1 and $X_1 \to \operatorname{Spec} R_1$ as above such that

a) $(X_1)_\eta \cong X_\eta \times_K K_1$
b) L_η extends to a relatively ample L_1 on X_1
c) There is an n_0 such that if $n \geqq n_0$ and $(X_1)_s$ is embedded by $\Gamma\big((L_1)_s^{\otimes n}\big)$, then its Chow form is semi-stable.

If one looks at the case where the generic fibre X_η is a surface of general type, it its natural to seek a best possible degeneration using the canonical ring. But then one comes up against another very hard *conjecture*[18] (see VIEHWEG [327]):

Finite generation of the relative canonical ring: Let $X \to \operatorname{Spec} R$ be a flat projective morphism, R a discrete valuation ring, X_η a surface of general type, X regular of dimension 3, and X_s reduced with normal crossings. Then is

$$\bigoplus_{n=0}^{\infty} \Gamma(\omega_{X/R}^{\otimes n})$$

a finitely generated R-algebra?

Returning, however, to $K3$ surfaces, it is important to tie together the analytic construction of compactifications of $\mathscr{K}_d$, realized as a locally

[18] Proved by KAWAMATA [653].

symmetric variety D/Γ_d, with these algebro-geometric descriptions of "good degenerations" of $K3$ surfaces. This can be understood in some respects via the limit of the Hodge structures and the mixed Hodge structure of the singular limit ([70], [111]). But a new twist appears: the families of $K3$ surfaces plus asymptotically stable singular limits appear to produce compactifications of D/Γ *intermediate* between the Baily-Borel one and the toroidal ones, i.e. dominating Baily-Borel and dominated by a suitable toroidal one. Some of these have been described by E. LOOIJENGA [714].

Appendix to Chapter 7

A. The tower of moduli spaces of abelian varieties [19]

The moduli spaces $\mathscr{A}_{g,d,n}$ described in the text are, for fixed g, all part of one "tower" of varieties and it is helpful to study this tower as a whole rather than describe the $\mathscr{A}_{g,d,n}$ one at a time. The links between the moduli spaces arise from the easy lemma (see [217], p. 234):

Lemma: For all polarized abelian varieties (X, λ), there is an isogeny $\pi: X \to Y$ and a principal polarization $\mu: Y \xrightarrow{\sim} \hat{Y}$ such that $\lambda = \hat{\pi} \circ \mu \circ \pi$:

$$\begin{array}{ccc} X & \xrightarrow{\lambda} & \hat{X} \\ {\scriptstyle\pi}\downarrow & & \uparrow{\scriptstyle\hat{\pi}} \\ Y & \xrightarrow[\mu]{\sim} & \hat{Y} \end{array}$$

Thus (X, λ) can be reconstructed from (Y, μ, H), where $H \subset Y$ is the finite subgroup $\mu^{-1}(\operatorname{Ker} \hat{\pi})$. Note that the degree of λ is the order of H. Therefore if $\mathscr{A}_{g,1}^{(d)}$ is the moduli space of principally polarized abelian varieties (Y, μ) plus subgroups $H \subset Y$ of order d, there is a surjective map:

$$\mathscr{A}_{g,1}^{(d)} \twoheadrightarrow \mathscr{A}_{g,d} \; (= \mathscr{A}_{g,d,1})$$

This, in many ways, reduces the study of the $\mathscr{A}_{g,d}$'s to that of the $\mathscr{A}_{g,1,n}$'s.

Within the variety $\mathscr{A}_{g,d}$, there is an open set which is much easier to handle. For every sequence $\delta_1, \ldots, \delta_g$ such that

$$\delta_1 | \delta_2 | \cdots | \delta_g$$

$$\prod_{i=1}^{g} \delta_i = d$$

[19] More recent references related to this section include [791], [792]. Others related to Chapters 6 and 7 in general include [345], [491], [633], [755], [783], [875].

let

$$\mathscr{A}_{g,\delta} = \left\{ \begin{array}{l} \text{open subscheme of } \mathscr{A}_{g,d} \text{ of pairs } (X, \lambda) \text{ such that} \\ \qquad \ker(\lambda) \cong \prod_1^g \mathbf{Z}/\delta_i\mathbf{Z} \times \prod_1^g \mu_{\delta_i} \end{array} \right\}$$

(Here μ_n = the group scheme of nth roots of 1). The $\mathscr{A}_{g,\delta}$'s are disjoint and exhaust all of $\mathscr{A}_{g,d}$ except for (X, λ)'s over a field k such that char $(k) \mid d$ and ker(λ) contains a subgroup isomorphic to α_p. The local structure of $\mathscr{A}_{g,\delta}$ is quite simple [75], [259]), and in particular all its components dominate Spec $\mathbf{Z}$. Moreover, over $\mathbf{C}$, it has the simple analytic description:

$$\mathscr{A}_{g,\delta} \times \operatorname{Spec} \mathbf{C} \cong \mathfrak{H}_g/\Gamma_\delta$$

$$\mathfrak{H}_g = \text{Siegel upper } \frac{1}{2} \text{ space}$$

$$= \{\Omega \mid \Omega = g \times g \text{ symm. cx. matrix, } \operatorname{Im} \Omega > 0\}$$

$$\Gamma_\delta = \{A \in GL(2g, \mathbf{Z}) \mid {}^tA J_\delta A = J_\delta\}$$

$$J_\delta = \left(\begin{array}{c|c} 0 & \begin{matrix} \delta_1 & & \\ & \ddots & \\ & & \delta_g \end{matrix} \\ \hline \begin{matrix} -\delta_1 & & \\ & \ddots & \\ & & -\delta_g \end{matrix} & 0 \end{array} \right)$$

hence is irreducible. And $\bigcup_\delta \mathscr{A}_{g,\delta}$ is *dense* in $\mathscr{A}_{g,d}$ (see [252], [222]). Therefore the closures $\bar{\mathscr{A}}_{g,\delta}$ are exactly the irreducible components of the scheme $\mathscr{A}_{g,d}$ over $\mathbf{Z}$. I would like to mention in passing that the local structure of $\mathscr{A}_{g,d}$ at points not in any $\mathscr{A}_{g,\delta}$ can be really very complicated. An effective machine for working out this local structure has been found by NORMAN, relying on CARTIER's "Courbes typiques" ([249], [250]). It is not yet known, however, whether $\mathscr{A}_{g,\delta} \times \operatorname{Spec}(k)$ is irreducible, where k is an algebraically closed field of characteristic p[20].

We can modify the concept of level n structure to accommodate ordinary abelian varieties in char. p where $p \mid n$: let $\mathscr{A}^*_{g,1,n}$ be the moduli space:

$$\mathscr{A}^*_{g,1,n} = \left\{ \begin{array}{l} \text{moduli space of triples } (X, \lambda, \alpha),\ X \text{ a } g\text{-dim. abelian} \\ \text{variety, } \lambda \text{ a principal polarization and } \alpha: X_n \xrightarrow{\sim} (\mathbf{Z}/n\mathbf{Z})^g \\ \times \mu_n^g \text{ a symplectic isomorphism} \end{array} \right\}$$

over the ground ring $\mathbf{Z}[\zeta_n]$, ζ_n a primitive n^{th} root of unity. As in the text, for $n > 3$, $\mathscr{A}^*_{g,1,n}$ is a fine moduli space. In fact, it is smooth over $\mathbf{Z}[\zeta_n]$

[20] Proved by CHAI [447], [448], FALTINGS [535], CHAI-FALTINGS [449].

and irreducible. The irreducibility follows from the analytic description:

$$\mathscr{A}^*_{g,1,n} \times \operatorname{Spec} \boldsymbol{C} \cong \mathfrak{H}_g/\Gamma_n$$

$$\Gamma_n = \{A \in \operatorname{Sp}(2g, \boldsymbol{Z}) \mid A \equiv I_{2g} (\operatorname{mod} n)\}.$$

The schemes $\mathscr{A}^*_{g,1,n}$ form a tower with respect to quasi-finite morphisms

$$\mathscr{A}^*_{g,1,mn} \to \mathscr{A}^*_{g,1,n}$$

given by $(X, \lambda, \alpha) \mapsto (X, \lambda, \operatorname{res}_{X_n} \alpha)$. In this situation, it is natural to enlarge the schemes $\mathscr{A}^*_{g,1,n}$:

$$\mathscr{A}_{g,1,n} = \text{normalization of } \mathscr{A}_{g,1,1} \text{ in the field } \boldsymbol{Q}(\zeta_n)\,(\mathscr{A}^*_{g,1,n}).$$

Note that $\mathscr{A}^*_{g,1,n}$ is an open dense subscheme of $\mathscr{A}_{g,1,n}$ and $\mathscr{A}^*_{g,1,n} = \mathscr{A}_{g,1,n}$ except over primes dividing n. The schemes $\mathscr{A}_{g,1,n}$ are normal and irreducible, they form a tower with respect to *finite* morphisms $\mathscr{A}_{nm} \to \mathscr{A}_n$, and are smooth over $\boldsymbol{Z}[\zeta_n]$ except at non-ordinary abelian varieties at primes dividing n.

The link between the $\mathscr{A}_{g,1,n}$'s and the $\mathscr{A}_{g,\delta}$'s is given by the morphism:

$$\mathscr{A}^*_{g,1,\delta_g} \to \mathscr{A}_{g,\delta}$$

$$(X, \lambda, \alpha) \mapsto (Y, \mu)$$

where Y is the étale cover of X defined by the diagram

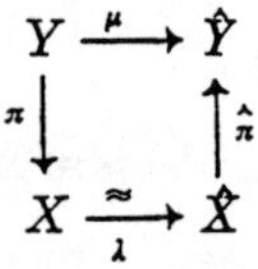

$$\begin{array}{ccc} Y & \xrightarrow{\mu} & \hat{Y} \\ {\scriptstyle \pi}\downarrow & & \uparrow{\scriptstyle \hat{\pi}} \\ X & \xrightarrow[\lambda]{\approx} & \hat{X} \end{array}$$

where

$$\operatorname{Ker}(\hat{\pi}) = \lambda\left(\alpha^{-1}\left((0) \times \prod_i \mu_{\delta_i}\right)\right)$$

$$\begin{array}{cccc} \downarrow & & & \\ \mathscr{A}_{g,1,\eta} & \supset \mathscr{A}^*_{g,1,n} & & \\ \downarrow & \searrow & \mathscr{A}_{g,1,\delta} & \\ \vdots & & & \subset \mathscr{A}_{g,n} \\ \vdots & \searrow & & \cup \\ \downarrow & & \mathscr{A}_{g,1,\delta'} & \\ \mathscr{A}_{g,1,1} & & & \end{array}$$

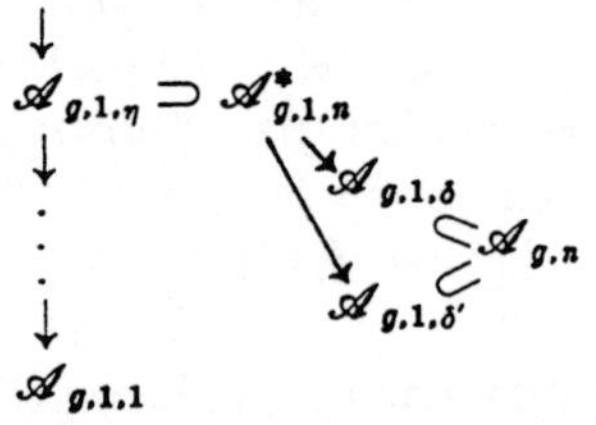

Summary of Moduli Spaces

The importance of the tower $\mathscr{A}_{g,1,n}$ is that it has even more structure, namely the extra morphisms given by the Hecke correspondences. In addition to the obvious automorphisms of $\mathscr{A}_{g,1,n}$ given by the fact that $\boldsymbol{Q}(\zeta_n)\,(\mathscr{A}_{g,1,n})$ is Galois over $\boldsymbol{Q}(\mathscr{A}_{g,1,1})$ with group $\operatorname{CSp}(2g, \boldsymbol{Z}/n\boldsymbol{Z})$

(CSp = symplectic similitudes), we have the "extra" morphisms

$$\hat{\mathscr{A}}^*_{g,1,n} \to \mathscr{A}_{g,1,1}$$
$$(X, \lambda, \alpha) \mapsto (Y, \mu)$$
$$Y = X/\alpha^{-1}\big((Z/nZ)^g \times (0)\big) \text{ and}$$

$$\begin{array}{ccc} X & \xrightarrow{n\lambda} & \hat{X} \\ {\scriptstyle \pi}\downarrow & & \uparrow{\scriptstyle \hat{\pi}} \\ Y & \xrightarrow[\mu]{} & \hat{Y} \end{array}$$

These morphisms make better sense when you pass to the inverse limit:

$$\hat{\mathscr{A}}_{g,1} = \varprojlim_n \mathscr{A}_{g,1,n}.$$

This certainly exists as a scheme, albeit a non-noetherian one, and CSp $(2g, \boldsymbol{A}_f)$ acts on it, where $\boldsymbol{A}_f$ is the ring of finite adéles (the restricted product of the $\boldsymbol{Q}_p$'s). This scheme and this group action are the most concise way of describing all the moduli spaces for abelian varieties of dimension g (compare [80]).

B. Theta functions and moduli of abelian varieties [21]

The classical theory of theta functions provides a direct and explicit approach to the moduli space of abelian varieties over $\boldsymbol{C}$. This approach, in fact, is valid over any field k or even over $\boldsymbol{Z}$. We will first explain the essential idea in the simplest case: start with an elliptic curve E with origin P over an algebraically closed field k. Fix a 4th root of 1, $i \in k$. Embed E in $\boldsymbol{P}^3$ by the linear system $\Gamma\big(E, \underline{o}_E(4P)\big)$. Then it is not hard to check that translation $T_x: E \to E$ by a point $x \in E$ extends to a projective transformation T'_x:

$$\begin{array}{ccc} E & \xrightarrow{T_x} & E \\ {\scriptstyle i}\downarrow & & \downarrow{\scriptstyle i} \\ \boldsymbol{P}_3 & \xrightarrow[T'_x]{} & \boldsymbol{P}^3 \end{array}$$

if and only if $4x = P$. The same holds for "reflections" $R_x: E \to E$ given by $R_x(y) = x - y$. Let $H \subset PGL(4, k)$ be the group of order 16 of translations T'_x. Purely projectively, we can characterize H and R_0:

$$H = \left\{ \Phi \in PGL(4, k) \;\middle|\; \begin{array}{l} \Phi(E) = E \text{ and either } \Phi = \text{id. or} \\ \Phi \text{ has no fixed points on } E \end{array} \right\}$$

$$R_0 = \left\{ \begin{array}{r} \text{the unique } \Phi \in PGL(4, k) \text{ s.t. } \Phi(E) = E, \\ \Phi(P) = P, \Phi^2 = \text{id.}, \Phi \neq \text{id.} \end{array} \right\}$$

Let $\mathscr{G}$ be the inverse image of H in $GL(4, k)$.

[21] More recent references related to this section include [377], [458], [533], [890].

The first main point is that there is a normal form for H and R_0:

H = group generated by

$$\begin{pmatrix} 0 & 1 & 0 & 0 \\ 1 & 0 & 0 & 0 \\ 0 & 0 & 0 & 1 \\ 0 & 0 & -1 & 0 \end{pmatrix} \qquad \begin{pmatrix} 0 & 0 & 1 & 0 \\ 0 & 0 & 0 & -i \\ 1 & 0 & 0 & 0 \\ 0 & -i & 0 & 0 \end{pmatrix}$$

call this A' call this B'

$$R_0 = \begin{pmatrix} 1 & 0 & 0 & 0 \\ 0 & 1 & 0 & 0 \\ 0 & 0 & 1 & 0 \\ 0 & 0 & 0 & -1 \end{pmatrix}$$

call this R_0'

in suitable coordinates. Elaborating on this, let H_0 be the subgroup of $PGL(4, k)$ generated by A' and B', and let $\mathscr{G}_0$ be its inverse image in $GL(4, k)$. Then it is very easy to check that $\mathscr{G}_0$ acts irreducibly on k^4 and this is its only irreducible representation in which the scalar matrices $\alpha \cdot I_4$ act by multiplication by α. In fact, $\mathscr{G}_0$ is a finite analog of the Heisenberg group and k^4 is the analog of the unique representation of the Heisenberg commutation relations (see [220], Ch. 1, §3). The second main point follows from this: there are only finitely many choices of coordinates in which H and R_0 have the above normal form. In fact, this normal form determines a pair of generators of H, hence of E_4, the points of E of order 4; these generators have a special property making them a "symplectic" basis of E_4 (see [212], p. 222). To get the normal form requires in fact the choice of an isomorphism Φ of $\mathscr{G}$ with $\mathscr{G}_0$ under which conjugation by R_0 goes to conjugation by R_0'. It can be shown that there are 192 such isomorphisms Φ, and that every symplectic basis of E_8 determines a Φ (see [209], p. 319). Thus the choice of a Φ is intermediate in complexity between the choice of a symplectic basis of E_4 and the choice of one for E_8: for this reason, the choice of a Φ is called a "level (4,8)-structure" on E.

After all this, though, we have in hand *coordinates* on the moduli space $\mathscr{A}_{1,(4,8)}$ of pairs (E, Φ), E an elliptic curve and Φ a level (4,8)-structure. Namely, given (E, Φ), we have a canonical projective embedding $i: E \to \mathbf{P}^3$. Then the coordinates of the image of the origin $i(P)$ are coordinates on the moduli space. In fact, $i(P)$ is fixed by R_0'. The fixed points of R_0' in $\mathbf{P}^3$ are a plane $\mathbf{P}^2$ with coordinates X_0, X_1, X_2 and the point (0, 0, 0, 1): it turns out that $i(P) \in \boldsymbol{P}^2$. The final result is that

$$(E, \Phi) \mapsto i(P)$$

gives an isomorphism:

$$\mathscr{A}_{1,(4,8)} = \begin{pmatrix}\text{moduli space}\\ \text{of pairs } (E,\Phi)\end{pmatrix} \xrightarrow{\approx} \left\{\begin{array}{l}\text{smooth curve } X_0^4 = X_1^4 + X_2^4 \text{ in } \boldsymbol{P}^2\\ \text{minus 12 points where}\\ X_0 = 0,\ X_1 = 0, \text{ or } X_2 = 0\end{array}\right\}$$

The reason this map is injective is that E and Φ can be immediately reconstructed from $i(P)$:

a) take the 16 points $\sum$ obtained by acting on $i(P)$ with H_0;

b) there is a pencil of quadrics in $\boldsymbol{P}^3$ containing $\sum$, and its base locus is $i(E)$.

The curve $\mathscr{A}_{1,(4,8)}$ is a covering of the moduli space $\mathscr{A}_1$ of "bare" elliptic curves of degree 96 (because there are 192 (4,8)-structures but the general E has an automorphism group of order 2).

These canonical coordinates are very simple to describe analytically. Over $\boldsymbol{C}$ the moduli space $\mathscr{A}_1$ may be described as the quotient:

$$H/PSL(2, \boldsymbol{Z})$$

$$H = \{\tau \in \boldsymbol{C} \mid \operatorname{Im} \tau > 0\}$$

$$PSL(2, \boldsymbol{Z}) \text{ acting by } \tau \mapsto \frac{a\tau + b}{c\tau + d}, \text{ for all}$$

$$\begin{pmatrix} a & b\\ c & d\end{pmatrix} \in SL(2, \boldsymbol{Z}).$$

The covering $\mathscr{A}_{1,(4,8)}$ is

$$H/\Gamma$$

$$\Gamma = \left\{\begin{pmatrix} a & b\\ c & d\end{pmatrix} \in SL(2,\boldsymbol{Z}) : 8 \mid b \text{ and } c,\ 4 \mid a - 1 \text{ and } d - 1\right\}.$$

The canonical embedding of the elliptic curve $E = \boldsymbol{C}/\boldsymbol{Z} + \boldsymbol{Z}\cdot\tau$ is given by

$$z \mapsto \left(\vartheta\begin{bmatrix}0\\0\end{bmatrix}(2z,\tau),\ \vartheta\begin{bmatrix}0\\ 1/2\end{bmatrix}(2z,\tau),\ \vartheta\begin{bmatrix}1/2\\0\end{bmatrix}(2z,\tau),\ \vartheta\begin{bmatrix}1/2\\ 1/2\end{bmatrix}(2z,\tau)\right)$$

where

$$\vartheta\begin{bmatrix}a\\b\end{bmatrix}(z,\tau) = \sum_{n\in\boldsymbol{Z}} e^{\pi i(n+a)^2\tau + 2\pi i(n+a)(z+b)}$$

is the standard theta function with characteristics a, b. Thus the image of the origin $z = 0$ has coordinates

$$\left(\vartheta\begin{bmatrix}0\\0\end{bmatrix}(0,\tau),\ \vartheta\begin{bmatrix}0\\ 1/2\end{bmatrix}(0,\tau),\ \vartheta\begin{bmatrix}1/2\\0\end{bmatrix}(0,\tau)\right)$$

and the map carrying τ to this point in P^2 embeds H/Γ in P^2. For details on this example, see the author's forthcoming book, *Tata lectures on theta functions* [220], Ch. 1, § 11.

This approach generalizes to higher dimensions and provides not only a construction but even an *explicit coordinatization* of a finite covering of $\mathscr{A}_{g,1}$, the moduli space of g-dimensional, principally polarized abelian varieties. Unfortunately, although these coordinates are explicit, their algebra is quite complicated and this explicit description is not easy to use, e. g., to compute invariants of $\mathscr{A}_{g,1}$

The coordinatization is based on the following generalization of the algebraic group $\mathscr{G}$ in the above example:

Definition: For all invertible sheaves L on abelian varieties X, let

$$H(L) = \{x \in X \mid T_x^* L \cong L\}$$

($T_x(y) = x + y$ is translation by x). In char. p, this is to be interpreted scheme-theoretically, i. e., $H(L)$ is the full subscheme whose R-valued points x are those such that L is invariant under translation by x, for all local rings R. Let

$$\mathscr{G}(L) = \{(\Phi, x) \mid x \in H(L) \text{ and } \Phi\colon T_x^* L \xrightarrow{\approx} L \text{ is an isomorphism}\}.$$

Elements of $\mathscr{G}(L)$ may be thought of as automorphisms of the pair (X, L) and as such they form a group. In fact, $\mathscr{G}(L)$ is an algebraic group, and is part of an exact sequence:

$$1 \to \mathbf{G}_m \to \mathscr{G}(L) \to H(L) \to 1$$

where $\mathbf{G}_m$ maps to $\mathscr{G}(L)$ by taking α to the autormorphism, mult. by α, of L. For details, see [212], § 23. Let $R_0(x) = -x$ be reflection. L is symmetric if $R_0^* L \cong L$. In this case, R_0 induces an outer automorphism ϱ of $\mathscr{G}(L)$ fitting into the diagram:

$$\begin{array}{ccccccccc} 1 & \to & \mathbf{G}_m & \to & \mathscr{G}(L) & \to & H(L) & \to & 1 \\ & & \downarrow{\scriptstyle \mathrm{id.}} & & \downarrow{\scriptstyle \varrho} & & \downarrow{\scriptstyle -1} & & \\ 1 & \to & \mathbf{G}_m & \to & \mathscr{G}(L) & \to & H(L) & \to & 1. \end{array}$$

We call L totally symmetric if there is an isomorphism $R_0^* L \xrightarrow{\sim} L$ which is the identity over the subscheme $X_2 \subset X$ of points of order 2: this is equivalent to L being $L^{\Delta}(\lambda)$ for some λ.

We can now introduce a moduli "level" $(n, 2n)$ intermediate between level n and level $2n$, or equivalently, a moduli space in the middle of the diagram:

$$\begin{array}{c} \mathscr{A}^*_{g,1,2n} \\ \downarrow \\ \mathscr{A}^*_{g,1,(2n,n)} \\ \downarrow \\ \mathscr{A}^*_{g,1,n} \end{array}$$

For simplicity, we assume that n is even. Then $\mathscr{A}^*_{g,1,(2n,n)}$ is the space of triples (X, λ, α) where

X = a g-dimensional abelian variety,

$\lambda: X \xrightarrow{\sim} \hat{X}$ is a principal polarization of X. Let $L = \left(L^{\Delta}(\lambda)\right)^{\otimes \frac{n}{2}}$

$\alpha: \mathscr{G}(L) \xrightarrow{\sim} \boldsymbol{G}_m \times (\boldsymbol{Z}/n\boldsymbol{Z})^g \times \mu_n^g$ is an isomorphism, such that

$\alpha(\varrho(g)) = \varrho_0(\alpha(g))$

$\varrho_0(\lambda, x, y) = (\lambda, -x, y^{-1})$

and the group law on the right is

$$(\lambda, x, y) \cdot (\lambda', x', y') = (\lambda\lambda' y^{x'}, x + x', y \cdot y').$$

The map from $\mathscr{A}^*_{g,1,(2n,n)}$ to $\mathscr{A}^*_{g,1,n}$ carries (X, λ, α) to (X, λ, Φ) where Φ is the map induced by α from $\mathscr{G}(L)/\boldsymbol{G}_m = H(L) = X_n$ to $[\boldsymbol{G}_m \times (\boldsymbol{Z}/n\boldsymbol{Z})^g \times \mu_n^g]/\boldsymbol{G}_m = (\boldsymbol{Z}/n\boldsymbol{Z})^g \times \mu_n^g$. For the map from $\mathscr{A}^*_{g,1,2n}$ to $\mathscr{A}^*_{g,1,(2n,n)}$, see [209], p. 319. Over $\boldsymbol{C}$,

$$\mathscr{A}^*_{g,1,(2n,n)} \times \operatorname{Spec} \boldsymbol{C} \cong \mathfrak{H}_g/\Gamma_{n,2n}$$

$$\Gamma_{n,2n} = \left\{ \begin{pmatrix} A & B \\ C & D \end{pmatrix} \middle| \begin{array}{l} A - I_n, D - I_n, B, C \equiv 0 \pmod{n} \\ \text{and the diagonals of } B, C \text{ divisible by } 2n \end{array} \right\}$$

See Igusa [147]).

Now the group $\mathscr{G}(L)$ acts in a natural way on $\Gamma(X, L)$. A basic result states that this action is irreducible and is the unique irreducible representation of $\mathscr{G}(L)$ in which every $\alpha \in \boldsymbol{G}_m$ acts by mult. by α. (See [209], § 1 and Sekiguchi [296]). The canonical coordinates on $\mathscr{A}^*_{g,1,(n,2n)}$ are defined most succinctly as follows:

i) Via α, $\boldsymbol{G}_m \times \boldsymbol{Z}/n\boldsymbol{Z} \times \mu_n^g \times$ acts on $\Gamma(X, L)$: let (λ, x, y) act by $U_{(\lambda,x,y)}$.

ii) There is a section $s \in \Gamma(X, L)$, unique up to scalars, such that

$$U_{(1,0,c)}s = s, \qquad \text{all } c \in \mu_n^g.$$

iii) Let $s \mapsto s(0)$ denote evaluation of sections at $0 \in X$.

We obtain a function

$$(\boldsymbol{Z}/n\boldsymbol{Z})^g \to k$$
$$x \mapsto \left(U_{(1,x,0)}(s)\right)(0)$$

unique up to multiplication by a constant, which is not identically zero.

iv) If $N = n^g - 1$ and the homogeneous coordinates in $\boldsymbol{P}^N$ are labelled by the elements of $(\boldsymbol{Z}/n\boldsymbol{Z})^g$, this defines a morphism:

$$\Theta: \mathscr{A}^*_{g,1,(n,2n)} \to \boldsymbol{P}^N.$$

The main theorem is that if $n \geq 4$, Θ is a closed immersion. A weaker form of this theorem was proven over $\boldsymbol{C}$ by Baily [51]; the full theorem over C when $4|n$ was proven by IGUSA [157], [160: Th. 4,8, Ch. 5]; in the general case, it is proven when $4|n$ in [209], and the essential steps to deal with the general case (including $n \geqq 4$, n odd) are in [211].

A proof can be given exactly as in the elliptic curve case:

i) $G_m \times (Z/nZ)^g \times \mu_n^g$ has au nique irreducible representation on k^N, $N = n^g$, in which $\alpha \in G_m$ acts by mult. by α.

ii) $\mathcal{G}(L)$ acts on $\Gamma(X, L)$ and, via α, this representation is the same as that on k^N. Hence X is embedded in P^{N-1} not just up to a projective transformation, but canonically. The point $p = \Theta((X, \lambda, \alpha))$ just defined is the image of $e \in X$ in this embedding.

iii) Given p, we reconstruct $X \subset P^{N-1}$ by taking the orbit $0(p)$ of p under $(Z/nZ)^g \times \mu_n^g$ and intersecting all quadrics $Q \subset P^{N-1}$ containing $0(p)$.

Thus p determines X, $\underline{o}_X(1)$ and an action of $G_m \times (Z/nZ)^g \times \mu_n^g$ on $\underline{o}_X(1)$ hence (X, λ, α).

Over C, Θ is the morphism defined analytically by

$$\Omega \mapsto \left(\ldots, \vartheta \begin{bmatrix} 0 \\ \alpha/n \end{bmatrix} (0, n\Omega), \ldots\right)_{\alpha \in (Z/nZ)^g}$$

where

$$\vartheta \begin{bmatrix} \alpha \\ \beta \end{bmatrix} (z, \Omega) = \sum_{n \in Z^g} e^{\pi i\, {}^t(n+\alpha)\Omega(n+\alpha) + 2\pi i\, {}^t(n+\alpha)(z+\beta)}$$

is Riemann's theta function in g variables. The theorem may then be restated over C in terms of Siegel modular forms (see [160: p. 222]) as saying that the ring of Siegel modular forms with respect to the congruence subgroup of level $(n, 2n)$ in $\mathrm{Sp}\,(2g, Z)$ is the integral closure in its fraction field of the ring of functions

$$C\left[\ldots, \vartheta \begin{bmatrix} 0 \\ \alpha/n \end{bmatrix} (0, n\Omega) \cdot \vartheta \begin{bmatrix} 0 \\ \beta/n \end{bmatrix} (0, n\Omega), \ldots\right]_{\alpha, \beta \in Z/nZ^g}$$

and that the conductor ideal has no zeroes for $\Omega \in \mathfrak{H}_g$. If $8|n$, one can even find an explicit set of homogeneous quartic equations — Riemann's theta relations — such that the image of Θ is an *open* part of the subscheme of P^N defined by these quartics (see [209], § 6).

Even in the char. p case, it is possible to reformulate these canonical coordinates as values of a type of theta function. We describe the theory of [209], § 8 to § 12. These functions are not functions on the universal covering space of X, but rather on the Tate group of X, defined as follows:

$V(X) =$ group of sequences $\{x_i \in X \mid i \geqq 1, p \nmid i\}$
such that $nx_{in} = x_i$, x_1 of finite order prime to p.

$T(X) = \{$subgroup of $V(X)$ of sequences where $x_1 = 0\}$.

These groups have the following normal form:

$$\begin{array}{ccccccccc} 0 \to & T(X) & \longrightarrow & V(X) & \longrightarrow & X_{\text{tor, prime to } p} & \to 0 \\ & \wr\| & & \wr\| & & \wr\| & \\ 0 \to & \Big(\prod\limits_{l \neq p} \mathbf{Z}_l\Big)^{2g} & \to & \Big(\prod\limits_{l \neq p}{}' \mathbf{Q}_l\Big)^{2g} & \to & \Big(\bigoplus\limits_{l \neq p} \mathbf{Q}_l/\mathbf{Z}_l\Big)^{2g} & \to 0 \end{array}$$

(where $\prod'$ is the restricted direct product usual in the theory of adeles). Via the canonical map

$$V(X) \xrightarrow{\pi} X$$

$$\{x_i\} \mapsto x_1$$

all invertible sheaves on X can be pulled back to $V(X)$.

Theorem: All symmetric line bundles $\boldsymbol{L}$ on X, pulled back to $V(X)$, can be canonically trivialized up to a constant, i.e., there is a canonical continuous isomorphism:

$$\pi^*\boldsymbol{L} \cong \boldsymbol{L}(0) \times V(X).$$

Corollary: If $\deg \boldsymbol{L} = 1$, the unique section s of $\boldsymbol{L}$ defines a continuous function $\vartheta\colon V(X) \to k$, canonical up to multiplication by a scalar.

Assume now char $(k) \neq 2$. One can prove that the functions ϑ that arise in this way satisfy

a) $\vartheta(x + a) = e_*\left(\frac{a}{2}\right) e\left(\frac{a}{2}, x\right) \vartheta(x)$, all $x \in V(X)$, $a \in T(X)$.

Here $e\colon V \times V \to k^*$ is the canonical multiplicative skew-symmetric form with values in roots of 1 determined by λ, and $e_*\colon \frac{1}{2}\, T(X)/T(X) \to \{\pm 1\}$ is the canonical multiplicative quadratic form determined by $\boldsymbol{L}$ (see [209], § 2).

b) $\vartheta(-x) = \pm\vartheta(x)$, the sign depending on the Arf invariant of e_*,

c) $\prod\limits_{i=1}^{4} \vartheta(x_i) = 2^{-g} \cdot \sum\limits_{\eta \in \frac{1}{2}T(X)/T(X)} e(y, \eta) \cdot \prod \vartheta\,(x_i + \eta + y)$, $y = \dfrac{x_1 + \cdots + x_4}{2}$

d) $\forall x \in V(X)$, $\exists\, \eta \in \frac{1}{2}\, T(X)$ such that $\vartheta(x + \eta) \neq 0$

e) Up to an elementary linear transformation whose coefficients are roots of 1, the set of values of ϑ on $\frac{1}{n}\, T(X)$ is equal to the set of values of the canonical coordinates Θ on the triple (X, L, α_n) (for any symmetric α_n defined on $\mathcal{G}(L^{n^2})$) The set of *all* values of ϑ on $V(X)$ is a canonical set of homogeneous coordinates on $\hat{\mathcal{A}}_{g,1}$ (over $\mathbf{Q}$, or over $\mathbf{Z}/p\mathbf{Z}$ if, in defining $\hat{\mathcal{A}}$, we consider only levels prime to p).

f) Over $\mathbf{C}$, if Ω is the period matrix for X, ϑ is essentially the function

$$(a_1, a_2) \mapsto e^{-\pi i^t a_1 a_2} \cdot \vartheta \begin{bmatrix} a_1 \\ a_2 \end{bmatrix} (0, \Omega), \qquad a_i \in \mathbf{Q}^g$$

g) Moreover, if we restrict ϑ to $V_2(X) \subset V(X)$, where

$$V_2(X) = \bigcup_n 2^{-n} T(X),$$

then all functions $f\colon V_2(X) \to k$ satisfying a, b, c and d arise from a unique principally polarized abelian variety.

A related algebraic theory of theta functions is due to Barsotti [53]. Instead of trivializing line bundles $\mathbf{L}$ on $V(X)$, he assumes the characteristic of the ground field k is 0 and considers the pull-back of line bundles on X by the canonical map:

$$U = \operatorname{Spec}(\underline{o}_{e,X}) \xrightarrow{\pi} X.$$

He proves that for every line bundle $\mathbf{L}$ on X, there is a canonical isomorphism

$$\pi^* \mathbf{L} \cong \underline{o}_U$$

up to multiplication by $e^{Q(x)}$, Q a quadratic function in the exponential coordinates on X (i.e. coordinates in which the group law is linear.)

He then goes on to characterize the formal power series in g variables which arise from rational sections of such $\mathbf{L}$, trivialized on U as above.

The essential idea behind this construction may be described as follows: let D be any divisor on X such that $(0) \notin \operatorname{Supp}(D)$. Let (U_α) be a covering of X such that there are local equations f_α of D in U_α. Then the trivialization

$$\underline{o}_X(D) \otimes \underline{o}_{e,X} \cong \underline{o}_{e,X}$$

is defined so that it carries $1 \in \underline{o}_{e,X}(D)$ to a power series ϑ with non-zero constant term such that for all invariant vector fields D_1, D_2 on X, $D_1 D_2 \log \vartheta$ is a rational function on X and for all α

$$(*) \qquad D_1 D_2 \log \vartheta - D_1 D_2 \log f_\alpha \in \Gamma(U_\alpha, \underline{o}_X)$$

When the ground field k is $\mathbf{C}$, the data given by the power series ϑ modulo $\vartheta \sim e^Q \vartheta$ turns out to be equivalent to the data given by the family of theta series with *pluri-harmonic* polynomial coefficients (see Freitag [110], Oda [255]; for the link, see [220], Ch. 4). Thus Barsotti's theory is, in essence, an algebraization of the theory of harmonic theta series.

In characteristic p, infinitesimal neighborhoods of e are p^n-torsion for high enough n, so one would expect that BARSOTTI's theory and my theory are related by considering trivializations of line bundles L in char. p on the p^n-torsion in X (or the p-divisible group of X). This is indeed the case and a start on this theory has been made by BARSOTTI [54] and CRISTANTE [76].

C. The covariant of Weierstrass points [22]

There is a direct approach in char. 0 to the construction of the moduli space of curves somewhat similar to the construction of the moduli space of abelian varieties carried out in Chapter 7. The idea is to replace the covariant of n^{2g} points of order n by a covariant formed from $g \cdot (g-1)^2 \cdot (2n-1)^2$ "n^{th} order Weierstrass points". We sketch here this method (which had in fact been planned as another Chapter of the original book). We assume throughout that the characteristic is 0.

Let $\Phi\colon C \to \boldsymbol{P}^n$ be any morphism such that $\Phi(C)$ spans $\boldsymbol{P}^n$. We define the set $HO(\Phi)$ of points of hyperosculation of C for Φ to be:

$$\left\{ x \in C \,\middle|\, \begin{array}{l} \exists \text{ hyperplane } H \subset \boldsymbol{P}^n \text{ such that} \\ \Phi^*H = (n+1)\cdot X + \mathfrak{A}, \text{ some effective cycle } \mathfrak{A} \end{array} \right\}$$

Thus if $n = 1$, $HO(\Phi)$ is the set of branch points of Φ. If $n = 2$ and Φ is an embedding, $HO(\Phi)$ is the set of points of inflection of $\Phi(C)$. In general, if Φ is an embedding, $HO(\Phi)$ is the set of points where the Frenet-Serret frame attached to $\Phi(C)$ is not defined because the top torsion is zero. And if Φ is the canonical map, then

$$\begin{aligned} HO(\Phi) &= \{x \in C \mid \exists \text{ 1-form } \omega \text{ such that } (\omega) = g\cdot x + \mathfrak{A} \\ &= \{x \in C \mid \dim |g\cdot x| \geqq 1\} \\ &= \text{set of Weierstrass points.} \end{aligned}$$

The first problem is to attach multiplicities to the set $HO(\Phi)$, making it into a divisor, in such a way that

i) if C and Φ vary in a smooth projective family

$$\begin{array}{ccc} \mathscr{C} & \xrightarrow{\Phi} & \boldsymbol{P}^n \times S \\ & \searrow \quad \swarrow & \\ & S & \end{array}$$

the divisors $HO(\Phi)_s$ fit together into a relative Cartier divisor $HO(\Phi) \subset \mathscr{C}$.

ii) the divisor class of $HO(\Phi)$ can be computed in terms of the canonical class K of C, $\Phi^{-1}(H)$ and n.

---

[22] More recent references related to this section include [464], [479], [480], [528], [751], [852], [877].

These results are essentially classical (e.g. HENSEL-LANDSBERG [138], p. 494) but modern treatments have been given by LAKSOV [181], HUBBARD [152], OGAWA [256], MOUNT-VILLAMAYOR [206]. A local equation for $HO(\Phi)$, both for one C and for a family of C's, is given by a suitable Wronskian, and globalizing this equation, one finds

$$HO(\Phi) \equiv \frac{n(n+1)}{2} K_C + (n+1)\,\Phi^*(H).$$

In particular, if $H = mK_C$, $m \geqq 2$, and Φ is defined by the complete linear system $|mK_C|$, then

$$HO(\Phi) \equiv (2m-1)^2 \frac{g(g-1)}{2} K_C.$$

The second problem is to bound the multiplicity with which any point can appear in a divisor $HO(\Phi)$, if Φ is defined by a complete linear system. In fact, no point occurs with multiplicity greater than g^2. This is proven in 2 steps: take $x \in C$ and take affine coordinates on $\boldsymbol{P}^n$ with $\Phi(x) = (0, \ldots, 0)$. In suitable coordinates, Φ is defined near x by

$$X_1 = u_1 t^{r_1}, \ldots, X_n = u_n t^{r_n}$$

where

$$0 < r_1 < \cdots < r_n,\ u_i(x) \neq 0,\ t \text{ a local coordinate on } C \text{ at } x.$$

Then a local calculation of the Wronskian shows that

$$\operatorname{mult}_x HO(\Phi) = \sum_{k=1}^{n} (r_k - k).$$

On the other hand, application of the Riemann-Roch theorem on C shows that $r_k = k$ if $k \leqq \deg \Phi^* H - 2g + 2$, while $r_{n-k} \leqq \deg \Phi^* H - k$ is obvious. This gives the bound.

With these results in hand, $\mathcal{M}_g$ is constructed exactly as $\mathcal{A}_g$ was in Chapter 7. Fix an integer μ such that

$$\mu \geqq \left[\frac{1}{2}\left(\sqrt{30g}+1\right)\right] + 1$$

and let

$$\mu' = g(g-1)^2 \cdot (2\mu - 1)^2.$$

Let H be the Hilbert scheme of Corollary 6.4 parametrizing tri-canonically embedded smooth curves of genus g. By associating the divisor of $HO(\Phi|\mu K|)$ of points of hyper-osculation for the map defined by $|\mu K|$ to each curve C, we obtain a $PGL(5g-5)$-equivariant morphism*

$$H \to \operatorname{Symm}^{\mu'}(\boldsymbol{P}^{5g-6}).$$

* Using the canonical morphism from the Hilbert scheme of finite subschemes of $\boldsymbol{P}^n$ to the symmetric powers of $\boldsymbol{P}^n$. This morphism was asserted to exist by GROTHENDIECK and was constructed recently by A. NEEMAN [770].

One checks that because the multiplicities in HO are at most g^2, and because of the lower bound for μ, HO, considered as a 0-cycle in $\mathbf{P}^{5g-5}$, is properly stable for the action of $PGL(5g-5)$. Therefore $H/PGL \times (5g-5)$ exists.

The difficulty in extending this proof to char. p lies in the possibility that $HO(\Phi) = C$, i.e., that every point $x \in C$ can be a point of hyperosculation. Thus there are plane curves such as $y = x^p$ for which every point is a point of inflection. I conjecture that if Φ is defined by a complete linear system $\Gamma(C, L)$ and the degree of L is large enough (e.g., $\deg L > k \cdot g$ for some small k) then $HO(\Phi) \subsetneq C$. If such a conjecture is true, the above construction extends to char. p without further difficulty[23].

Another interesting fact is that when the ground field is $\mathbf{C}$, the sets $HO(\Phi)$ tend to be rather uniformly distributed on C in the complex topology. Fix any L of positive degree and let

$$H_n = HO(\Phi_n)$$

$$\Phi_n = \text{morphism defined by } \Gamma(C, L^n).$$

Then Olsen [258] proved that $\bigcup_{n=1}^{\infty} H_n$ is dense on C. In fact, if μ_n is the singular measure on C, supported on H_n, such that for all x in H_n,

$$\mu_n = \frac{\text{mult of } x \text{ in } H_n}{\deg H_n},$$

then I have proven (unpublished) that

$$\frac{1}{n} \sum_{k=1}^{n} \mu_k$$

converges weakly to the "Bergmann measure"

$$\frac{1}{g\sqrt{-1}} \sum_{i=1}^{g} \omega_i \wedge \overline{\omega}_i, \quad \{\omega_i\} \text{ orthonormal basis of } \Gamma(C, \Omega^1)$$

on C.

D. The Schottky problem[24]

Following the notation of Ch. 7, § 4, we wish to discuss the morphism

$$j: \mathcal{M}_g \to \mathcal{A}_{g,1,1}$$

where $\mathcal{M}_g$ is the coarse moduli space of curves, $\mathcal{A}_{g,1,1}$ that of principally

[23] This conjecture is false but weaker versions of it are true, as was discovered by Neeman [767].

[24] More recent references related to this section include [349], [353], [354], [384], [470], [492–494], [516–518], [536], [560], [562], [594], [754], [758], [759], [850], [911–913].

polarized abelian varieties, and j is given by

$$C \mapsto Jac\ C = \text{Jacobian of } C.$$

As described in the text, j is injective on geometric points, and its image is locally closed. More precisely, the results of MATSUSAKA and HOYT ([22], [151]) show that if

$\bar{\mathcal{M}}_g$ = complete moduli space of stable curves, as in Appendix 5C

and

$\mathcal{M}_g^*$ = open subset of $\bar{\mathcal{M}}_g$ of stable curves C whose components are non-singular and which are connected together in a tree

then for all stable C,

$$Pic^\tau(C) \text{ is complete} \Leftrightarrow [C] \in \mathcal{M}_g^*,$$

hence j extends to a *proper* morphism j^*:

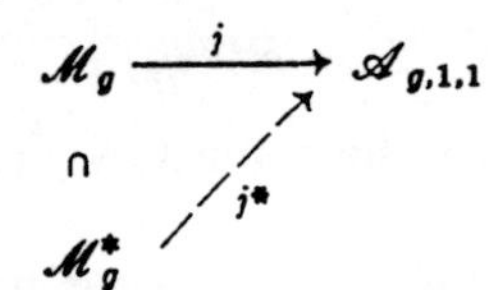

The image of j^* is precisely the locus of products $\prod_{i=1}^k J_i$ of jacobians, polarized by the product of Θ-polarizations $\lambda_i: J_i \to \hat{J}_i$ of the factors. This is the locus called Z in the text. As the theta divisor Θ (the unique effective divisor representing the polarization) in $\prod J_i$ is

$$\sum_{i=1}^k (J_1 \times \cdots \times \Theta_i \times \cdots \times J_k)$$

the open subset $Z_0 \subset Z$ of true jacobians is just the subset whose theta divisors are irreducible.

OORT-STEENBRINK [260] have analyzed whether or not Z_0 is normal: they prove that it is normal over $\mathbf{Q}$, i.e.,

$$j \times \text{Spec}\,(\mathbf{Q}): \mathcal{M}_g \times \text{Spec}\,(\mathbf{Q}) \to Z_0 \times \text{Spec}\,(\mathbf{Q})$$

is an isomorphism. However, in char. 2, it is definitely *not* an isomorphism.

The problem which has attracted the most interest is to find equations or other characterizations of the locus Z inside $\mathcal{A}_{g,1,1}$. This is known as the Schottky problem. The author gave a survey of work on this question in [216], Ch. 4. In a rough sort of way, one can divide the known approaches into 4 groups. The first is based on the lemma:

Lemma: Let J be the jacobian of C, Θ its theta-divisor, and Φ: $C \to J$ the canonical map. If Θ_a denotes the translate of Θ by $a \in J$, then

$$\left\{\begin{array}{l} \Theta \cap \Theta_a = W_1 \cup W_2 \text{ where} \\ W_1 \subset \Theta_b,\ W_2 \subset \Theta_c, \text{ some} \\ b, c \in J, \text{ distinct from } 0, a \end{array}\right\} \Leftrightarrow \{a = \Phi(x) - \Phi(y) \text{ for some } x, y \in C\}$$

This break-up of $\Theta \cap \Theta_a$ has many implications, and is used in most of the known proofs of Torelli's theorem. The lemma is more or less equivalent to a remarkable identity for the theta function $\vartheta(z)$ of J:

For any 4 distinct points $x, y, u, v \in C$, there are constants c_1, c_2 such that

$$(*)\ \vartheta(z)\cdot\vartheta\left(z+\int_{u+v}^{x+y}\omega\right) = c_1\vartheta\left(z+\int_u^x\omega\right)\cdot\vartheta\left(z+\int_v^y\omega\right) + c_2\vartheta\left(z+\int_v^x\omega\right)\vartheta\left(z+\int_u^y\omega\right).$$

(See FAY [99], p. 34). Here $z \in C^g$ and ω is the (suitably normalized) vector of holomorphic 1-forms on C. Letting the 4 points x, y, u, v approach one another in different ways, (*) gives rise to many differential identities on ϑ (FAY [99], pp. 20–29). When all 4 come together, it finally collapses to a non-linear partial differential equation, called the KADOMTSEV-PETVIASHVILI equation (or K–P equation) after the mathematicians who introduced it in hydrodynamics [167]. In Fay's form, it states that (for suitable constants c_1, c_2)

$$(**)\ \frac{1}{6}D_1^4\log\vartheta + (D_1^2\log\vartheta + c_1)^2 - \frac{1}{3}D_1D_3\log\vartheta + \frac{1}{2}D_2^2\log\vartheta = c_2$$

where

$$D_i = \sum_{j=1}^{g} a_{ij}\,\partial/\partial z_j$$

and for some $x \in C$, the differentials of 1st kind have local expansions in a coordinate t near x:

$$\omega_j = \left(a_{1j} + a_{2j}t + \frac{1}{2}a_{3j}t^2 + \cdots\right)dt.$$

Taking D_1^2, you get the K–P equation itself:

$$D_1\left(\frac{1}{6}D_1^3u + 2u\cdot D_1u - \frac{1}{3}D_3u\right) + \frac{1}{2}D_2^2u = 0$$

where $u = D_1^2\log\vartheta + c_1$. NOVIKOV has conjectured that the existence of constant partial differential operators D_1, D_2, D_3 such that ϑ satisfies (**) is equivalent to J being a jacobian[25]. In this direction, it has turned out that the theta functions of jacobians, especially hyperelliptic jacobians and generalized jacobians associated to singular limit curves, give special but important solutions of many of the non-linear PDE's of mathematical physics. For details, see KRIČEVER [178], NOVIKOV et al. [93], and the author's lectures [219].

[25] This conjecture was proved by SHIOTA [850].

The second approach to the Schottky problem is based on the fact that "Θ is doubly of translation type", i.e., Θ is the locus of points $\Phi(x_1) + \cdots + \Phi(x_{g-1})$, $x_i \in C$. Applying this representation near a point $z \in \Theta$ and its mirror image $-z \in \Theta$ gives 2 distinct local representations of the hypersurface Θ near z as the pointwise vector sum of $(g-1)$ germs of analytic curves. Besides the classical work of LIE and WIRTINGER [332], this has been studied by SAINT-DONAT [286] and GRIFFITHS [120], who simplify and greatly extend the older results. In particular, SAINT-DONAT gives a proof of the Torelli theorem based on these ideas.

A third approach is based on the fact that for Jacobians J of dimension g, Θ has a singular locus of dimension $g-4$ or $g-3$ (according as the curve is non-hyperelliptic of hyperelliptic). The main result here is ANDREOTTI and MAYER's theorem [344] that if

$$\mathcal{S} = \left\{ \begin{array}{l} \text{principally polarized abel. var. } X, \Theta \text{ such that} \\ \Theta \text{ has a singular locus of codim.} \leqq 4 \end{array} \right\} \subset \mathcal{A}_{g,1,1}$$

then Z is one of the components of $\mathcal{S}$. Using the theory of Prym varieties, BEAUVILLE [57] showed that when $g = 4$, $\mathcal{S}$ has *two* components: generically on Z, Θ has exactly 2 singularities of the form $\pm P$, while on the other component, Θ has only one singularity at a point of order 2. Both of thes eare 9-dimensional subvarieties of the 10-dimensional $\mathcal{A}_{g,1,1}$. DONAGI [88] showed that when $g = 5$, $\mathcal{S}$ has *five* components: Z of dimension 10, the 11-dimensional locus of products $X^1 \times X^4$, and 3 8-dimensional varieties of Prym's of curves C of genus 6 covering elliptic curves doubly.

The fourth approach was used by SCHOTTKY, and is based on Prym varieties. Let $\pi: \hat{C} \to C$ be an étale double covering, let $\iota: \hat{C} \to \hat{C}$ be the map interchanging sheets, and let J, $\hat{J}$ be the jacobians. Then $Prym(\hat{C}/C)$ is the *odd* part of $\hat{J}$: it can be defined either as the quotient $\hat{J}/J$ or as the subvariety $\{\iota x - x) \mid x \in \hat{J}\}$ of $\hat{J}$. It turns out to be a principally polarized abelian variety. If $g =$ genus C, then by Hurwitz's formula, $2g - 1 =$ genus $\hat{C}$, hence $g - 1 = \dim Prym$. The theory of these abelian varieties has been developed by FAY [99], myself [215] and BEAUVILLE [57]. One application of this theory is to construct in a useful way abelian varieties which are not jacobians. Thus whereas the jacobians of curves of genus g form (for $g \geqq 2$) a $(3g-3)$-dimensional locus Z inside the $\frac{g(g+1)}{2}$-dimensional $\mathcal{A}_{g,1,1}$, the $Prym(\hat{C}/C)$'s for curves C of genus $g+1$ and their limits give us (for $g \geqq 5$) a $3g$-dimensional locus $Z_{\text{Prym}} \subset \mathcal{A}_{g,1,1}$. In particular, whereas

$$Z = \mathcal{A}_{g,1,1} \text{ only if } g = 1, 2 \text{ or } 3$$

we have

$$Z_{\text{Prym}} = \mathcal{A}_{g,1,1} \quad \text{if } g = 4 \text{ or } 5 \text{ too.}$$

Note that $Z \subset Z_{\mathrm{Prym}}$ because if C of genus $g+1$ approaches a curve C_0 obtained from glueing 2 points on a curve C_1 of genus g, then some of its coverings $\hat{C} \to C$ approach curves $\hat{C}_0$ obtained from glueing 2 copies of C_1 at 2 pairs of points, and then $Prym(\hat{C}_0/C_0) \cong Jac\ C_1$. This approach to $\mathscr{A}_4$ and $\mathscr{A}_5$ gives the analysis of $\mathscr{S}$ mentioned above. Moreover, if $\mathscr{R}_g$ is the moduli space of étale double covers $\{(\pi, \hat{C}, C)\}$ where $g = $ genus C, we may study the morphism:

$$\mathscr{R}_g \to Z_{\mathrm{Prym}} \subset \mathscr{A}_{g-1,1,1}.$$

Donagi-Smith [90] have shown that

$$\mathscr{R}_6 \to \mathscr{A}_{5,1,1}$$

is a covering of degree 27, and Donagi [89] has shown that $\mathscr{R}_6$, hence $\mathscr{A}_{5,1,1}$, is a unirational variety.

The application made by Schottky-Jung of this theory, however, is based on the map

$$j': \mathscr{R}_g \to \mathscr{A}_{g,1,1} \times \mathscr{A}_{g-1,1,1}$$

which takes the point representing $(\pi, \hat{C}, C)$ to the pair of points representing $Jac(C)$, $Prym(\hat{C}/C)$. If Ω_1, Ω_2 are the $g \times g$ and $(g-1) \times (g-1)$ period matrices of these 2 abelian varieties, then there are very simple polynomials in the theta constants $\vartheta \begin{bmatrix} a \\ b \end{bmatrix} (0, \Omega_1)\ \vartheta \begin{bmatrix} c \\ d \end{bmatrix} (0, \Omega_2)$ that vanish on the image of j'. Moreover, if Z' is the subset* of $\mathscr{A}_g \times \mathscr{A}_{g-1}$ defined by these polynomials, $\mathrm{Im}(j')$ is one of the components of Z'. For details, see [98], [99], [215].[26]

* To be precise, $Z' \subset \mathscr{A}_{g,4,8} \times \mathscr{A}_{g-1,4,8}$ because the theta constants involve half-integer characteristics $\begin{bmatrix} a \\ b \end{bmatrix}, \begin{bmatrix} c \\ d \end{bmatrix}$.

[26] A totally new way of looking at the Schottky locus is due to Buser-Sarnak [438], who show that the whole locus of Jacobians is "very close to the boundary" of $\mathscr{A}_g$ when g is large.

References

[1] AHLFORS, L.: Analytic Functions. Princeton **1960**, p. 45.
[2] ANDREOTTI, A.: On a theorem of Torelli. Am. J. Math., **80**, 801 (1958).
[3] AUSLANDER, M., and D. BUCHSBAUM: Homological Dimension in local rings. Trans. Am. Math. Soc. **85**, 390 (1957).
[4] BAILY, W.: On the moduli of Jacobian varieties. Annals of Math. **71**, 303 (1960).
[5] CARTAN, H.: Séminaire 13. Paris 1960/61.
[6] CARTIER, P.: Séminaire Sophus Lie 1. Paris 1954/55.
[7] —: Isogenies and duality of abelian varieties. Annals of Math. **71**, 315 (1960).
[8] CHEVALLEY, C.: Séminaire 1. Paris 1956/1958.
[9] —: Séminaire 2. Paris 1958.
[10] —: Sur la théorie de la variété de Picard. Am. J. Math. **82**, 435 (1960).
[11] GROTHENDIECK, A.: Eléments de géométrie algébrique, (with J. DIEUDONNÉ). Publ. de l'Inst. des Hautes Etudes Scientifiques. Referred to as EGA.
[12] —: Séminaire de géométrie algébrique, 1960/61. Springer Lecture Notes **224** (1971). Referred to as SGA.
[13] —: Fondements de la géometrie algébrique, collected Bourbaki talks. Paris 1962.
[14] HILBERT, D.: Über die vollen Invariantensysteme. Math. Annalen **42**, 313 (1893).
[15] HIRONAKA, H.: An example of a non-kählerian deformation. Annals of Math. **75**, 190 (1962).
[16] IWAHORI, N., and M. MATSUMOTO: On some Bruhat decompositions and the structure of Hecke rings of p-adic Chevalley groups, Publ. I. H. E. S. **25**, 5–48 (1965).
[17] KODAIRA, K., and D. C. SPENCER: On deformations of complex analytic structures. Annals of Math **67**, 328 (1958).
[18] —, —: A theorem of completeness of characteristic systems of complete continuous systems. Am. J. Math. **81**, 477 (1959).
[19] KOIZUMI, S.: On specialization of the Albanese and Picard varieties. Mem. Coll. Sci., Kyoto **32**, 371 (1960).
[20] KOSTANT, B.: Lie group representations on polynomial rings. Am. J. Math., **85**, 327–404 (1963).
[21] LANG, S.: Abelian varieties. Interscience, N. Y., 1959.
[22] MATSUSAKA, T.: On a theorem of Torelli. Am. J. Math. **80**, 784 (1958).
[23] —: On a characterization of a Jacobian variety. Mem. Coll. Sci., Kyoto **32**, 1 (1959).
[24] — and D. MUMFORD: Two fundamental theorems on deformations of polarized varieties. Am. J. Math., **86**, 668 (1964).
[25] MUMFORD, D.: An elementary theorem in geometric invariant theory. Bull. Am. Math. Soc. **67**, 483 (1961).
[26] NAGATA, M.: Invariants of a group in an affine ring. J. Math. Kyoto Univ. **3**, 369 (1964).
[27] NAGATA, M.: Note on orbit spaces. Osaka Math. J. **14**, 21 (1962).

[28] —: Complete reducibility of rational representations of a matrix group. J. Math. Kyoto Univ. **1**, 87 (1961).

[29] Nishi, M.: Some results on abelian varieties. Nat. Sci. Report Ochanomizu Univ. **9**, 1 (1958).

[30] —: The Frobenius theorem and the duality theorem on an abelian variety. Mem. Coll. Sci., Kyoto **32**, 333 (1959).

[31] Palais, R.: On the existence of slices for actions of non-compact Lie groups. Annals of Math. **73**, 295 (1961).

[32] Rosenlicht, M.: Toroidal algebraic groups. Proc. Am. Math. Soc. **12**, 984 (1961).

[33] Severi, F.: Geometria dei sistemi algebrici sopra una superficie. Vol. 2. Rome 1959.

[34] Tits, J.: Théorème de Bruhat et sous-groupes paraboliques. C. R. Acad. Sci., Paris **254**, 2910 (1962).

[35] Weil, A.: Zum Beweis des Torellischen Satzes. Nachr. der Akad. Wissen. Göttingen, 1957, p. 33.

[36] Zariski, O.: Algebraic surfaces. Chelsea, N. Y. 1948.

[37] Bialynicki-Birula, A.: On homogeneous affine spaces of linear algebraic groups. Am. J. Math. **85**, 577 (1963).

[38] Samuel, P.: Méthodes d'algèbre abstraite en géométrie algébrique. Ergebnisse, No. 4. Berlin-Göttingen-Heidelberg: Springer 1955.

[39] Narasimhan, M. S., and C. S. Seshadri: Stable bundles and unitary bundles on a compact Riemann Surface, Proc. Nat. Acad. Sci. USA, **52**, 207 (1964).

[40] Mumford, D.: Lectures on curves on an algebraic surface. Annals of Math. Studies, **59** (1966).

[41] Adamovich, O. A., and Y. O. Galovina: Simple linear Lie groups with free algebras of invariants. Proc. of Yaroslavl Univ., **1979**, p. 3.

[42] Altman, A., and S. Kleiman: Compactification of the Picard Scheme I, II Adv. in Math. **35**, 50 (1980) and Am. J. Math. **101**, 10 (1979).

[43] Arakelov, S.: Families of algebraic curves with fixed degeneracies. Izv. Akad. Nauk **35**, (1971).

[44] Artin, M.: On Azumaya algebras and finite-dimensional representations of rings. J. of Algebra **11**, 532 (1969).

[45] —: Lectures on Deformations of Singularities. Tata Institute Lecture Notes **54** (1976), distr. by Springer-Verlag.

[46] —: Algebraic spaces. Yale Math. Monographs 3, Yale Univ. Press, 1971, New Haven.

[47] — and G. Winters: Degenerate fibres and stable reduction of curves. Topology **10**, 373 (1971).

[48] Ash, A., D. Mumford, M. Rapoport and Y. Tai: Smooth compactification of locally symmetric varieties. Math.-Sci. Press, 53 Jordan Rd., Brookline, Mass., 1975.

[49] Atiyah, M.: Geometry of Yang-Mills Fields. Scuola Normale Sup., Pubblic. della classe di Scienze (Lezioni Fermiane), Pisa, 1979.

[50] — and R. Bott: The Yang-Mills equations over Riemann surfaces. Philos. Trans. Roy. Soc. London Ser. A **308**, 523 (1982).

[51] Baily, W.: On the moduli of abelian varieties with multiplications. J. Math. Soc. Japan **15**, (1963).

[52] — and A Borel: Compactification of arithmetic quotients of bounded symmetric domains. Annals of Math. **84**, 442 (1966).

[53] Barsotti, I.: Considerazioni sulle funzioni theta. 1st. Naz. di Alta Mat. Symp. Math. **3** 247 (1970).

[54] BARSOTTI, I.: Theta functions in positive characteristics. Astérisque **63**, 5 (1979).

[55] BARTH, W.: Moduli of vector bundles on the projective plane. Inv. Math. **42**, 63 (1977).

[56] — and K. HULEK: Monds and moduli of vector bundles. Manuscripta Math. **25**, 323 (1978),

[57] BEAUVILLE, A.: Prym Varieties and the Schottky problem. Inv. Math. **41**, 149 (1977).

[58] BERNSTEIN, I., and GEL'FAND, I. and PONOMAREV, V.: Coxeter functors and Gabriel's theorem, Russian Math. Surveys, **28**, 17 (1973).

[59] BIRKES, D.: Orbits of linear algebraic groups. Annals of Math. **93**, 459 (1971).

[60] BOGOMOLOV, F. A.: Holomorphic tensors and vector bundles on projective varieties, Ivz. Akad. Nauk., **42**, (1978) (transl. Math. USSR Izv., **13**, 499 (1979)).

[61] BOREL, A.: Linear algebraic groups. Lecture notes in Math., 1968, Benjamin Press.

[62] — and N. WALLACH: Continuous cohomology, discrete subgroups and representations of reductive groups. Annals of Math. Studies **94**, Princeton Univ. Press, 1980.

[63] BRENNER, S.: Endomorphism algebras of vector spaces with distinguished sets of subspaces, J. Alg., **6**, 100 (1967).

[64] BURNS, D., and M. RAPOPORT: On the Torelli problem for Kählerian K3 surfaces. Annales Ec. Norm. Sup. **8**, 235 (1975).

[65] BURNS, D., and J. WAHL: Local contributions to global deformations Inv. Math. **26**, 67 (1974).

[66] BYRNES, C.: On the control of certain deterministic infinite-dimensional systems by algebro-geometric techniques. Am. J. Math. **100**, 1333 (1978).

[67] — and P. L. FALB: Applications of algebraic geometry in system theory. Am. J. Math. **101**, 337 (1979).

[68] — and N. HURT: On the moduli of linear dynamical systems. Adv. in Math.: Studies in Analysis **4**, (1978).

[69] CATTANI, E., and A. KAPLAN: Polarized mixed Hodge structures and the local monodromy of a variation of Hodge structures. Inv. Math. **67**, 101 (1982).

[70] CARLSON, J., E. CATTANI and A. KAPLAN: Mixed Hodge structures and compactifications of Siegel's space. Proc. of Angiers Conference Noordhof, 1980.

[71] CATANESE, F.: The moduli and global period mapping of surfaces with $K^2 = p_g = 1$: a counterexample to the global Torelli problem. Compositio Math. **41**, 401 (1980).

[72] CHAKIRIS, K.: Counterexamples to global Torelli for certain simply connected surfaces. Bull. AMS **2**, 297 (1980).

[73] CLEMENS, C.: Degeneration of Kähler manifolds. Duke Math. J. **44**, 215 (1977).

[74] CLINE, E., B. PARSHALL, L. SCOTT and W. VAN DER KALLEN: Rational and generic cohomology. Inv. Math. **39**, 143 (1977).

[75] CRICK, S.: Local moduli of abelian varieties. Am J. Math. **97**, 851 (1975).

[76] CRISTANTE, V.: Theta functions and Barsotti-Tate groups. Ann. Scuola Norm. Sup. Pisa Cl. Sci. (4) **7**, 181 (1980).

[77] DE CONCINI, C. and C. PROCESI: A characteristic-free approach to invariant theory, Adv. in Math. **21**, 330 (1976).

[78] DELIGNE, P.: Théorie de Hodge I. Actes du Congrés Int. des Math., Nice, 1, 425 (1970).

[79] —: Théorie de Hodge II, III. Publ. I. H. E. S. 40, 5 (1971); 44, 5 (1975).

[80] —: Variétiés de Shimura (cont. of Travaux de Shimura), Proc. Symp. in Pure Math. 33, Automorphic Forms, 247 (1979), part 2; and Sém. Bourbaki, exp. 389, in Springer Lecture Notes 244 (1971).

[81] —: Travaux de Griffiths. Sém. Bourbaki, exp. 376, in Springer Lecture Notes 180 (1971).

[82] — and D. MUMFORD: The irreducibility of the space of curves of given genus. Publ. I. H. E. S. 36, 75 (1969).

[83] DESALE, U. V., and S. RAMANAN: Classification of vector bundles of rank 2 on hyperelliptic curves. Inv. Math. 38, 161 (1976).

[84] —, —: Poincaré polynomials of spaces of stable bundles. Math. Annalen 216, 233 (1975).

[85] DIEUDONNÉ, T., and CARREL, T.: Invariant theory, old and new, Advances in Math., 4, (1970).

[86] DLAB, V. and C. M. RINGEL: Indecomposable representations of graphs and algebras, Memoirs A. M. S., 173 (1976).

[87] DLAB, V. and GABRIEL, P.: Representations of algebras, Springer Lecture notes, 488 (1974).

[88] DONAGI, R.: The tetragonal construction. Bull. Amer. Math. Soc. 4, (1981).

[89] —: The unirationality of $\mathscr{A}_5$. Ann. Math. (2) 119, 269 (1984).

[90] — and R. SMITH: The structure of the Prym map. Acta Math. 145, (1981).

[91] DOUBILET, P., G. C. ROTA and J. STEIN: On the foundations of combinational theory IX, Studies in Appl. Math., 103, 185 (1974).

[92] DRINFELD, V. G.: Langland's conjecture for GL(2) over functional fields. Proc. Int. Congress of Math. (Helsinki, 1978), p. 565—54; Acad. Sci. Fennica, Helsinki, 1980.

[93] DUBROVIN, V., V. MATVEEV, and S. NOVIKOV: Non-linear equations of Korteweg-de Vries type, finite-zone linear operators and abelian manifolds. Russian Math. Surveys 31 59 (1976).

[94] EDMUNDS, J.: Minimum partitions of a matroid into independent subsets. Jour. Res., Nat. Bur. Standards 69B, 67 (1965).

[95] ÉLASHVILI, A. G.: Canonical form and stationary subalgebras of points of general position for simple linear Lie groups. Funct. Anal. Appl. 6, 44 (1972).

[96] — and E. B. VINBERG: Classification of trivectors of ninedimensional space. Trudy Sem. Vekt. Analizu 18, 197 (1978).

[97] FALB, P.: Linear systems and feedback. Lecture Notes, Control group, Lund Univ., 1974.

[98] FARKAS, H., and H. RAUCH: The kinds of theta constants and period relations on a Riemann surface. Proc. Nat. Acad. USA 62, (1969).

[99] FAY, J.: Theta functions on Riemann surfaces. Springer Lecture Note 352 (1973).

[100] FOGARTY, J.: Truncated Hilbert functors. J. Reine und Angew. Math. 234, 65 (1969).

[101] —: Kähler differentials and Hilbert's fourteenth problem for finite groups. Amer. Jour. Math. 102 1159 (1980).

[102] —: Geometric quotients are algebraic schemes. Adv. in Math. 48, 166 (1983).

[103] FORMANEK, E.: The center of the ring of 4×4 generic matrices. J. Algebra **62**, 304 (1980).
[104] —: The center of the ring of 3×3 generic matrices. Linear and Multilinear Alg. **7**, (**1979**).
[105] — and C. PROCESI: Mumford's conjecture for the general linear group. Adv. in Math. **19**, **292** (**1976**).
[106] FORSTER, O., A. HIRSCHOWITZ and M. SCHNEIDER: Type de scindage générlaisé pour les fibrés stables. Vector Bundles and Differential Equations, Berkhauser, Boston (**1980**).
[107] FOSSUM, R. and B. IVERSON: On Picard groups of algebraic fibre spaces. J. Pure Appl. Math. **3**, **269** (**1973**).
[108] FREITAG, E.: Der Körper der Siegelschen Modulfunktionen, Abh. Math. Sem. Univ. Hamburg, **47**, 25 (**1978**).
[109] —: Der Kodaira dimension von Körpern automorphes Funktionen. Crelle **296**, 162 (**1977**).
[110] FREITAG, E.: Thetareihen mit harmonischen Koeffizienten zur Siegelschen Modulgruppe, Math. Ann. **254**, 27 (**1980**).
[111] FRIEDMAN, R.: Hodge theory, degenerations and the global Torelli problem, Ph. D. Thesis, Harvard University, **1981**.
[112] GABRIEL, P.: Unzerlegbare darstellungen I, II, Manuscr. Math. **6**, 71 (**1972**) and Symp. Math. Inst. Naz. Alta Mat. **11**, 81 (**1973**).
[113] GARDNER, R. B.: The Fundamental Theorem of Vector Relative Invariants. J. of Alg. **36**, **314** (**1975**).
[114] GEL'FAND, I. M.: Remarks on the classification of a pair of commuting linear transformations in a finite-dimensional vector space. Funct. Anal. Appl. **3**, **325** (**1969**).
[115] — and V. A. PONOMAREV: Problems of linear algebra and classification of quadruples of subspaces of a finite-dimensional vector space. Colloq. Math. Soc. János Bolyai, 5: Hilbert Space Operators, Tihany, Hungary, **1970**.
[116] GIESEKER, D.: Global moduli for surfaces of general type. Inv. Math. **43**, **233** (**1977**).
[117] —: On the moduli of vector bundles on an algebraic surface. Ann. of Math. **106**, 45 (**1977**).
[118] —: Some applications of Geometric Invariant Theory to moduli problems. Proc. of I. C. M. Helsinki 525 (**1978**).
[119] —: Lectures on moduli of curves. Tata Institute Lecture Notes on Math., Springer-Verlag (1982).
[120] GRIFFITHS, P.: Variations on a Theorem of Abel. Inv. Math. **35**, **321** (**1976**).
[121] —: Periods of integrals on algebraic manifolds, I, II, III. Amer. J. Math. **90**, 568, 805 (1968) and Publ. I. H. E. S. **38**, 125 (1970).
[122] —: Periods of integrals on algebraic manifolds; summary of main results and discussion of open problems. Bull. Amer. Math. Soc. **76**, 228 (1970).
[123] — and J. HARRIS: Principles of Algebraic Geometry. Wiley-Interscience, **1978**.
[124] — and W. SCHMID: Recent developments in Hodge theory, in Discrete Subgroups of Lie groups, Oxford Univ. Press, **1973**.
[125] GROSSHANS, F.: Observable groups and Hilbert's fourteenth problem. Amer. J. Math. **95** **229** (**1973**).
[126] —: Subgroups satisfying the codimension 2 condition, to appear.
[127] GUREVICH, G. B.: Classification des trivecteurs ayant le rang huit. Dokl. Akad. Nauk SSR **2**, 355 (**1935**).

[128] HABOUSH, W. J.: Reductive groups are geometrically reductive. Annals of Math. **102**, 67 (1975).

[129] —: Homogenous vector bundles and reductive subgroups of reductive algebraic groups. Am. J. Math. **100**, **1123** (**1978**).

[130] HARDER, G.: Eine bemerkung zu einer arbeit von P. E. Newstead. J. für Math. **242**, **16** (**1970**).

[131] — and D. KAZHDAN: Automorphic forms on GL_2 over function fields (after V. G. Drinfeld). Proc. of Symp. on Pure Math. **83** (part 2). Automorphic Forms, 357 (1975).

[132] — and M. NARASIMHAN: On the cohomology groups of moduli spaces of vector bundles on curves. Math. Annalen **212**, **215** (**1975**).

[133] HARTSHORNE, R.: Algebraic vector bundles on projective spaces: a problem list. Topology **18**, **117** (**1979**).

[134] HAYASHIDA, T.: A class number associated with the product of an elliptic curve with itself. J. Math, Soc. of Japan **20**, **26** (**1968**).

[135] — and M. NISHI: Existence of curves of genus 2 on a product of 2 elliptic curves. J. Meth. Soc. Japan **17**, **1** (**1965**).

[136] HAZEWINKEL, M.: Moduli and canonical forms for linear dynamical systems III. **1976** Ames Conference on Geometric Control Theory, Math.-Sci. Press **291** (**1977**).

[137] — and R. E. KALMAN: On invariants, canonical forms and moduli for linear constant finite-dimensional dynamical systems. Lecture Notes Econ.-Math. System Theory, Springer-Verlag **131** (**1976**).

[138] HENSEL, K. and G. LANDSBERG: Theorie der Algebraischen Funktionen einer Variablen, Leipzig 1902, reprint Chelsea Publ. Co., **1965**.

[139] HERMANN, R. and C. MARTIN: Applications of algebraic geometry to systems theory: the McMillan degree and Kronecker indices. SIAM J. Control and Optimization **16**, **743** (**1978**).

[140] HESSELINK, W.: Uniform instability in reductive groups, J. Reine Angew. Math. **303**, 74 (1978).

[141] HIRZEBRUCH, F.: Hilbert modular surfaces. L'Ens. Math. (**1973**)

[142] HOCHSTER, M.: Rings of invariants of tori, Cohen-Macaulay rings generated by monomials, and polytopes. Ann. of Math. **96**, **318** (**1972**).

[143] — and J. EAGON: Cohen-Macaulay rings, invariant theory and the generic perfection of determinantal loci. Amer. J. Math. **93**, **1020** (**1971**).

[144] — and J. ROBERTS: Rings of invariants of reductive groups acting on regular rings on Cohen-Macaulay. Adv. in Math. **13**, **115** (**1974**).

[145] HORIKAWA, E.: Surjectivity of the period map of K3 surfaces of degree 2. Math. Annalen **228**, **113** (**1977**).

[146] —: Algebraic surfaces of general type with small c_1^2, I–V. Annals of Math. **104**, 357 (1976); Inv. Math. **37**, 121 (1976); Inv. Math. **47**, 209 (1978); Inv. Math. **50**, 103 (1969); J. Fac. Sci. Univ. Tokyo Sect. IA Math. **28**, 745 (1981).

[147] —: On the periods of Enriques surfaces I, II. Math. Annalen **234**, **73** (**1978**) and **235**, **217** (**1978**).

[148] HORROCKS, G.: A construction for locally free sheaves. Topology **7**, **117** (**1968**).

[149] —: Vector bundles on the punctured spectrum of a local ring. Proc. London Math. Soc. **14**, **689** (**1964**).

[150] — and D. MUMFORD: A rank 2 vector bundle on $\boldsymbol{P}^4$ with 15000 symmetries. Topology **12**, 63 (1973).

[151] HOYT, W.: On products and algebraic families of jacobian varieties. Annals of Math. **77**, (**1963**).

[152] HUBBARD, J.: Sur les sections analytiques de la courbe universelle de Teichmüller, Memoirs A. M. S., **166**, **1976**.
[153] HULEK, K.: Stable rank-2 vector bundles on $\mathbf{P}_2$ with c_1 odd. Math. Ann. **242**, 241 (1979).
[154] —: On the classification of stable rank r vector bundles over the projective plane. Progress in Math. 7, Birkhauser-Boston, Inc. (1981).
[155] HUMPHREYS, J.: On the hyperalgebra of a semi-simple algebraic group. Contributions to Algebra, Academic Press, N. Y., **203** (1977).
[156] — and J. JANTZEN: Blocks and indecomposable modules for semi-simple algebraic groups. J. Algebra **54**, **494** (1978).
[157] IGUSA, J.-I.: On the graded ring of theta constants I, II. Am. J. Math. **86**, **219** (1964) and **88**, **221** (1966).
[158] —: A desingularization problem in the theory of Siegel modular functions. Math. Annalen **168**, 228 (1967).
[159] —: On certain representations of semi-simple algebraic groups and the arithmetic of the corresponding invariants. Inv. Math. **12**, 62 (1971).
[160] —: Theta functions, Springer-Verlag, NY, **1972**.
[161] —: Geometry of absolutely admissible representations. Number Theory, Algebraic Geometry and Commutative Algebra, Kinokunüya, Tokyo, 1973.
[162] —: On the ring of modular forms of degree 2 over **Z**. Amer. J. Math. **101**, 149 (1979).
[163] KAC, V. G.: Concerning the question of description of the orbit space of a linear algebraic group. Uspekhi Math. Nauk **30**, **173** (1975).
[164] —: Some remarks on nilpotent orbits. J. of Algebra (1980).
[165] —: Infinite root systems, representations of graphs and invariant theory. Inv. Math. **56**, **57** (1980).
[166] KAC, V. G., V. POPOV and E. VINBERG: Sur les groupes algébriques dont l'algèbre des invariants est libre. Comptes Rendus de l'Acad. Sci **283 A**, 875 (1976).
[167] KADOMTSEV, B., and V. PETVIASHVILI: On the stability of solitary waves in weakly dispersing media. Soviet Physics Doklady **15**, **539** (1971).
[168] KALMAN, R.: Kronecker invariants and feedback. Ordinary Differential Equations, L. Weiss, ed., Academic Press, **1972**.
[169] KAS, A.: Ordinary double points and obstructed surfaces. Topology **16**, **51** (1977).
[170] KEMPF, G.: Some quotient varieties have rational singularities. Mich. Math. J. **24**, **347** (1977).
[171] —: Instability in invariant theory. Annals of Math. **108**, **299** (1978).
[172] —: Hochster-Roberts' theorem in invariant theory. Mich. Math. J. **26**, **19** (1979).
[173] —: Some quotient surfaces are smooth, Michigan Math. J. **27**, 295 (1980).
[174] —, F. KNUDSEN, D. MUMFORD, and B. SAINT-DONAT: Toroidal Embeddings I, Springer Lecture Notes **339**, **1973**.
[175] — and L. NESS: The length of vectors in representation spaces. Algebraic Geometry, Proceedings, Copenhagen **1978**, Lect. Notes in Math. **732**, **233** (1979).
[176] KNUDSEN, F.: The projectivity of the moduli space of stable curves, I, II and III. Math. Scand. **39**, 19 (1976), **52**, 161 (1983) and **52**, 200 (1983).
[177] KNUTSON, D.: Algebraic Spaces, Springer Lecture Notes **203** (1971).
[178] KRIČEVER, I. M.: Methods of Algebraic Geometry in the Theory of Non-linear Equations. Russian Math. Surveys **32**, 185 (1977).

[179] KULIKOV, V. S.: Degenerations of K3 surfaces. Izv. Akad. Nauk USSR translation **11**, 957 (1577).

[180] LAKSHMIBAI, V., C. MUSILI and C. S. SESHADRI: Geometry of G/P. Bull. Amer. Math. Soc. **1**, 432 (1979).

[181] LAKSOV, D.: Weierstrass points on curves. Young tableaux and Schur functions in algebra and geometry (Torun, 1980) Astérisque 87–88 (1981).

[182] LANG, S.: Diophantine Geometry. Interscience Press, **1964**.

[183] LANGTON, S.: Valuative criteria for families of vector bundles on algebraic varieties. Annals of Math. **101**, 88 (1975).

[184] LEW, John S.: Reducibility of matrix polynomials and their traces. J. of Appl. Math. and Physics **18**, 289 (1967).

[185] LICHTENBAUM, S., and M. SCHLESSINGER: The cotangent complex of a morphism. Trans. Amer. Math. Soc. **128**, 41 (1967).

[186] LIEBERMAN, D., and D. MUMFORD: Matsusaka's big theorem, in Algebraic Geometry — Arcata 1974. Proc. Symp. in Pure Math. **29**, AMS, 1975.

[187] LUNA, D.: Sur les orbites fermés des groupes algébriques reductifs. Invent. Math. **16**, 1 (1972).

[188] —: Slices Étales. Bull. Soc. Math. France, Mémoire **33**, 81 (1973).

[189] —: Adhérences d'orbit et invariants. Invent. Math. **29**, 231 (1975).

[190] — and R. RICHARDSON: A generalization of the Chevalley restriction theorem. Duke Math. Jour. **46**, 487 (1979).

[191] — and T. VUST: Une théoreme sur les orbites affines des groupes algébriques semi-simples. Ann. Scuol. Norm. Pisa **27**, 527 (1973).

[192] MARUYAMA, M.: Stable vector bundles on an algebraic surface. Nagoya Math. J., **58**, (1975).

[193] —: Moduli of stable sheaves I. J. Math. Kyoto **17**, 91 (1977).

[194] —: Moduli of stable sheaves II. J. Math. Kyoto **18**, 557 (1978).

[195] —: Boundedness of semi-stable sheaves of small ranks. Nagoya Math. J. **78**, 65 (1980).

[196] MATSUSAKA, T.: Polarized varieties with a given Hilbert polynomial. Am. J. Math. **94**, 1027 (1972).

[197] — and D. MUMFORD: Two fundamental theorems on deformations of polarized varieties. Am. J. Math. **86**, 668 (1964).

[198] MATSUSHIMA, Y.: Espaces homogènes de Stern des groupes de Lie complexes. Nagoya Math. Jour. **16**, 205 (1960).

[199] MEHTA, V., and C. S. SESHADRI: Moduli of vector bundles on curves with parabolic structures, Math. Annalen **248** (1980).

[200] MILLSON, J., and M. RAGHUNATHAN: Geometric construction of the cohomology of arithmetic groups. Proc. Indian Acad. Sci. Math. Sci. **90**, 103 (1981).

[201] MIRANDA, R.: On the stability of pencils of cubic curves, Am. J. Math., **102**, 1177 (1980).

[202] MIYATA, T., and M. NAGATA: Note on semi-reductive groups. J. Math. Kyoto **3**, 379 (1964).

[203] MOIŠEZON, B.: The algebraic analog of compact complex spaces with a sufficiently large field of meromorphic functions, I, II, III. Izv. Akad. Nauk **33** (1969) (translation in Math. USSR — Izv. **3**, 167, 305 (1969).

[204] MORRISON, D.: Semistable degenerations of Enriques' and hyperelliptic surfaces, Harvard Ph. D. Thesis, Feb. 1980.

[205] MORRISON, I.: Projective stability of ruled surfaces. Inv. Math. **56**, 269 (1980).

[206] MOUNT, K., and O. VILLAMAYOR: Weierstrass points as singularities of maps in arbitrary characteristic. J. Alg. **31**, 343 (1974).
[207] MUMFORD, D.: Picard groups of moduli problems. Arith. Alg. Geom., Harper and Row, 1963.
[208] —: *The boundary of moduli schemes* and *Further comments on boundary points*, unpublished lecture notes distributed at the AMS Summer Institute, Woods Hole, 1964.
[209] —: On the equations defining abelian varieties, I, II, III. Inv. Math. **1** (1966) and **3** (1967).
[210] —: Abelian quotients of the Teichmüller modular group, J. d'Anal. Math. **18**, 227 (1967).
[211] —: Varieties defined by quadratic equations. Quest. sulle var. alg., Edizioni Cremonese, Roma, 1969.
[212] —: Abelian Varieties. Tata Studies in Math., Oxford Univ. Press (1970), second edition (1975).
[213] —: Introduction to the theory of moduli. Algebraic Geometry, Oslo 1970, F. Oort, ed., Wolters-Noordhoff 171.
[214] —: Analytic construction of degenerating abelian varieties. Comp. Math. **24**, 239 (1972).
[215] —: Prym Varieties I. Contributions to Analysis, Academic Press, N. Y., 1974, p. 325.
[216] —: Curves and their Jacobians. Univ. of Mich. Press, Ann Arbor (1975).
[217] —: A new approach to compactifying locally symmetric varieties. Discrete subgroups of Lie groups, Oxford Univ. Press, 1975.
[218] —: Stability of projective varieties. Lectures given at the I. H. E. S. (1976), L'Ens. Math. **24** (1977).
[219] —: Theta functions and applications to non-linear differential equations. Les Presses de l'Univ. de Montréal, to appear.
[220] —: Tata lectures on theta functions I, II. Progress in Math. **28** and **43**, Birkhäuser (1983 and 1984).
[221] —: Hilbert's 14th problem — the finite generation of subrings such as rings of invariants, in A. M. S. Proc. Symp in Pure Math., **28**, 431 (1976).
[222] —: Bi-extensions of formal groups, in Proc. Bombay Colloq. Alg. Geom., Oxford Univ. Press (1969).
[223] —, and P. NEWSTEAD: Periods of a moduli space of bundles on curves. Am. J. Math. **90**, 1201 (1968).
[224] NAGATA, M.: On the fourteenth problem of Hilbert. Amer. J. Math. **81**, 766 (1959).
[225] —: Some questions on rational group actions. Alg. Geom., Oxford Univ. Press, Bombay, 1969.
[226] NAKAMURA, I.: Relative compactification of the Néron model. Complex Anal. and Alg. Geom., Iwanami-Shoten, Tokyo, 1977.
[227] NAMIKAWA, Y.: A new compactification of the Siegel space and degeneration abelian varieties, I, II. Math. Ann. **221**, 97, 201 (1976).
[228] —: Toroidal degeneration of abelian varieties, I., II. Part I in Complex Anal. and Alg. Geom., Iwanami-Shoten, Tokyo, 1977; Part II in Math. Ann. **245**, 117 (1979)
[229] —: Toroidal compactification of Siegel spaces, Springer Lecture Notes **812** (1980).
[230] NARASIMHAN, M. S., and M. V. NORI: Polarizations on an abelian variety. Proc. Indian Acad. Sci. Math. Sci. **90**, 125 (1981).

[231] — and S. Ramanan: Vector bundles on curves. Proc. Int. Colloq. on Alg. Geom., Oxford Univ. Press, Bombay, 1968.

[232] —, —: Moduli of vector bundles on a compact Riemann surface. Annals of Math. —, 14 (1969).

[233] —, —: Deformations of the moduli space of vector bundles on curves. Annals of Math. **101**, 391 (1975).

[234] —, —: Generalized Prym varieties as fixed points. J. Ind. Math. Soc. **39**, 1 (1975).

[235] Narasimhan, M. and S. Ramanan: Geometry of Hecke cycles I, in C. P. Ramanujam-a tribute, Springer-Verlag (1978).

[236] — and C. S. Seshadri: Holomorphic vector bundles on a compact Riemann surface. Math. Annalen **155**, 69 (1964).

[237] —, —: Stable and unitary vector bundles on a compact Riemann surface. Annals of Math. **82**, 540 (1965).

[238] — and R. R. Simha: Manifolds with ample canonical class. Inv. Math. **5**, 120 (1968).

[239] Nazarova, L. A.: Representation of a tetrad. Math. USSR-Izv. **1**, 1305 (1967).

[240] —: Representations of quivers of infinite type. Math. USSR-Izv. **7**, 749 (1973).

[241] Ness, L.: Mumford's numerical function and stable projective surfaces. Algebraic geometry (Copenhagen, 1978) Lecture Notes in Math. **732**, Springer (1979).

[242] Newstead, P. E.: Topological properties of some space of stable bundles. Topology **6**, 241 (1967).

[243] —: Stable bundles of rank 2 and odd degree over a curve of genus 2. Topology **7**, 205 (1968).

[244] —: Characteristic classes of stable bundles of rank 2 over an algebraic curve. Trans. Am. Math. Soc. **169**, 337 (1972).

[245] —: A non-existence theorem for families of stable bundles. J. London Math. Soc. **6**, 259 (1973).

[246] —: Rationality of moduli spaces of stable bundles. Math. Annalen **215**, 251 (1975). (Correction in **249**, 281 (1980).)

[247] —: Introduction to moduli problems and orbit spaces. Tata Inst. Lecture Notes, Springer-Verlag, 1978.

[248] —: Invariants of pencils of binary cubics. Math. Proc. Camb. Phil. Soc. **89**, 201 (1981).

[249] Norman, P.: An algorithm for computing local moduli of abelian varieties. Annals of Math. **101**, 499 (1975).

[250] —: Intersection of components of the moduli space of abelian varieties. J. pure appl. alg. **13**, 105 (1978).

[251] —: Quadratic obstructions to extending polarizations on abelian varieties. Proc. AMS, to appear.

[252] — and F. Oort: Moduli of abelian varieties. Annals of Math. **112**, 413 (1980).

[253] Oda, Tadao: On Mumford's conjecture concerning reducible rational representations of algebraic linear groups. J. Math. Kyoto Univ. **3**, 275 (1963).

[254] — and C. S. Seshadri: Compactification of the generalized Jacobian variety. Trans. AMS **253**, 1 (1979).

[255] Oda, Takayaki: On theta series of quadratic forms and modular forms adjunct to spherical functions. Deformation theory of linear differential equations and a new viewpoint for extension of abelian function theory (Kyoto, 1980), Kyoto University Press (1980).

[256] Ogawa, R. H.: On the points of Weierstrass in dimensions greater than one. Trans. AMS **184**, 401 (1973).

[257] Okonek, M. Schneider, and H. Spindler: Vector bundles on complex projective spaces. Progress in Math. **3**, Birkhäuser, Boston (1980).

[258] Olsen, B.: On higher Weierstrass points. Annals of Math. **95**, 357 (1972).

[259] Oort, F.: Finite group schemes, local moduli for abelian varieties and lifting problems. Algebraic Geometry, Oslo 1970, Wolters-Noordhoff, Groningen, 1972.

[260] Oort, F. and J. Steenbrink: The local Torelli problem for algebraic curves, Proc. of Angiers Conference, Noordhof, 1980.

[261] Parthasarathy, R.: Holomorphic forms in $\Gamma\backslash G/K$ and Chern classes. Topology **21**, 157 (1982).

[262] Pearcy, C.: A complete set of unitary invariants for 3×3 complex matrices. Trans. AMS **104**, 425 (1962).

[263] Persson, U.: On degenerations of algebraic surfaces. Memoirs of the A.M.S., v. 11, no. 189 (1977).

[264] — and H. Pinkham: Degeneration of surfaces with trivial canonical bundle, Annals of Math., **112**, (1980).

[265] Popov, A. M.: Irreducible simple linear Lie groups with finite stationary subgroups of general position. Funct. Anal. Appl. **9**, 346 (1976).

[266] —: Stationary subgroups of general position for certain actions of simple Lie groups. Funct. Anal. Appl. **10**, 239 (1977).

[267] Popov, V. L.: Stability criteria for the actions of a semisimple group on an algebraic manifold. Izv. Akad. Nauk. SSSR, Ser. Math. **4**, 527 (1970).

[268] —: Reperesentations with a free module of covariants. Funct. Anal. Appl. **10**, 242 (1977).

[269] —: On Hilbert's theorem on invariants. Dokl. Akad. Nauk. **249**, 551 (1979).

[270] Popp, H.: On the moduli of algebraic varieties I, II, III. Inv. Math. **22**, 1 (1973); Comp. Math. **28**, 51 (1974) and **31**, 237 (1975).

[271] —: Modulräume algebraischer mannigfaltigkeiten, Springer Lecture Notes **412**, 219 (1974).

[272] Platetskii-Shapiro, I., and I. Safarevic: A Torell theorem for albebraic surfaces of type K3. Izv. Akad. Nauk. **35** (1971), transl. Math. USSR Izv. **5**, 547 (1971).

[273] Procesi, C.: Non-commutative affine rings. Atti Acc. Naz. Lincei **8**, 239 (1967).

[274] —: The invariant theory of $n\times n$ matrices, Adv. in Math. **19**, 306 (1976).

[275] —: Young diagrams, standard monomials and invariant theory, Proc. Int. Cong. Math., Helsinki (1978).

[276] Ramanan, S.: The moduli spaces of vector bundles over an algebraic curve. Math. Annalen **200**, 69 (1973).

[277] —: Orthogonal and spin bundles over hyperelliptic curves. Proc. Indian Acad. Sci. Math. Sci. **90**, 151 (1981).

[278] —: Vector bundles on algebraic curves. Proc. Int. Cong. Math., Helsinki, 1978.

[279] Rasmyslev, J.: Trace identities of matrix full algebras over a field of char. 0. Izv. Akad. Nauk. **8**, 727 (1974).

[280] Reid, M.: Bogomolov's theorem $c_1^2 \leqq 4c_2$. Proc. Int. Symp. Alg. Geom. (Kinokuniya, Tokyo, 1978), p. 623.

[281] Richardson, R.: Affine coset spaces of reductive groups. Bull. Lond. Math. Soc. **9**, 38 (1977).

[282] RIVLIN, R. S.: Viscoelastic fluids. In: Frontiers of Research in Fluid Dynamics, G. Temple, R. Seeger, eds., Wiley (Intersience), New York 1965.
[283] ROSENLICHT, M.: Some rationality questions on algebraic groups. Annal. Mat. Pura Appl. **43**, 25 (1957).
[284] —: A remark on quotient spaces. An Acad. Brasil **35**, 487 (1963).
[285] ROUSSEAU, G.: Immeubles sphériques et théorie des invariants. C. R. Acad. Sci. Paris **286**, 247 (1978).
[286] SAINT-DONAT, B.: Varieties de translation et theoreme de Torelli. Comptes Rendus de l'Acad. Sci. **280**, 1611 (1975).
[287] SATAKE, I.: On the compactification of the Siegel space. J. Ind. Math. Soc. **20**, 259 (1956).
[288] SATO, M., and T. KIMURA: A classification of irreducible pre-homogeneous vector spaces and their relative invariants. Nagoya Math. J. **65**, 1 (1977).
[289] SCHLESSINGER, M.: Functors on Artin rings. Trans. AMS **130**, 208 (1968).
[290] SCHMID, W.: Variations of Hodge structure: the singularities of the period mapping. Inv. Math. **22**, 211 (1973).
[291] SCHOUTEN, J. A.: Klassifizierung der alternierenden Größen dritten Grades in 7. Dimension. Rend Cir. Mat. Palermo **55** (1931).
[292] SCHWARZ, G.: Representations of simple Lie groups with regular rings of invariants. Inv. Math. **49**, 167 (1978).
[293] —: Representations of simple Lie groups with a free module of covariants. Inv. Math. **50**, 1 (1878).
[294] SCHWARZENBERGER, R.: Vector bundles on the projective plane. Proc. London Math. Soc. **11**, 623 (1961).
[295] SEILER, W. K.: Quotientenprobleme in der Invariantentheorie. Diplomarbeit 1977, Univ. Karlsruhe.
[296] SEKIGUCHI, T.: On projective normality of abelian varieties I, II. J. Math. Soc. Japan **28**, 307 (1976) and **29**, 709 (1977).
[297] —: On the cubics defining abelian varieties. J. Math. Soc. Japan **30**, 703 (1978).
[298] SESHADRI, C. S.: On a theorem of Weitzenböck in invariant theory, Mem. College Sci Kyoto **1** (1962).
[299] —: Space of unitary vector bundles on a compact Riemann surface. Annals of Math. **85**, 303 (1967).
[300] —: Mumford's conjecture for GL(2) and applications. Proc. Int. Colloq. on Alg. Geom., Oxford University Press, 347 (1968).
[301] —: Moduli of π-vector bundles over an algebraic curve. Proc. C. I. M. E., Varenna, 1969.
[302] —: Quotient spaces modulo reductive algebraic groups and applications to moduli of vector bundles on algebraic curves. Actes du Congrès Int. des Math., Nice, 1970, p. 479.
[303] —: Quotient spaces modulo reductive algebraic groups. Annals of Math. **95**, 511 (1972).
[304] —: Theory of moduli. Proc. Symp. in Pure Math. **29**: Alg. Geom., AMS, 263 (1975).
[305] —: Geometric reductivity over an arbitrary base. Adv. in Math. **26**, 225 (1977).
[306] —: Moduli of vector bundles with parabolic structure. Bull. of the AMS **83**, 124 (1977).
[307] —: Desingularization of the moduli varieties of vector bundles on curves. Proc. Int. Symp. Alg. Geom. Kyoto, 1977, Kinokuniya, Tokyo, 155 (1978).

[308] SHAH, J.: Surjectivity of the period map in the case of quartic surfaces and sextic double planes. Bull. AMS **82**, 716 (1976).
[309] —: Stability of local rings of dimension 2. Proc. Nat. Acad. Sci. **75**, 4085 (1978).
[310] —: Insignificant limit singularities of surfaces and their mixed Hodge structure. Annals of Math. **109**, 497 (1979).
[311] —: A complete moduli space for K3 surfaces of degree 2. Annals of Math. **112**, (1980).
[312] —: Degeneration of K3-surfaces of degree 4. Trans. Amer. Math. Soc. **263**, 271 (1981).
[313] SHIODA, T.: On the graded ring of invariants of binary octavics. Amer. J. Math. **89**, 1022 (1967).
[314] SMITH, G. F.: A complete set of unitary invariants for N 3×3 complex matrices. Tensor **21**, 273 (1970).
[315] SPECHT, W.: Zur theorie der matrizen II. Jahresberichten der Deutsche Math. Ver. **50**, 19 (1940).
[316] SPENCER, A. J. M.: Theory of Invariants. "Continuum physics", vol. I, ed. by C. Eringen, Academic Press, 1971.
[317] STEENBRINK, J.: Limits of Hodge structures. Inv. Math. **31**, 229 (1976).
[318] STEINBERG, R.: Representations of algebraic groups. Nagoya Math. J. **22**, 33 (1963).
[319] SUMIHIRO, H.: Equivariant completion I and II. J. Math. Kyoto **14**, 1 (1974) and **15**, 573 (1975).
[320] TAKEUCHI, M.: A note on geometrically reductive groups. J. Fac. Sci. Univ. Tokyo **20**, 387 (1973).
[321] TJURIN, A. N.: Analog of Torelli's theorem for 2-dimensional bundles over algebraic curves of arbitrary genus. Izv. Akad. Nauk **33**, 1149 (1969) (translation: **3**, 1081 (1969)).
[322] —: Analogs of Torelli's theorem for multi-dimensional vector bundles over an arbitrary curve. Izv. Akad. Nauk **34** (1970) (translation: **4**, 343 (1970)).
[323] —: The geometry of the Fano surface of a non-singular cubic 3-fold and Torelli theorems for Fano surfaces and cubics. Izv. Akad. Nauk **35**, 498 (1971).
[324] —: The geometry of moduli of vector bundles. Uspekhi Mat. Nauk **29**, 57 (1974).
[325] —: Finite-dimensional bundles on infinite varieties. Izv. Akad. Nauk **40**, 1248 (1976).
[326] TODOROV, A.: A construction of surfaces with $p_g = 1$, $q = 0$ and $2 \leqq K^2 \leqq 8$. Inv. Math. **63**, 287 (1981).
[327] VIEHWEG, E.: Klassifikationstheorie algebraischer Varietäten der Dimension drei. Compositio Math. **41**, 361 (1980).
[328] VUST, T.: Sur la théorie classique des invariants, Comm. Math. Helv., **52**, 259 (1977).
[329] WAHL, J.: Deformations of plane curves with nodes and cusps. Am. J. Math. **96**, 529 (1974).
[330] WEIL, A.: Généralisation des fonctions abéliennes. J. Math. pures et appl. **17**, 47 (1938).
[331] WEYL, H.: The Classical Groups. Princeton Univ. Press, 1946.
[332] WIRTINGER, W.: Lie's Translations mannigfaltigkeiten und Abelsche Integrale. Monatsh. für Math. und Phys. **46**, 384 (1938).
[333] ZARISKI, O.: Interprétations algébro-géométriques de quatorzième probleme de Hilbert. Bull. Sci. Math. France **78**, 155 (1954).

[334] BOGOMOLOV, F.: Holomorphic tensors and vector bundles on projective varieties. Izv. Akad. Nauk SSR **42**, 1227 (1978).
[335] GUILLEMIN, V. and S. STERNBERG: Geometric quantization and multiplicities of group representations. Invent. Math. **67**, 515 (1982).
[336] —: Convexity properties of the moment mapping I, II. Invent. Math. **67**, 491 (1982) and **77**, 533 (1984).
[337] HARRIS, J. and D. MUMFORD: On the Kodaira dimension of the moduli space of curves. Invent. Math. **67**, 23 (1982).
[338] KAMBAYASHI, T.: Projective representation of algebraic linear groups of transformations. Amer. J. Math. **88**, 199 (1966).
[339] MARTENS, H.: A new proof of Torelli's theorem. Annals of Math. **78** (1963).
[340] RAMANAN, S. and A. RAMANATHAN: Some remarks on the instability flag. Tohoku Math. J. (2) 36, 269 (1984).
[341] RAYNAUD, M.: Faisceaux amples sur les schemas en groupes et les espaces homogenes. Springer Lecture Notes **119** (1970).
[342] DOLGACHEV, I.: Cohomologically insignificant degenerations of algebraic varieties. Comp. Math. **42**, 279 (1981).
[343] FOGARTY, J.: Invariant Theory. Benjamin (1969).
[344] ANDREOTTI, A. and A. MAYER: On period relations for Abelian integrals on algebraic curves. Ann. Scu. Norm. Sup. Pisa (1967).

New references

[345] ADLER, M. and P. VAN MOERBEKE: The intersection of four quadrics in P^6, abelian surfaces and their moduli. Math. Ann. **279**, 25 (1987).
[346] AKBULUT, S. and J. MCCARTHY: Casson's invariant for oriented homology spheres. Math. Notes **36**, Princeton University Press (1990).
[347] AKHIEZER, D.: Modality and complexity of the actions of reductive groups. Russian Math. Surveys **43**, 157 (1988).
[348] AKYILDIZ, E.: On the G_m decomposition of G/P. J. Fac. Sci. Karadeniz Tech. Univ. **4**, 1 (1981).
[349] ARBARELLO, E.: Fay's trisecant formula and a characterization of Jacobian varieties. Algebraic geometry (Bowdoin, 1985), Proc. Sympos. Pure Math. **46**, Vol. 1, 49 (1988).
[350] — and M. CORNALBA: A few remarks about the variety of plane curves of given degree and genus. Ann. Sci. Ecole Norm. Sup. **16**, 467 (1983).
[351] — and M. CORNALBA: The Picard groups of the moduli spaces of curves. J. Diff. Geom. **34**, 839 (1987).
[352] —, M. CORNALBA, P. GRIFFITHS, and J. HARRIS: Geometry of algebraic curves I. Springer (1985).
[353] — and C. DE CONCINI: On a set of equations characterizing Riemann matrices. Ann. Math. (2) 120, 119 (1984).
[354] — and C. DE CONCINI: Another proof of a conjecture of S. P. Novikov on periods of abelian integrals on Riemann surfaces. Duke Math. J. **54**, 163 (1987).
[355] — and E. SERNESI: The equation of a plane curve. Duke Math. J. **46**, 469 (1979).
[356] ARMS, J., R. CUSHMAN, and M. GOTAY: A universal reduction procedure for Hamiltonian group actions. The geometry of Hamiltonian systems (T. Ratiu, editor), MSRI publications **20**, Springer (1991).

[357] ARMS, J., M. GOTAY, and G. JENNINGS: Geometric and algebraic reduction for singular momentum maps. Advances in Math. **79**, 43 (1990).

[358] ARMS, J., J. MARSDEN, and V. MONCRIEF: Symmetry and bifurcations of moment mappings. Comm. Math. Phys. **78**, 455 (1981).

[359] ATIYAH, M.: Convexity and commuting Hamiltonians. Bull. London Math. Soc. **23**, 1 (1982).

[360] —: Angular momentum, convex polyhedra and algebraic geometry. Proc. Edinburgh Math. Soc. **26**, 121 (1983).

[361] —: The moment map in symplectic geometry. Durham Symposium on global Riemannian geometry **43**, Ellis Horwood (1984).

[362] —: Instantons in two and four dimensions. Comm. Math. Phys. **93**, 437 (1984).

[363] —: New invariants for 3 and 4 dimensional manifolds. The mathematical heritage of Hermann Weyl, Proc. Symp. Pure Math. **48**, 285 (1988).

[364] — and R. BOTT: The moment map and equivariant cohomology. Topology **23**, 1 (1984).

[365] —, V. DRINFELD, N. HITCHIN, and Y. MANIN: Construction of instantons. Phys. Letters **A 65**, 185 (1978).

[366] — and N. HITCHIN: Low-energy scattering of non-Abelian monopoles. Phil. Trans. R. Soc. London **315**, 459 (1985).

[367] — and N. HITCHIN: The geometry and dynamics of magnetic monopoles. Princeton Univ. Press (1988).

[368] — and J. JONES: Topological aspects of Yang-Mills theory. Comm. Math. Phys. **61**, 97 (1978).

[369] — and A. PRESSLEY: Convexity and loop groups. Progress in Math. **36**, 33 (1983).

[370] AUDIN, M.: The topology of torus actions on symplectic manifolds. Progress in Math. **93**, Birkhäuser, 1991.

[371] BALAJI, V.: Cohomology of certain moduli spaces of vector bundles. Proc. Indian Acad. Sci. Math. Sci. **98**, 1 (1988).

[372] —: Intermediate Jacobian of some moduli spaces of vector bundles on curves. Amer. J. Math. **112**, 611 (1990).

[373] — and C. SESHADRI: Cohomology of a moduli space of vector bundles. The Grothendieck Festschrift, Vol. I, Progress in Math. **86**, **87**, Birkhäuser, 1990.

[374] BALLICO, E.: Stable rationality for the variety of vector bundles over an algebraic curve. J. London Math. Soc. (2) 30, 21 (1984).

[375] — and P. OLIVERIO: Stability of pencils of plane quartic curves. Alti Accad. Naz. Lincei Rend. Cl. Sci. Fis. Mat. Natur (8) 74, 234 (1983).

[376] BANICA, C. and N. MANOLACHE: Rank 2 stable vector bundles on $P^3(C)$ with Chern classes $c_1 = -1$ and $c_2 = 4$. Math. Z. **190**, 315 (1985).

[377] BARSOTTI, I.: A new look for thetas. Theta functions (Bowdoin, 1987) Amer. Math. Soc. (1989).

[378] BARTH, W. : Irreducibility of the space of mathematical instanton bundles with rank 2. Math. Ann. **258**, 81 (1981/2).

[379] —: Lectures on mathematical instanton bundles. Gauge theories: fundamental interactions and rigorous results, Progress in Physics **5**, Birkhäuser, 1982.

[380] BASS, H. and W. HABOUSH: Linearizing certain reductive group actions. Trans. Amer. M. Soc. **292**, 463 (1985).

[381] BEAUVILLE, A.: Variétés kählériennes dont la première classe de Chern est nulle. J. Diff. Geom. **18**, 755 (1983).

[382] —: Variétés kählériennes compactes avec $c_1 = 0$. Astérisque **126**, 181 (1985).

[383] —: Fibrés de rang 2 sur un courbe, fibré déterminant et fonctions théta I, II. Bull. Soc. Math. France **116**, 431 (1988) and **119**, 259 (1991).

[384] — and O. DEBARRE: Une relation entre deux approches du problème de Schottky. Inv. Math. **86**, 195 (1986).

[385] —, M. NARASIMHAN, and S. RAMANAN: Spectral curves and the generalised theta divisor. J. Reine Angew. Math. **398**, 169 (1989).

[386] BEILINSON, A., J. BERNSTEIN, P. DELIGNE, and O. GABBER: Faisceaux pervers. Astérisque **100** (1982).

[387] BERLINE, N. and M. VERGNE: Zéros d'un champ de vecteurs et classes caractéristiques équivariantes. Duke Math. J. **50**, 539 (1983).

[388] BERTIN, J.: Symetries des fibrés vectoriels sur P^n et nombre d'Euler. Duke Math. J. **49**, 807 (1982).

[389] BERTRAM, A.: Moduli of rank 2 vector bundles, theta divisors, and the geometry of curves in projective space. J. Diff. Geom. **35**, 429 (1992).

[390] —: A partial verification of the Verlinde formulae for vector bundles of rank 2. Harvard preprint.

[391] —: Stable pairs and stable parabolic pairs. Harvard preprint.

[392] —: Generalised $SU(2)$ theta functions. Harvard preprint.

[393] — and B. FEINBERG: On stable rank 2 bundles with canonical determinant and many sections. Harvard preprint.

[394] — and A. SZENES: Hilbert polynomials of moduli spaces of rank 2 vector bundles II. To appear in Topology.

[395] BESSENROD, C. and L. LEBRUYN: Stable rationality of certain PGL_n-quotients. Inv. Math. **104**, 179 (1991).

[396] BHOSLE, U.: Moduli of orthogonal and spin bundles over hyperelliptic curves. Comp. Math. **51**, 15 (1984).

[397] —: Moduli of parabolic G-bundles. Bull. Amer. Math. Soc. **20**, 45 (1989).

[398] —: Parabolic vector bundles on curves. Ark. Mat. **27**, 15 (1989).

[399] —: Pencils of quadrics and hyperelliptic curves in characteristic two. J. Reine Angew. Math. Math. **407**, 75 (1990).

[400] —: Generalised parabolic sheaves on an integral projective curve. Proc. Indian Acad. Sci. Math. Sci. **102**, 23 (1992).

[401] — and A. RAMANATHAN: Moduli of parabolic G-bundles on curves. Math. Z. **202**, 161 (1989).

[402] BIALYNICKI-BIRULA, A.: Some theorems on actions of algebraic groups. Ann. Math. **98**, 480 (1973).

[403] —: On actions of SL(2) on complete algebraic varieties. Pacific J. Math. **86**, 53 (1980).

[404] —, J. CARRELL, P. RUSSELL, and D. SNOW (eds.): Group actions and invariant theory. Proceedings of the 1988 Montreal Conference, CMS Conference Proceedings **10**, Amer. Math. Soc. (1989).

[405] — and A. SOMMESE: Quotients by C^* and SL(2,C) actions. Trans. Amer. Math. Soc. **279**, 773 (1983).

[406] — and A. SOMMESE: Quotients by $C^* \times C^*$ actions. Trans. Amer. Math. Soc. **289**, 519 (1985).

[407] — and J. SWIECICKA: Complete quotients by algebraic torus actions. Group actions and vector fields (Vancouver, 1981), Lecture Notes in Math. **956**, Springer, 1982.

[408] — and J. SWIECICKA: Generalized moment functions and orbit spaces. Amer. J. Math. **109**, 229 (1987).

[409] — and J. SWIECICKA: On complete orbit spaces of SL(2) actions. Colloq. Math. **55**, 229 (1988).

[410] — and J. SWIECICKA: A reduction theorem for the existence of good quotients. Amer. J. Math. **113**, 189 (1991).

[411] BIFET, E.: On complete symmetric varieties. Adv. Math. **80**, 225 (1990).

[412] BOGOMOLOV, F.: Unstable vector bundles and curves on surfaces. Proc. Int. Cong. Math. (Helsinki, 1978) Acad. Sci. Fennica (1980).

[413] — and KATSYLO, P.: Rationality of some quotient varieties. Math. USSR Sbornik **54**, 571 (1986).

[414] BOTT, R.: Nondegenerate critical manifolds. Ann. Math. **60**, 248 (1954).

[415] —: On some recent interactions between mathematics and physics. Canad. Math. Bull. **28**, 129 (1985).

[416] —: On E. Verlinde's formula in the context of stable bundles. Intern. J. Modern Phys. **A 6**, 2847 (1991).

[417] —: Stable bundles revisited. Surveys in differential geometry 1, 1. Lehigh University, Bethlehem PA (1991).

[418] BOURGUIGNON, J.-P. and H. B. LAWSON: Yang-Mills theory: its physical origins and differential geometric aspects. Seminar on Differential Geometry (editor S.-T. Yau), Ann. Math. Studies **102**, Princeton (1982).

[419] BOUTOT, J.-F.: Singularités rationelles et quotients par les groupes réductifs. Inv. Math. **88**, 65 (1987).

[420] BOYER, C. and B. MANN: Homology operations on instantons. J. Diff. Geom. **28**, 423 (1988).

[421] BOYER, C., J. HURTUBISE, B. MANN, and R. MILGRAM: The topology of instanton moduli spaces I: the Atiyah-Jones conjecture. To appear in Ann. Math.

[422] BRADLOW, S.: Special metrics and stability for holomorphic bundles with global sections. J. Diff. Geom. **33**, 169 (1991).

[423] — and G. DASKALOPOULOS: Moduli of stable pairs for holomorphic bundles over Riemann surfaces I and II. International J. Math. **2**, 477 (1991).

[424] BRAMBILA PAZ, L.: Existence of certain universal extensions. Algebraic geometry and complex analysis (Pátzcuaro, 1987), Lecture Notes in Math. **1414**, Springer, 1989.

[425] —: Algebras of endomorphisms of semistable vector bundles of rank 3 over a Riemann surface. J. Alg. **123**, 414 (1989).

[426] —: Moduli of endomorphisms of semistable vector bundles over a compact Riemann surface. Glasgow Math. J. **32**, 1 (1990).

[427] BRION, M.: Invariants de plusieurs formes binaires. Bull. Soc. Math. France **110**, 429 (1982).

[428] —: Invariants d'un sous-groupe unipotent maximal d'un groupe semisimple. Ann. Inst. Fourier (Grenoble) **33**, 1 (1983).

[429] —: Sur l'image de l'application moment. Lecture Notes in Math. **1296**, Springer, 1987.

[430] —: Groupe de Picard et nombres caractéristiques des variétés sphériques. Duke Math. J. **58**, 397 (1989).

[431] —: Cohomologie équivariante des points semi-stables. C. R. Acad. Sci. Paris (I) 311, 281 (1990).

[432] —: Cohomologie équivariante des points semi-stables. J. Reine Angew. Math. **421**, 125 (1991).

[433] — and D. LUNA: Sur la structure locale des variétés sphériques. Bull. Soc. Math. France **115**, 211 (1987).
[434] — and C. PROCESI: Action d'un tore dans une variété projective. Operator algebras, unitary representations, enveloping algebras and invariant theory (Paris, 1989), Progress in Math. **92**, 509, Birkhäuser, 1990.
[435] BRUN, J. and A. HIRSCHOWITZ: Droites de saut des fibrés stables de rang élevé sur P_2. Math. Z. **181**, 171 (1982).
[436] BUCHDAHL, N.: Hermitian Einstein connections and stable vector bundles over compact complex surfaces. Math. Ann. **280**, 625 (1988).
[437] BURNOL, J.-F.: Remarques sur la stabilité en arithmétique. Preprint, Ecole Polytechnique.
[438] BUSER, P. and P. SARNAK: On the period matrix of a Riemann surface of large genus. Preprint.
[439] CAPORASO, L.: On a compactification of the universal Picard variety over the moduli space of stable curves. Harvard Ph.D. thesis (1993).
[440] CARRELL, J. and M. GORESKY: A decomposition theorem for the integral homology of a variety. Inv. Math. **73**, 367 (1983).
[441] CARRELL, J. and A. SOMMESE: $\mathbf{C}^*$ actions. Math. Scand. **43**, 49 (1978).
[442] CARRELL, J. and A. SOMMESE: Some topological aspects of $\mathbf{C}^*$ actions on compact Kähler manifolds. Comment. Math. Helvetici **54**, 567 (1979).
[443] CATANESE, F.: Moduli of surfaces of general type. Algebraic geometry — open problems (Ravello, 1982), Lecture Notes in Math. **997**, Springer, 1983.
[444] —: Moduli and classification of irregular Kaehler manifolds (and algebraic varieties) with Albanese general type fibrations. Inv. Math. **104**, 263 (1991).
[445] —: Chow varieties, Hilbert schemes and moduli spaces of general type. J. Alg. Geom. **1**, 561 (1992).
[446] CATTANI, E., A. KAPLAN, and W. SCHMID: Degeneration of Hodge structures. Ann. Math. (2) 123, 457 (1986).
[447] CHAI, C.: Compactification of Siegel moduli schemes. London Math. Soc. Lecture Notes **107**, Cambridge (1985).
[448] —: Arithmetic compactifications of the Siegel moduli space. Proc. A.M.S. Symp. Pure Math. **49**, 19 (1989).
[449] — and G. FALTINGS: Degeneration of abelian varieties. Springer, 1990.
[450] CHANG, M.: Stable rank 2 bundles on P^3 with $c_1 = 0$, $c_2 = 4$, $\alpha = 1$. Math. Z. **184**, 487 (1983).
[451] —: Stable rank 2 reflexive sheaves on P^3 with large c_3. J. Reine Angew. Math. **343**, 99 (1983).
[452] — and Z. RAN: Unirationality of the moduli spaces of curves of genus 11, 13 (and 12). Inv. Math. **76**, 41 (1984).
[453] — and Z. RAN: The Kodaira dimension of the moduli space of curves of genus 15. J. Diff. Geom. **24**, 205 (1986).
[454] — and Z. RAN: On the slope and Kodaira dimension of $\overline{\mathcal{M}}_g$ for small g. J. Diff. Geom. **34**, 267 (1991).
[455] CHARNEY, R. and R. LEE: Moduli space of stable curves from a homotopy viewpoint. J. Diff. Geom. **20**, 185 (1984).
[456] CILIBERTO, C. and G. VAN DER GEER: Subvarieties of the moduli space of curves parametrizing Jacobians with non-trivial endomorphisms. Amer. J. Math. **114**, 551 (1991).

[457] CILIBERTO, C., G. VAN DER GEER, and M. TEIXIDOR I BIGAS: On the number of parameters of curves whose Jacobians possess non-trivial endomorphisms. J. Alg. Geom. **1**, 215 (1992).
[458] CLEMENS, C.: Double solids. Adv. Math. **47**, 107 (1983).
[459] COHEN, R. COHEN, B. MANN, and R. MILGRAM: The topology of rational functions and divisors of surfaces. Acta Math. **166**, 163 (1991).
[460] CORLETTE, K.: Flat G-bundles with canonical metrics. J. Diff. Geom. **28**, 361 (1988).
[461] CORNALBA, M.: Systèmes pluricanoniques sur l'espace des modules des courbes et diviseurs de courbes k-gonales (d'apres Harris et Mumford). Seminar Bourbaki 1983/84, Astérisque 121–122, 7 (1985).
[462] —: On the projectivity of the moduli space of curves. Univ. of Pavia preprint.
[463] — and J. HARRIS: Divisor classes associated to families of stable varieties, with applications to the moduli space of curves. Ann. Sci. Ecole Norm. Sup. (4) 21, 455 (1988).
[464] CUKIERMAN, F. and L. FONG: On higher Weierstrass points. Duke Math. J. **62**, 179 (1991).
[465] DANILOV, V: The geometry of toric varieties. Russ. Math. Surv. **33**, 97 (1978).
[466] DASKALOPOULOS, G.: The topology of the space of stable bundles on a compact Riemann surface. To appear in J. Diff. Geom.
[467] — and K. UHLENBECK: An application of transversality to the topology of the space of stable bundles. Preprint (1992).
[468] — and R. WENTWORTH: Local degeneration of the moduli space of vector bundles and factorization of rank two theta functions I. To appear in Math. Ann.
[469] — and R. WENTWORTH: Geometric quantization for the moduli space of vector bundles with parabolic structure. To appear in Duke Math. J.
[470] DEBARRE, O.: Trisecant lines and Jacobians. J. Alg. Geom. **1**, 5 (1992).
[471] DECKER, W.: Über den Modul-Raum für stabile 2-Vectorbündel über P_3 mit $c_1 = -1$, $c_2 = 2$. Manuscripta Math. **42**, 211 (1983).
[472] —: Monads and cohomology modules of rank 2 vector bundles. Comp. Math. **76**, 7 (1990).
[473] DE CONCINI, C. and C. PROCESI: Symmetric functions, conjugacy classes and the flag variety. Inv. Math. **64**, 203 (1981).
[474] —: Complete symmetric varieties. Invariant theory (Montecatini, 1982) Lecture Notes in Math. **996**, Springer, 1983.
[475] —: Complete symmetric varieties II. Algebraic groups and related topics (Kyoto/Nagoya, 1983) Adv. Stud. Pure Math. 6, North-Holland, 1985.
[476] DE CONCINI, C. and T. SPRINGER: Betti numbers of complete symmetric varieties. Geometry today (Rome, 1984) Progress in Math. **60**, Birkhäuser, 1985.
[477] DEL CENTINA, A. and S. RECILLAS: On a property of the Kummer variety and a relation between two moduli spaces of curves. Algebraic geometry and complex analysis (Pátzcuaro, 1987), Lecture Notes in Math. **1414**, Springer, 1989.
[478] DIAZ, S.: A bound on the dimensions of complete subvarieties of $\mathcal{M}_g$. Duke Math. J. **51**, 405 (1984).
[479] —: Tangent spaces in moduli via deformations with applications to Weierstrass points. Duke Math. J. **51**, 905 (1984).

[480] —: Exceptional Weierstrass points and the divisor on the moduli space that they define. Mem. Amer. Math. Soc. **56**, 327 (1985).

[481] DIMCA, A.: On the algebraic structures and the automorphism groups of the nonsingular projective hypersurfaces. Rev. Roumaine Math. Pures Appl. **24**, 545 (1979).

[482] DIXMIER, J.: Sur les invariants des formes binaires. C. R. Acad. Sci. Paris Ser. I Math. **292**, 987 (1981).

[483] —: Séries de Poincaré et systèmes de parametres pour les invariants des formes binaires. Acta. Math. **45**, 151 (1983).

[484] —: Quelques résultats et conjectures concernant les séries de Poincaré des invariants des formes binaires. Dubreil and Malliavin Seminar (Paris, 1983–84), Lecture Notes in Math. **1146**, Springer, 1985.

[485] —: On the projective invariants of quartic plane curves. Adv. in Math. **64**, 279 (1987).

[486] —: Quelques aspects de la théorie des invariants. Gaz. Math. **43**, 39 (1990).

[487] — and D. LAZARD: Le nombre minimum d'invariants fondamentaux pour les formes binaires de degré 7. Portugal. Math. **43**, 377 (1985/86).

[488] — and M. RAYNAUD: Sur le quotient d'une variété algébrique par un groupe algébrique. Adv. Math. Suppl. Stud. **7A**, 327 (1981).

[489] DOLGACHEV, I.: Rationality of fields of invariants. Algebraic geometry, Bowdoin (Brunswick, Maine, 1985) Proc. Sympos. Pure Math. **46**, 3, Amer. Math. Soc. (1987).

[490] — and Y. HU: Variation of quotients in geometric invariant theory and applications to moduli problems. Michigan preprint (1992).

[491] DONAGI, R.: The unirationality of $\mathscr{A}_5$. Ann. Math. (2) 119, 269 (1984).

[492] —: Non-Jacobians in the Schottky loci. Ann. Math. **126**, 193 (1987).

[493] —: Big Schottky. Inv. Math. **89**, 569 (1987).

[494] —: The Schottky problem. Theory of moduli (Montecatini, 1985) Lecture Notes in Math. **1337**, 84, Springer, 1988.

[495] DONALDSON, S.: A new proof of a theorem of Narasimhan and Seshadri. J. Diff. Geom. **18**, 269 (1983).

[496] —: An application of gauge theory to four-dimensional topology. J. Diff. Geom. **18**, 279 (1983).

[497] —: Instantons and geometric invariant theory. Comm. Math. Phys. **93**, 453 (1984).

[498] —: Nahm's equations and the classification of monopoles. Comm. Math. Phys. **93**, 453 (1984).

[499] —: Anti self-dual Yang-Mills connections over complex algebraic surfaces and stable vector bundles. Proc. London Math. Soc. (3) 50, 1 (1985).

[500] —: Connections, cohomology and the intersection forms of 4- manifolds. J. Diff. Geom. **24**, 275 (1986).

[501] —: Irrationality and the h-cobordism conjecture. J. Diff. Geom. **26**, 141 (1987).

[502] —: The orientation of Yang-Mills moduli spaces and 4-manifold topology. J. Diff. Geom. **26**, 397 (1987).

[503] —: The geometry of 4-manifolds. Proc. Int. Congress Math. (Berkeley, 1986) 43, Amer. Math. Soc. (1987).

[504] —: Infinite determinants, stable bundles and curvature. Duke Math. J. **54**, 231 (1987).

[505] —: Polynomial invariants for smooth 4-manifolds. Topology **29**, 257 (1990).
[506] —: Gluing techniques in the cohomology of moduli spaces. Oxford preprint (1992).
[507] —: Gauge theory and four-manifold topology. Oxford preprint (1992).
[508] — and P. KRONHEIMER: The geometry of four-manifolds. Oxford Univ. Press, 1990.
[509] DOSTAGLOU, S. and D. SALAMON: Self-dual instantons and holomorphic curves. Warwick preprint (1992).
[510] DOUADY, A. and J.-L. VERDIER: Les equations de Yang-Mills. Astérisque 71–72 (1980).
[511] DREZET J.-M.: Cohomologie des variétés de modules de hauteur nulle. Math. Ann. **281**, 43 (1988).
[512] —: Groupe de Picard des variétés de modules de faisceaux semi-stables sur $P_2(C)$. Ann. Inst. Fourier (Grenoble) **38**, 105 (1988).
[513] —: Variétés de modules extrémales de faisceaux semi-stables sur $P_2(C)$. Math. Ann. **290**, 727 (1991).
[514] — and M. NARASIMHAN: Groupe de Picard des variétés de modules de fibrés semi-stables sur les courbes algébriques. Inv. Math. **97**, 53 (1989).
[515] — and C. SESHADRI: Fibrés vectoriels sur les courbes algébriques. Astérisque **96** (1982).
[516] DUBROVIN, B.: Theta functions and nonlinear equations. Russian Math. Surveys **36**, 11 (1981).
[517] —: The Kadomcev-Petviasvili equation and the relation between the periods of holomorphic differentials on Riemann surfaces. Math. USSR Izv. **19**, 285 (1982).
[518] DUCROT, F.: Fibré déterminant et courbes relatives. Bull. Soc. Math. France **118**, 311 (1990).
[519] DUISTERMAAT, J. and G. HECKMAN: On the variation in the cohomology of the symplectic form of the reduced phase space. Inv. Math. **69**, 259 (1982) and **72**, 153 (1983).
[520] DUMA, A. and W. RADKTE: Über die Dimension des singularen Ortes des Modulraumes M^g. Manuscripta Math. **45**, 147 (1984).
[521] DUMA, E.: On SL(2) actions without 3-dimensional orbits. Colloq. Math. **58**, 233 (1990).
[522] EDGE, W.: Geometry related to the key del Pezzo surface and the associated mapping of plane cubics. Proc. Roy. Soc. Edinburgh Sect. **A 107**, 75 (1987).
[523] EDIDIN, D.: The codimension-two homology of the moduli space of stable curves is algebraic. Duke Math. J. **67**, 241 (1992).).
[524] —: Brill-Noether theory in codimension two. To appear in J. Alg. Geom.
[525] EIN, L. and R. HARTSHORNE: Restriction theorems for stable rank 3 vector bundles on P^n. Math. Ann. **259**, 541 (1982).
[526] EISENBUD, D. and J. HARRIS: Limit linear series, the irrationality of M_g, and other applications. Bull. Amer. Math. Soc. **10**, 277 (1984).
[527] —: The Kodaira dimension of the moduli space of curves of genus 23. Inv. Math. **90**, 359 (1987).
[528] —: Existence, decomposition and limits of some Weierstrass points. Inv. Math. **87**, 495 (1987).
[529] —: Progress in the theory of complex algebraic curves. Bull. Amer. Math. Soc. **21**, 205 (1989).

[530] EISENBUD, D. and A. VAN DE VEN: On the variety of smooth rational space curves with given degree and normal bundle. Inv. Math. **67**, 89 (1982).

[531] ELLINGSRUD, G.: Sur l'irreductibilité du module des fibrés stables sur P^2. Math. Zeitschrift **182**, 189 (1983).

[532] — and S. STROMME: On the Chow ring of a geometric quotient. Ann. Math. (2) 130, 159 (1989).

[533] ERCOLANI, N. and H. MCKEAN: Geometry of KDV: Abel sums, Jacobi variety and theta function in the scattering case. Inv. Math. **99**, 483 (1990).

[534] FABER, C.: Chow rings of moduli spaces of curves I, II. Ann. Math. **132**, 331 and 421 (1990).

[535] FALTINGS, G.: Arithmetische Kompaktifizierung des Modulsraums der abelschen Varietäten. Springer Lect. Notes in Math. **1111**, 321 (1985).

[536] FARKAS, H.: On Fay's trisecant formula. J. Analyse Math. **44**, 205 (1984).

[537] FAUNTLEROY, A.: Invariant theory for linear algebraic groups I and II. Comp. Math. **55**, 63 (1985) and **68**, 23 (1988).

[538] —: On the moduli of curves on rational ruled surfaces. Amer. J. Math. **109**, 417 (1987).

[539] —: Quasi-projective orbit spaces for linear algebraic group actions. Invariant theory (Denton, Texas, 1986) Contemp. Math. **88**, 399. Amer. Math. Soc. (1989).

[540] —: G.I.T. for general algebraic groups. Group actions and invariant theory (Montreal, 1988) CMS Conf. Proc. **10**, 45. Amer. Math. Soc. (1989).

[541] FINTUSHEL, R. and R. STERN: SO(3) connections and the topology of 4-manifolds. J. Diff. Geom. **20**, 523 (1984).

[542] FINTUSHEL, R. and R. STERN: Pseudo-free orbifolds. Ann. Math. (2) 122, 335 (1985).

[543] FINTUSHEL, R. and R. STERN: Definite 4-manifolds. J. Diff. Geom. **28**, 133 (1988).

[544] FLOER, A.: An instanton invariant for 3-manifolds. Comm. Math. Phys. **118**, 215 (1989).

[545] FOSSUM, R.: Invariant theory, representation theory, commutative algebra – ménage a trois. Dubreil and Malliavin Seminar (Paris, 1980) Lecture Notes in Math. **867**, Springer, 1981.

[546] FREED, D. and K. UHLENBECK: Instantons and four-manifolds. M.S.R.I. Publications **1**, Springer, 1984.

[547] FRIED, M. and R. BIGGERS: Moduli spaces of covers and the Hurwitz monodromy group. J. Reine Angew. Math. **335**, 87 (1982) and correction **340**, 213 (1983).

[548] FRIEDLANDER, E.: Homology using Chow varieties. Bull. Amer. Math. Soc. **20**, 49 (1989).

[549] FRIEDMAN, R.: Base change, automorphisms and stable reduction for type III K3 surfaces. The birational geometry of degenerations (Cambridge, Massachusetts, 1981) Progress in Math. **29**, 277. Birkhäuser, 1983.

[550] —: Rank two vector bundles over regular elliptic surfaces. Inv. Math. **96**, 283 (1989).

[551] —, B. MOISHEZON and J. MORGAN: On the C^∞ invariance of the canonical class of certain algebraic surfaces. Bull. Amer. Math. Soc. **17**, 357 (1987).
[552] — and J. MORGAN: On the diffeomorphism types of certain algebraic surfaces I, II. J. Diff. Geom. **27**, 297 (1988).
[553] — and J. MORGAN: Algebraic surfaces and 4-manifolds: some conjectures and speculations. Bull. Amer. Math. Soc. **18**, 1 (1988).
[554] — and J. MORGAN: Complex versus differentiable classification of algebraic surfaces. Topology conference (Athens, Georgia, 1987). Topology and its Applications **32**, 135 (1989).
[555] FROHMAN, C.: Unitary representations of knot groups. To appear in Topology.
[556] — and A. NICAS: An intersection homology invariant for knots in a rational homology 3-sphere. To appear in Topology.
[557] FULTON, W.: On the irreducibility of the moduli spaces of curves. Inv. Math. **67**, 87 (1982)
[558] —: On nodal curves. Algebraic geometry – open problems (Ravello, 1982) Lecture Notes in Math. **997**, 146. Springer, 1983.
[559] FUTAKI, A.: The Ricci curvature of symplectic quotients of Fano manifolds. Tohoku Math. J. **39**, 329 (1987).
[560] VAN GEEMEN, B: Siegel modular forms vanishing on the moduli space of curves. Inv. Math. **78**, 329 (1984).
[561] GELFAND, I., R. M. GORESKY, R. MACPHERSON, and V. SERGANOVA: Combinatorial geometries, convex polyhedra and Schubert cells. Adv. Math. **63**, 301 (1987).
[562] GERRITZEN, L. and F. HERRLICH: The extended Schottky space. J. Reine Angew. Math. **389**, 190 (1988).
[563] —: Stable n-pointed trees of projective lines. Nederl. Akad. Wetensch. Indag. Math. **50**, 131 (1988).
[564] GIESEKER, D.: On a theorem of Bogomolov on Chern classes of stable bundles. Amer. J. Math. **101**, 79 (1979).
[565] —: Geometric invariant theory and applications to moduli problems. Invariant theory (Montecatini, 1982) Lecture Notes in Math. **996**, 45. Springer, 1983.
[566] —: A degeneration of the moduli space of stable bundles. J. Diff. Geom. **19**, 173 (1984).
[567] —: A construction of stable bundles on an algebraic surface. J. Diff. Geom. **27**, 137 (1988).
[568] — and J. LI: Irreducibility of moduli of rank 2 vector bundles on algebraic surfaces. To appear in J. Diff. Geom.
[569] — and I. MORRISON: Hilbert stability of rank two bundles on curves. J. Diff. Geom. **19**, 1 (1984).
[570] GINSBURG, V.: Equivariant cohomology and Kähler geometry. Funkts. Anal. Priloj. **21**, 271 (1987).
[571] GOLDMAN, W.: The symplectic nature of fundamental groups of surfaces. Adv. Math. **54**, 200 (1984).
[572] —: Invariant functions on Lie groups and Hamiltonian flows of surface group representations. Inv. Math. **85**, 263 (1986).
[573] —: Representations of fundamental groups of surfaces. Lecture Notes in Math. **1167**, 95 (1985).
[574] GÓMEZ-MONT, X. and G. KEMPF: Stability of meromorphic fields in projective spaces. Comment. Math. Helv. **64**, 462 (1989).

[575] GONZALEZ DIEZ, G. and W. HARVEY: On complete curves in moduli space. Math. Proc. Cambridge Philos. Soc. **110**, 461 (1991).
[576] GORESKY, M. and R. MACPHERSON: Intersection homology theory. Topology **19**, 135 (1980).
[577] GORESKY, M. and R. MACPHERSON: Intersection homology II. Inv. Math. **71**, 77 (1983).
[578] GRAVESEN, J.: On the topology of spaces of holomorphic maps. Acta Math. **162**, 249 (1988).
[579] GRIFFITHS, P.: Topics in transcendental algebraic geometry. Ann. Math. Studies **106**, Princeton (1984).
[580] — and J. HARRIS: Principles of algebraic geometry. Wiley (1978).
[581] GROISSER, D: The geometry of the moduli space of $\mathbf{CP}^2$ instantons. Inv. Math. **99**, 393 (1990).
[582] — and T. PARKER: The geometry of the Yang-Mills moduli space for definite manifolds. J. Diff. Geom. **29**, 499 (1989).
[583] GROMOV, M.: Partial differential relations. Ergebnisse der Math., Springer, 1986.
[584] GROSS, D.: Compact quotients by C^* actions. Pacific J. Math. **114**, 149 (1984).
[585] GROSSHANS, F.: The variety of points which are not semistable. Illinois J. Math. **26**, 138 (1982).
[586] —: The invariants of unipotent radicals of parabolic subgroups. Inv. Math. **73**, 1 (1983).
[587] —: Hilbert's fourteenth problem for non-reductive groups. Math. Z. **193**, 95 (1986).
[588] —: Separated orbits for certain nonreductive subgroups. Manuscripta Math. **62**, 205 (1988).
[589] —: Contractions of the actions of reductive algebraic groups in arbitrary characteristic. Inv. Math. **107**, 127 (1992).
[590] GUILLEMIN, V. and S. STERNBERG: Convexity properties of the moment mapping, I and II. Inv. Math. **67**, 491 (1982) and **77**, 533 (1984).
[591] —: A normal form for the moment map. Differential geometric methods in mathematical physics, Reidel, Dordrecht, 1984.
[592] —: Symplectic techniques in physics. Cambridge Univ. Press, 1984.
[593] —: Birational equivalence in the symplectic category. Inv. Math. **97**, 485 (1989).
[594] GUNNING, R.: Some curves in Abelian varieties. Inv. Math. **66**, 377 (1982).
[595] HARER, J.: The second homology group of the mapping class group of an orientable surface. Inv. Math. **72**, 221 (1983).
[596] —: Stability of the homology of the mapping class groups of orientable surfaces. Ann. Math. (2) 121, 215 (1985).
[597] —: The virtual cohomology dimension of the mapping class group of an orientable surface. Inv. Math. **84**, 157 (1986).
[598] —: The cohomology of the moduli space of curves. Theory of moduli (Montecatini, 1985) Lecture Notes in Math. **1337**, 138. Springer, 1988.
[599] —: The third cohomology group of the moduli space of curves. Duke Math. J. **63**, 25 (1991).
[600] —: The fourth cohomology group of the moduli space of curves. Preprint.
[601] — and D. ZAGIER: The Euler characteristic of the moduli space of curves. Inv. Math. **85**, 457 (1986).

[602] HARRIS, J.: Recent work on $\mathcal{M}_g$. Proc. Int. Cong. Math. (Warsaw, 1983) I, 719, Elsevier, 1984.
[603] —: On the Kodaira dimension of the moduli space of curves II. Inv. Math. **75**, 437 (1984).
[604] —: Families of smooth curves. Duke Math. J. **51**, 409 (1984).
[605] —: On the Severi problem. Inv. Math. **84**, 445 (1986).
[606] —: Curves and their moduli. Algebraic geometry, Bowdoin (Brunswick, Maine, 1985) Proc. Sympos. Pure Math. **46**, 99, Amer. Math. Soc. (1987).
[607] — and I. MORRISON: Slopes of effective divisors on the moduli space of curves. Inv. Math. **99**, 321 (1990).
[608] HARTSHORNE, R.: Four years of algebraic vector bundles: 1975–1979. Journées de Geometrie Algebrique d'Angers 1979, Sijthoff & Noordhoff, 1980.
[609] —: Stable reflexive sheaves. Math. Ann. **254**, 121 (1980).
[610] —: Stable reflexive sheaves II. Inv. Math. **66**, 165 (1982).
[611] — and A. HIRSCHOWITZ: Cohomology of a general instanton bundle. Ann. Sci. Ecole Norm. Sup. (4) 15, 365 (1982).
[612] — and I. SOLS: Stable rank 2 vector bundles on P^3 with $c_1 = -1$, $c_2 = 2$. J. Reine Angew. Math. **325**, 145 (1981).
[613] HAZEWINKEL, M.: A partial survey of the uses of algebraic geometry in systems and control theory. Symposia math. XXIV (INDAM, Rome, 1979) Acad. Press, 1981.
[614] —: Lectures on invariants, representations and Lie algebras in systems and control theory. Dubreil and Malliavin algebra seminar (Paris, 1982), Lecture Notes in Math. **1029**, Springer, 1983.
[615] HEINZNER, P.: Geometric invariant theory on Stein spaces. Math. Ann. **289**, 631 (1991).
[616] HELMKE, U.: Topology of the moduli space for reachable linear dynamical systems. Math. Systems Theory **19**, 155 (1986).
[617] HERNÁNDEZ, R.: Varieties of cuspidal curves in P^r. Math. Ann. **285**, 593 (1989).
[618] HESSELINK, W.: Desingularizations of varieties of nullforms. Inv. Math. **55**, 141 (1979).
[619] —: Characters of the nullcone. Math. Ann. **252**, 179 (1980).
[620] —: Concentration under action of algebraic groups. Dubreil and Malliavin algebra seminar (Paris, 1979/1980), Lecture Notes in Math. **867**, Springer, 1981.
[621] HIRSCHOWITZ, A.: Sur la restriction des faisceaux semi-stables. Ann. Sci. Ecole Norm. Sup. (4) 14, 199 (1981).
[622] — and M. NARASIMHAN: Fibrés de 't Hooft speciaux et applications. Enumerative geometry and classical algebraic geometry (Nice, 1981) Progress in Math. **24**, Birkhäuser, 1981.
[623] HITCHIN, N.: Monopoles and geodesics. Comm. Math. Phys. **83**, 579 (1982).
[624] —: Metrics on moduli spaces. Proc. Lefschetz Centennial Conference, Mexico City 1984, Amer. Math. Soc. (1986).
[625] —: The self-duality equations on a Riemann surface. Proc. London Math. Soc. **55**, 59 (1987).
[626] —: Stable bundles and integrable systems. Duke Math. J. **54**, 91 (1987).
[627] —, A. KARLHEDE, U. LINDSTROM, and M. ROCEK: Hyperkähler metrics and supersymmetry. Comm. Math. Phys. **108**, 535 (1987).

[628] —: Lie groups and Teichmüller space. Topology **31**, 449 (1992).
[629] HOCHSCHILD, G.: The structure of Lie groups. Holden-Day, San Francisco, London, Amsterdam, 1965.
[630] HOPPE, H.: Modulräume stabiler Vecktorraumbündel vom Rang 2 auf rationalen Regelflächen. Math. Ann. **264**, 227 (1983).
[631] — and H. SPINDLER: Modulräume stabiler 2-Bündel auf Regelflächen. Math. Ann. **249**, 127 (1980).
[632] HU, Y.: The geometry and topology of quotient varieties of torus actions. Duke Math. J. **68**, 151 (1992).
[633] HULEK, K., C. KAHN, and S. WEINTRAUB: Singularities of the moduli spaces of certain abelian surfaces. Comp. Math. **79**, 231 (1991).
[634] HULEK, K. and J. LE POTIER: Sur l'espace de modules des faisceaux semistables de rang 2, de classes de Chern $(0,3)$ sur P_2. Ann. Inst. Fourier (Grenoble) **39**, 251 (1989).
[635] IITAKA, S.: Algebra and geometry – classification and enumeration of algebraic varieties (in Japanese). Sugaku **29**, 334 (1978).
[636] ISHII, S.: Global moduli of equisingular complete curves. J. Algebra **78**, 255 (1982).
[637] —: Moduli space of polarized del Pezzo surfaces and its compactification. Tokyo J. Math. **5**, 289 (1982).
[638] —: Chow instability of certain projective varieties. Nagoya Math. J. **92**, 39 (1983).
[639] ITOH, M.: On the moduli space of anti-self-dual Yang-Mills connections on Kähler surfaces. Publ. Res. Inst. Math. Sci. **19**, 15 (1983).
[640] —: Geometry of Yang-Mills connections over a Kähler surface. Proc. Japan Acad. Ser. A Math. Sci. **59**, 431 (1983).
[641] JEFFREY, L. C. and J. WEITSMAN: Bohr-Sommerfeld orbits in the moduli space of flat connections and the Verlinde formula. IAS preprint IASSNS-HEP-91/82.
[642] JEFFREY, L. C. and J. WEITSMAN: Toric structures on the moduli space of flat connections on a Riemann surface: volumes and the moment map. IAS preprint IASSNS-HEP-92/95.
[643] JURKIEWICZ, J.: A remark on the orbit spaces under multiplicative group actions. Colloq. Math. **54**, 67 (1987), with correction in Colloq. Math. **57**, 361 (1989).
[644] —: On some reductive group actions on affine space. Group actions and invariant theory (Montreal, 1988) CMS Conf. Proc. **10**, Amer. Math. Soc., 1989.
[645] —: On the linearization of actions of linearly reductive groups. Comment. Math. Helv. **64**, 508 (1989).
[646] KAC, V.: Infinite root systems, representations of graphs and invariant theory II. J. Algebra **78**, 141 (1982).
[647] —: Root systems, representations of quivers and invariant theory. Invariant theory (Montecatini, 1982) Lecture Notes in Math. **996**, Springer, 1983.
[648] —: On geometric invariant theory for infinite-dimensional groups. Algebraic groups (Utrecht, 1986) Lecture Notes in Math. **1271**, Springer, 1987.
[649] KAMBAYASHI, T. and P. RUSSELL: On linearizing algebraic torus actions. J. Pure Appl. Algebra **23**, 243 (1982).
[650] KATSYLO, P.: Rationality of the moduli of hyperelliptic curves. Izv. Akad. Nauk. SSSR **48**, 705 (1984).

[651] —: Rationality of moduli varieties of plane curves of degree $3k$ (in Russian). Mat. Sb. **136** (178), 377 (1988).
[652] —: Rationality of the variety of moduli of curves of genus 5 (in Russian). Mat. Sb. **182**, 457 (1991).
[653] KAWAMATA, Y.: On the finiteness of generators of a pluricanonical ring for a 3-fold of general type. Amer. J. Math. **106**, 1503 (1984).
[654] KEEL, S.: Intersection theory of moduli space of stable n-pointed curves of genus zero. Trans. Amer. Math. Soc. **330**, 345 (1992).
[655] KEMPF, G.: Computing invariants. Lecture Notes in Math. **1278**, 81, Springer, 1987.
[656] —: Stability in representations. J. Pure Appl. Algebra **52**, 51 (1988).
[657] —: Rank g Picard bundles are stable. Amer. J. Math. **112**, 397 (1990).
[658] —, F. KNUDSEN, D. MUMFORD, and B. SAINT-DONAT: Toroidal embeddings. Lecture Notes in Math. **339**, Springer, 1973.
[659] KIRWAN, F.: Cohomology of quotients in symplectic and algebraic geometry. Math. Notes **31**, Princeton University Press, 1984.
[660] —: Convexity properties of the moment mapping III. Inv. Math. **77**, 547 (1984).
[661] —: Partial desingularisations of quotients of nonsingular varieties and their Betti numbers. Ann. Math. **122**, 41 (1985).
[662] —: On spaces of maps from Riemann surfaces to Grassmannians and applications to the cohomology of moduli of vector bundles. Ark. Math. **24**, 221 (1986).
[663] —: On the homology of compactifications of moduli spaces of vector bundles over a Riemann surface. Proc. London Math. Soc. (3) 53, 237 (1986) and **65**, 474 (1992).
[664] —: Rational intersection homology of quotient varieties. Inv. Math. **86**, 471 (1986).
[665] —: Rational intersection homology of quotient varieties II. Inv. Math. **90**, 153 (1987).
[666] —: Intersection homology and torus actions. J. Amer. Math. Soc. **1**, 385 (1988).
[667] —: The topology of reduced phase spaces of the motion of vortices on a sphere. Physica **D 30**, 99 (1988).
[668] —: Moduli spaces of degree d hypersurfaces in P_n. Duke Math. J. **58**, 39 (1989).
[669] —: The cohomology rings of moduli spaces of bundles over Riemann surfaces. J. Amer. Math. Soc. **5**, 853 (1992).
[670] —: Geometric invariant theory and the Atiyah-Jones conjecture. To appear in the proceedings of the Sophus Lie Memorial Conference, Oslo 1992.
[671] — and R. Lee: The cohomology of moduli spaces of $K3$ surfaces of degree 2, I. Topology **28**, 495 (1989).
[672] — and R. Lee: The cohomology of moduli spaces of $K3$ surfaces of degree 2, II. Proc. London Math. Soc. (3) 58, 559 (1989).
[673] —, R. LEE and S. WEINTRAUB: Quotients of the complex ball by discrete groups. Pacific J. Math. **130**, 115 (1987).
[674] KNOP, F.: Der kanonische Modul eines Invariantenrings. J. Algebra **127**, 40 (1989).
[675] KOLLÁR, J.: Projectivity of complete moduli. J. Diff. Geom. **32**, 235 (1990).

[676] — and S. MORI: Classification of three-dimensional flips, J. Amer. Math. Soc. **5**, 533 (1990).

[677] KONARSKI, J.: A pathological example of an action of k^*. Group actions and vector fields (Vancouver, 1981) Lecture Notes in Math. **956**, Springer, 1982.

[678] —: Properties of projective orbits of actions of affine algebraic groups. Group actions and vector fields (Vancouver, 1981) Lecture Notes in Math. **956**, Springer, 1982.

[679] KONTSEVICH, M.: Intersection theory on the moduli spaces of curves and the matrix Airy function. Preprint (1991).

[680] KORAS, M. and P. RUSSELL: On linearizing "good" C^*-actions on C^3. Group actions and invariant theory (Montreal, 1988) CMS Conf. Proc. 10, Amer. Math. Soc. (1989).

[681] KORAS, M. and P. RUSSELL: Codimension 2 torus actions on affine n-space. Group actions and invariant theory (Montreal, 1988) CMS Conf. Proc. 10, Amer. Math. Soc. (1989).

[682] KOSAREW, S. and C. OKONEK: Global moduli spaces and simple holomorphic bundles. Publ. Res. Inst. Math. Sci. **25**, 1 (1989).

[683] KOTSCHICK, D.: Moduli of vector bundles with odd c_1 on surfaces with $q = p_g = 0$. Amer. J. Math. **114**, 297 (1992).

[684] KOUVIDAKIS, A.: The Picard group of the universal Picard varieties over the moduli space of curves. J. Diff. Geom. **34**, 839 (1991).

[685] KRAFT, H.: Conjugacy classes and Weyl group representations. Young tableaux and Schur functions in algebra and geometry (Torun, 1980), Asterisque **87–88**, 191 (1981).

[686] —:Geometrische Methoden in Invariantentheorie.Vieweg Verlag (1984).

[687] —: Geometric methods in representation theory. Representations of algebras (Puebla, 1980) Lecture Notes in Math. **944**, Springer, 1982.

[688] —: Algebraic group actions on affine spaces. Geometry today (Rome, 1984) Progress in Math. **60**, Birkhäuser, 1985.

[689] —: Closures of conjugacy classes in G_2. J. of Algebra **126**, 424 (1989).

[690] —, T. Petrie and J. Randall: Quotient varieties. Adv. Math. **74**, 145 (1989).

[691] — and C. PROCESI: Closures of conjugacy classes of matrices are normal. Inv. Math. **53**, 227 (1979).

[692] — and C. PROCESI: Graded morphisms of G-modules. Ann. Inst. Fourier (Grenoble) **37**, 161 (1987).

[693] — and G. SCHWARZ: Reductive group actions on affine space with one-dimensional quotient. Group actions and invariant theory (Montreal, 1988) CMS Conf. Proc. 10, Amer. Math. Soc., 1989.

[694] —, P. SLODOWY and T. SPRINGER: Algebraische Transformationsgruppen und Invariantentheorie. DMV Seminar Bd. 13, Birkhäuser, 1989.

[695] KRONHEIMER, P.: The construction of ALE spaces as hyperkähler quotients. J. Diff. Geom. **29**, 665 (1989).

[696] —: A hyper-Kählerian structure on co-adjoint orbits of a semisimple complex group. J. London Math. Soc. (2) 42, 193 (1990).

[697] KULIKOV, V. and P. KURCHANOV: Complex algebraic manifolds: periods of integrals, Hodge structures. Current problems in mathematics, fundamental directions (Russian) Vol. 36, 280, 5. Itogi Nauki i Tekhniki, Akad. Nauk. SSSR, Vsesoyuz Inst. Naukn. i Tekhn. Inform., Moscow, 1989.

[698] LAUDAL, O. and G. PFISTER: Local moduli and singularities. Lecture Notes in Math. **1310**, Springer, 1988.

[699] LAUMON, G.: Un analogue global du cone nilpotent. Duke Math. J. **57**, 647 (1988).

[700] LAWSON, B.: Algebraic cycles and homotopy theory. Ann. Math. (2) 129, 253 (1989).

[701] LAX, R.: Gap sequences and moduli in genus 4. Math. Z. **175**, 67 (1980).

[702] LAZSLO, Y.: Une théorème de Riemann pour les diviseurs théta sur les espaces de modules de fibrés stables sur une courbe. Duke Math. J. **64**, 333 (1991).

[703] LE BRUYN, L.: Some remarks on rational matrix invariants. J. Algebra **118**, 487 (1988).

[704] — and C. PROCESI: Semisimple representations of quivers. Trans. Amer. Math. Soc. **317**, 585 (1990).

[705] LE POTIER, J.: Sur l'espace de modules des fibrés de Yang et Mills. Mathematics and physics (Paris, 1979/1982) Progress in Math. **37**, Birkhäuser, Boston, 1983.

[706] — : Variétés de modules de faisceaux semi-stables de rang élevé sur P_2. Algebraic geometry, Bowdoin (Brunswick, Maine, 1985) Proc. Sympos. Pure Math. **46**, Amer. Math. Soc. (1987).

[707] LEE, R. and S. WEINTRAUB: Moduli spaces of Riemann surfaces of genus two with level structures, I. Trans. Amer. Math. Soc. **310**, 217 (1988).

[708] LENSSEN, M: The singularities of the moduli space of stable bundles and its compactifications. Chapters 2,3, Oxford D. Phil. thesis (1992).

[709] LI, Y.: Spectral curves, theta divisors and Picard bundles. Internat. J. Math. 2, 525 (1991).

[710] LIN, T.: Some recent developments in the Popov-Pommerening conjecture. Group actions and invariant theory (Montreal, 1988), Canad. Math. Soc. Proc. **10**, 207, Amer. Math. Soc. (1989).

[711] LOJASIEWICZ, S.: Une propriété topologique des sous-ensembles analytiques réels. Actes du Colloque International du Centre National de la Recherche Scientifique 117 (Paris, 1962), 87 CRNS (1963).

[712] — : Ensembles semi-analytiques. IHES Notes (1965).

[713] LONSTED, K.: The singular points on the moduli spaces for smooth curves. Math. Ann. **266**, 397 (1984).

[714] LOOIJENGA, E.: New compactification of locally symmetric varieties. Algebraic geometry (Vancouver, 1984), Amer. Math. Soc. Conf. Proc. **6**, 341 (1986).

[715] LUNA, D. and Th. VUST: Plongements d'espaces homogenes. Comment. Math. Helv. **58**, 186 (1983).

[716] LÜBKE, M. and C. OKONEK: Stable bundles on regular elliptic surfaces. J. Reine Angew. Math. **378**, 32 (1987).

[717] MAGID, A.: Separated G_a-actions. Proc. Amer. Math. Soc. **76**, 35 (1979).

[718] — : Equivariant completions of rings with reductive group action. J. Pure Appl. Algebra **49**, 173 (1987).

[719] MANN, B. and R. J. MILGRAM: Some spaces of holomorphic maps to complex Grassmann manifolds. J. Diff. Geom. **33**, 301 (1991).

[720] MARGERIN, C.: Fibrés stable et métriques d'Hermite-Einstein. Bourbaki Seminar 683, Astérisque **152–153** (1987).

[721] MARLE, C.-M.: Normal forms generalizing action-angle coordinates for Hamiltonian actions of Lie groups. Lett. Math. Phys. **7**, 55 (1983).

[722] —: Modèle d'action Hamiltonienne d'un groupe de Lie sur une variété symplectique. Rendiconti del Seminario Matematico Torino **43**, 227 (1985).

[723] MARSDEN, J. and A. WEINSTEIN: Reduction of symplectic manifolds with symmetry. Rep. Math. Phys. **5**, 121 (1974).

[724] MARUYAMA, M.: The theorem of Grauert-Mulich-Spindler. Math. Ann. **255**, 317 (1981).

[725] —: The rationality of the moduli spaces of vector bundles of rank 2 on P^2. Algebraic geometry (Sendai, 1985) Adv. Stud. Pure Math. **10**, North-Holland, 1987.

[726] —: On a compactification of a moduli space of stable vector bundles on a rational surface. Algebraic geometry and commutative algebra I, Kinokuniya, 1988.

[727] — and G. TRAUTMANN: On compactifications of the moduli space of instantons. Int. J. Math. **1**, 431 (1990).

[728] — and G. TRAUTMANN: Limits of instantons. Int. J. Math. **3**, 213 (1992).

[729] MATSUSAKA, T.: On polarized varieties of dimension 3, I, II and III. Amer. J. Math. **101**, 212 (1979), **103**, 357 (1981) and **104**, 449 (1982).

[730] MCDUFF, D.: Examples of simply-connected symplectic non-Kählerian manifolds. J. Diff. Geom. **20**, 267 (1984).

[731] —: Examples of symplectic structures. Inv. Math. **89**, 13 (1987).

[732] MEHTA, V. and A. RAMANATHAN: Semistable sheaves on projective varieties and their restriction to curves. Math. Ann. **258**, 213 (1982).

[733] —: Restriction of stable sheaves and representations of the fundamental group. Inv. Math. **77**, 163 (1984).

[734] MESEGUER, J. and I. SOLS: Faisceaux semi-stables de rang 2 sur P^3. C. R. Acad. Sci. Paris Ser. I Math. **298**, 525 (1984).

[735] MESTRANO, N.: Conjecture de Franchetta forte. Inv. Math. **87**, 365 (1987).

[736] MILGRAM, R. and R. PENNER: Riemann's moduli spaces and the symmetric groups. Preprint.

[737] MILLER, E.: The homology of the mapping class group. J. Diff. Geom. **24**, 1 (1986).

[738] MIRANDA, R.: The moduli of Weierstrass fibrations over P^1. Math. Ann. **255**, 379 (1981).

[739] MIRÓ ROIG, R.: Some moduli spaces for rank 2 stable reflexive sheaves on $\mathbf{P}^3$. Trans. Amer. Math. Soc. **299**, 699 (1987).

[740] —: Chern classes of rank 3 stable reflexive sheaves. Math. Ann. **276**, 291 (1987).

[741] —: Stable rank 3 reflexive sheaves on P^3 with extremal c_3. Math. Z. **196**, 537 (1987).

[742] — and G. TRAUTMANN: The moduli scheme $M(0,2,4)$ over $\mathbf{P}_3$. Kaiserslautern preprint, 1992.

[743] MIYANISHI, M.: On group actions. Recent progress of algebraic geometry in Japan, Math. Studies **73**, North-Holland, 1983.

[744] MIYAOKA, Y.: On the Chern numbers of surfaces of general type. Inv. Math. **42**, 225 (1987).

[745] MONG, K.: On the reducibility of moduli spaces of stable 2-bundles over an elliptic surface. Math. Z. **208**, 667 (1991).

[746] Moore, R. and R. Wardelmann: $PGL(4)$ acts transitively on $M(-1,2)$. J. Reine Angew. Math. **346**, 48 (1984).
[747] Morgan, J.: Comparison of the Gieseker compactification and the Uhlenbeck compactification of Moduli space. Preprint (1991).
[748] Mori, S.: Flip theorem and existence of minimal models. J. Amer. Math. Soc. **1**, 117 (1988).
[749] — and S. Mukai: The uniruledness of the moduli space of curves of genus 11. Algebraic geometry (Tokyo/Kyoto, 1982) Lecture Notes in Math. **1016**, Springer, 1983.
[750] Morita, S.: Characteristic classes of surface bundles. Bull. Amer. Math. Soc. (N.S.) **11**, 386 (1984).
[751] Morrison, I. and H. Pinkham: Galois-Weierstrass points and Hurwitz characters. Ann. Math. (2) 124, 591 (1986).
[752] Mukai, S.: Symplectic structure of the moduli space of sheaves on an abelian or $K3$-surface. Inv. Math. **77**, 101 (1984).
[753] —: Fourier functor and its application to the moduli of bundles on an abelian variety. Algebraic geometry (Sendai, 1985) Adv. Stud. Pure Math. **10**, North-Holland, 1987.
[754] Mulase, M.: Cohomological structure in soliton equations and Jacobian varieties. J. Diff. Geom. **19**, 403 (1984).
[755] Mumford, D.: On the Kodaira dimension of the Siegel modular variety. Algebraic geometry – open problems (Ravello, 1982) Lecture Notes in Math. **997**, Springer, 1983.
[756] —: Towards an enumerative geometry of the moduli space of curves. Arithmetic and geometry II, Birkhäuser, 1983.
[757] —: Proof of the convexity theorem. Amer. J. Math. **106**, 1281 (1984).
[758] Munoz-Porras, J.: Geometric characterizations of Jacobians and the Schottky equations. Proc. Symp. Pure Math. **49** (1989).
[759] —: On the Schottky-Jung relations. Univ. of Salamanca preprint.
[760] Narasimhan, M. and T. Ramadas: Geometry of SU(2) gauge fields. Comm. Math. Phys. **67**, 121 (1979).
[761] —: Factorisation of generalised theta functions. Preprint (1992).
[762] Narasimhan, M. and G. Trautmann: Compactification of $M(0,2)$. Vector bundles on algebraic varieties (Bombay, 1984), Oxford Univ. Press, 1987.
[763] —: Compactification of $M_{P_3}(0,2)$ and Poncelet pairs of conics. Pacific J. Math. **145**, 245 (1990).
[764] —: The Picard group of the compactification of $M_{P_3}(0,2)$. J. Reine Angew. Math. **422**, 21 (1991).
[765] Naruki, I.: Cross ratio variety as a moduli space of cubic surfaces. Proc. London Math. Soc. (3) 45, 1 (1982).
[766] —: On the moduli space $M(0,4)$ of vector bundles. J. Math. Kyoto Univ. **27**, 723 (1987).
[767] Neeman, A.: Weierstrass points in characteristic p. Inv. Math. **75**, 359 (1984).
[768] —: The topology of quotient varieties. Ann. Math. **122**, 419 (1985).
[769] —: Analytic questions in geometric invariant theory. Invariant theory (Denton, Texas, 1986), Contemp. Math. **88**, Amer. Math. Soc. (1989).
[770] —: Zero cycles in $\mathbf{P}^n$. Adv. Math. **89**, 217 (1991).
[771] Ness, L.: A stratification of the null cone via the moment map. Amer. J. Math. **106**, 1281 (1984).

[772] NEWSTEAD, P.: The fundamental group of a moduli space of bundles on P^3. Topology **19**, 419 (1980).
[773] —: Covariants of pencils of binary cubics. Proc. Roy. Soc. Edinburgh Sect. **A 91**, 181 (1981/82).
[774] —: On the cohomology and the Picard group of a moduli space of bundles on P_3. Quart. J. Math. Oxford Ser. (2) 33, 349 (1982).
[775] —: Pencils on conic bundles. J. London Math. Soc. (2) 27, 19 (1983).
[776] —: On the relations between characteristic classes of stable bundles of rank 2 over an algebraic curve. Bull. Amer. Math. Soc. **10**, 292 (1984).
[777] —: Geography of stable bundles. Liverpool preprint (1992).
[778] NITSURE, N.: Cohomology of the moduli of parabolic vector bundles. Proc. Indian Acad. Sci. **95**, 61 (1986).
[779] —: Topology of conic bundles. J. London Math. Soc. (2) 35, 18 (1987).
[780] —: Cohomology of desingularization of moduli space of vector bundles. Comp. Math. **69**, 309 (1989).
[781] —: Moduli space of semistable pairs on a curve. Proc. London Math. Soc. (3) 62, 275 (1991).
[782] ODA, T.: Convex bodies and algebraic geometry. Springer, 1988.
[783] O'GRADY, K.: On the Kodaira dimension of moduli spaces of abelian surfaces. Comp. Math. **72**, 121 (1989).
[784] OKONEK, C.: Homotopiegruppen des Modulraumes $M_{P_2}(-1, c_2)$. Math. Ann. **258**, 253 (1981/82).
[785] —: Reflexive Garben auf P^4. Math. Ann. **260**, 211 (1982).
[786] —: Moduli extremer reflexiver Garben auf $\mathbf{P}^n$. J. Reine Angew. Math. **338**, 183 (1983).
[787] —, M. SCHNEIDER and H. SPINDLER: Vector bundles on complex projective spaces. Birkhäuser, 1980.
[788] — and H. SPINDLER: Stabile reflexive Garben vom Rang 3 auf P^3 mit kleinen Chernklassen. Math. Ann. **264**, 91 (1983).
[789] — and H. SPINDLER: Die Modulräume ${}^3M^{st}\mathbf{P}^3(-2, 3, c_3)$. Math. Ann. **267**, 365 (1984).
[790] — and A. VAN DE VEN: Stable vector bundles and differentiable structures on certain elliptic surfaces. Inv. Math. **86**, 357 (1986).
[791] OORT, F.: Abelian varieties: moduli and lifting properties. Algebraic geometry (Copenhagen, 1978) Lecture Notes in Math. **732**, Springer, 1979.
[792] —: Coarse and fine moduli spaces of algebraic curves and polarized abelian varieties. Sympos. Math. XXIV (INDAM, Rome, 1979) Acad. Press, 1981.
[793] PANYUSHEV, D.: On orbit spaces of finite and linear groups. Izv. Akad. Nauk SSSR Ser. Mat. **46**, 95 (1982), English translation Math. USSR-Izv. 20 (1983).
[794] PARKER, T.: A Morse theory for equivariant Yang-Mills. Duke Math. J. **66**, 337 (1992).
[795] PENNER, R.: Universal constructions in Teichmüller theory. Adv. Math. **98**, 143 (1993).
[796] —: The Poincaré dual of the Weil-Petersson Kähler 2-form. Preprint.
[797] PETRIE, T. and J. RANDALL: A topological view of algebraic actions on complex affine space. Algebra and topology (Korea, 1986) **272**, Korea Inst. Tech., 1987.

[798] POMMERENING, K.: Invarianten unipotenter Gruppen. Math. Z. **176**, 359 (1981).

[799] —: Ordered sets with the standardizing property and straightening laws for algebras of invariants. Adv. Math. **63**, 271 (1987).

[800] —: Invariants of unipotent groups (a survey). Lecture Notes in Math. **1278**, 8, Springer, 1987.

[801] POPOV, V.: On Hilbert's theorem on invariants. Dokl. Akad. Nauk SSSR **249**, 551 (1979), English translation Soviet Math. Dokl. 20 (1979).

[802] —: The constructive theory of invariants. Izv. Akad. Nauk SSSR Ser. Mat. **45**, 1100 (1981), English translation Math. USSR-Izv. 19 (1982).

[803] —: Constructive invariant theory. Young tableaux and Schur functions in algebra and geometry (Torun, 1980), Astérisque 87–88, 303 (1981).

[804] —: A finiteness theorem for representations with free algebra of invariants. Izv. Akad. Nauk SSSR Ser. Mat. **46**, 347 (1982), English translation Math. USSR-Izv. 20 (1983).

[805] —: Syzygies in invariant theory. Izv. Akad. Nauk SSSR Ser. Mat. **47**, 544 (1983), English translation Math. USSR-Izv. 22 (1984).

[806] —: Modern developments in invariant theory. Proc. Int. Cong. Math. (Berkeley, 1986), Amer. Math. Soc. (1987).

[807] —: On actions of G_a on A^n. Algebraic groups (Utrecht, 1986) Lecture Notes in Math. **1271**, Springer, 1987.

[808] —: Closed orbits of Borel subgroups (in Russian). Mat. Sb. (N.S.) **135** (177), 385 (1988).

[809] —: Groups, generators, syzygies and orbits in invariant theory. Transl. Math. Monographs **100**, Amer. Math. Soc. (1992).

[810] PROCESI, C. and G. SCHWARZ: Inequalities defining orbit spaces. Inv. Math. **81**, 539 (1985).

[811] QIN, Z.: Birational properties of moduli spaces of stable locally free rank-2 sheaves on algebraic surfaces. Manuscripta Math. **72**, 163 (1991).

[812] —: Stable rank-2 bundles on simply-connected elliptic surfaces. Duke Math. J. **67**, 557 (1992).

[813] QUILLEN, D.: Determinants of Cauchy-Riemann operators over a Riemann surface. Funct. Anal. Appl. **14**, 31 (1985).

[814] RAMADAS, T., I. SINGER, and J. WEITSMAN: Some comments on Chern-Simons gauge theory. Comm. Math. Phys. **126**, 409 (1989).

[815] RAMANAN, S. and A. RAMANATHAN: Some remarks on the instability flag. Tohoku Math. J. (2) 36, 269 (1984).

[816] RAMANATHAN, A.: Moduli for principal bundles. Algebraic geometry (Copenhagen, 1978) Lecture Notes in Math. **732**, Springer, 1979.

[817] — and S. SUBRAMANIAN: Einstein Hermitian connections on principal bundles and stability. J. Reine Angew. Math. **390**, 21 (1988).

[818] RAN, Z.: Families of plane curves and their limits: Enriques' conjecture and beyond. Ann. Math. **130**, 121 (1989).

[819] REICHSTEIN, Z.: Stability and equivariant maps. Inv. Math. **96**, 349 (1989).

[820] —: A functorial interpretation of the ring of matrix invariants. J. Algebra **136**, 439 (1991).

[821] REID, M.: What is a flip? Warwick preprint (1992).

[822] REIDER, I.: Vector bundles of rank 2 and linear systems on algebraic surfaces. Ann. Math. (2) 127, 309 (1988).

[823] RICHARDSON, R.: Irreducible components of the nullcone. Invariant theory (Denton, Texas, 1986), Contemp. Math. **88**, Amer. Math. Soc. (1989).

[824] —: Conjugacy classes of n-tuples in Lie algebras and algebraic groups. Duke Math. J. **57**, 1 (1988).

[825] — and P. SLODOWY: Minimum vectors for real reductive groups. J. London Math. Soc. (2) 42, 409 (1990).

[826] ROUSSEAU, G.: Instabilité dans les espaces vectoriels. Algebraic surfaces (Orsay, 1976-1978) Lecture Notes in Math. **868**, Springer, 1981.

[827] —: Instabilité dans les fibrés vectoriels (d'apres Bogomolov). Algebraic surfaces (Orsay, 1976–1978) Lecture Notes in Math. **868**, Springer, 1981.

[828] SALAMON, S.: Quaternionic Kähler manifolds. Inv. Math. **67**, 143 (1982).

[829] SAITO, M. and S. ZUCKER: Classification of nonrigid families of $K3$ surfaces and a finiteness theorem of Arakelov type. Math. Ann. **289**, 1 (1991).

[830] SCHNEIDER, M.: Chernklassen semi-stabiler Vektorraumbündel vom Rang 3 auf dem komplex-projektiven Raum. J. Reine Angew. Math. **315**, 211 (1980).

[831] —: Vector bundles and submanifolds of projective space: nine open problems. Algebraic geometry, Bowdoin (Brunswick, Maine, 1985) Proc. Sympos. Pure Math. **46**, Amer. Math. Soc. (1987).

[832] SCHUBERT, D.: A new compactification of the moduli space of curves. Comp. Math. **78**, 297 (1991).

[833] SCHUMACHER, G.: Construction of the coarse moduli space of compact polarized Kähler manifolds. Math. Ann. **264**, 81 (1983).

[834] —: Moduli of polarized Kähler manifolds.. Math. Ann. **269**, 137 (1984).

[835] —: On the geometry of moduli spaces. Manuscripta Math. **50**, 229 (1985).

[836] SCHWARZ, G.: On classical invariant theory and binary cubics. Ann. Inst. Fourier (Grenoble) **37**, 191 (1987).

[837] —: The topology of algebraic quotients. Topological methods in algebraic transformation groups (New Brunswick, 1988) Progress in Math. **80**, Birkhäuser (1989).

[838] —: Exotic algebraic group actions. C. R. Acad. Sci. Paris Ser. I Math. **309**, 89 (1989).

[839] SEKIGUCHI, T.: How coarse the coarse moduli spaces for curves are! Algebraic geometry and commutative algebra II, 693, Kinokuniya, 1988.

[840] SEPPÄLÄ, M.: Complex algebraic curves with real moduli. J. Reine Angew. Math. **387**, 209 (1988).

[841] —: Real algebraic curves in the moduli space of complex curves. Comp. Math. **74**, 259 (1990).

[842] —: Moduli space of stable real algebraic curves. Ann. Sci. Ec. Norm. Sup. (4) 24, 519 (1991).

[843] — and R. SILHOL: Moduli spaces for real algebraic curves and real abelian varieties. Math. Z. **201**, 151 (1989).

[844] SERNESI, E.: Unirationality of the variety of moduli of curves of genus twelve. Ann. Scuola Norm. Sup. Pisa Cl. Sci. (4) 8, 405 (1981).

[845] SESHADRI, C.: Fibrés vectoriels sur les courbes algébriques. Astérisque **96**, Soc. Math. France (1982).

[846] SHATZ, S: The decomposition and specialization of algebraic families of vector bundles. Comp. Math. **33**, 163 (1977).

[847] SHEPHERD-BARRON, N.: The rationality of some moduli spaces of plane curves. Comp. Math. **67**, 51 (1988).

[848] — Apolarity and its applications. Inv. Math. **97**, 433 (1989).

[849] —: Invariant theory for S_5 and the rationality of M_6. Comp. Math. **70**, 13 (1989).

[850] SHIOTA, T.: Characterization of Jacobian varieties in terms of soliton equations. Inv. Math. **83**, 333 (1986).

[851] SILHOL, R.: Compactifications of moduli spaces in real algebraic geometry. Inv. Math. **107**, 151 (1992).

[852] SILVERMAN, J. and J. VOLOCH: Multiple Weierstrass points. Comp. Math. **79**, 123 (1991).

[853] SIMPSON, C.: Constructing variations of Hodge structure using Yang-Mills theory, with applications to uniformisation. J. Amer. Math. Soc. **1**, 867 (1989).

[854] SINGHOF, W. and G. TRAUTMANN: On the topology of the moduli space $M(0,2)$ of stable bundles of rank 2 on $\mathbf{P}_3$. Quart. J. Math. Oxford (2) 41, 335 (1990).

[855] SJAMAAR, R.: Singular orbit spaces in Riemannian and symplectic geometry. Ph. D. thesis, Utrecht (1990).

[856] — and E. LERMAN: Stratified symplectic spaces and reduction. Ann. Math. **134**, 375 (1991).

[857] SLODOWY, P.: Simple singularities and simple algebraic groups. Lecture Notes in Math. **815**, Springer, 1980.

[858] ŚNIATYCKI, J. and A. WEINSTEIN: Reduction and quantization for singular momentum mappings. Lett. Math. Phys. **7**, 155 (1983).

[859] SNOW, D.: Unipotent actions on affine space. Topological methods in algebraic transformation groups (New Brunswick, 1988), Progress in Math. **80**, Birkhäuser, 1989.

[860] —: Reductive group actions on Stein spaces. Math. Ann. **259**, 79 (1982).

[861] SOBERON-CHAVEZ, S.: Rank 2 vector bundles over a complex quadric surface. Quart. J. Math. (2) 36, 159 (1985).

[862] SOURIAU, J.: Structures des systemes dynamiques. Dunod, Paris (1970).

[863] SPINDLER, H.: Ein Satz über die Einschrängkung holomorpher Vektorbündel auf $\mathbf{P}_n$ mit $c_1 = 0$ auf Hyperebenen. J. Reine Angew. Math. **327**, 93 (1981).

[864] —: Die Modulräume stabiler 3-Bündel auf P_3 mit den Chernklassen $c_1 = 0$, $c_3 = c_2^2 - c_2$. Math. Ann. **256**, 133 (1981).

[865] —: Holomorphe Vektorbündel auf P_n mit $c_1 = 0$ und $c_2 = 1$. Manuscripta Math. **42**, 171 (1983).

[866] — and G. TRAUTMANN: Special instanton bundles on P_{2N+1}, their geometry and their moduli. Math. Ann. **286**, 559 (1990).

[867] SPRINGER, T.: Trigonometric sums, Green functions of finite groups and representations of Weyl groups. Inv. Math. **36**, 173 (1976).

[868] —: Invariant theory. Lecture Notes in Math. **585**, Springer, 1977.

[869] STEENBRINK, J. and S. ZUCKER: Variation of mixed Hodge structure I. Inv. Math. **80**, 489 (1985).

[870] STERK, H.: Compactifications of the period space of Enriques surfaces. Math. Z. **207**, 1 (1991).

[871] STROMME, S.: Ample divisors on fine moduli spaces on the projective plane. Math. Z. **187**, 405 (1984).

[872] SUNDARAM, N.: Special divisors and vector bundles. Tohoku Math. J. **39**, 175 (1987).
[873] SUNDARARAMAN, D.: Moduli, deformations and classifications of compact complex manifolds. Research Notes in Math. **45**, Pitman (1980).
[874] SZENES, A.: Verification of Verlinde's formulas for $SU(2)$. Internat. Math. Res. Notices **93** (1991).
[875] TAI, Y.: On the Kodaira dimension of the moduli space of abelian varieties. Inv. Math. **68**, 425 (1982).
[876] — : Pluricanonical differentials on the Siegel modular variety. Inv. Math. **68**, 425 (1982).
[877] TAKIGAWA, N.: Weierstrass points on compact Riemann surfaces with nontrivial automorphisms. J. Math. Soc. Japan **33**, 235 (1981).
[878] TAN, L.: On the Popov-Pommerening cojecture for groups of type A_n. Proc. Amer. Math. Soc. **106**, 611 (1989).
[879] TANNENBAUM, A.: Invariance and system theory: algebraic and geometric aspects. Lecture Notes in Math. **845**, Springer, 1981.
[880] TAUBES, C.: Path-connected Yang-Mills moduli spaces. J. Diff. Geom. **19**, 337 (1984).
[881] — : A framework for Morse theory for the Yang-Mills functional. Inv. Math. **94**, 327 (1988).
[882] — : The stable topology of self-dual moduli spaces. J. Diff. Geom. **29**, 162 (1989).
[883] TEIXIDOR I BIGAS, M. and L. TU: Theta divisors for vector bundles. Preprint.
[884] — : Brill-Noether theory for vector bundles of rank 2. Tohoku Math. J. **43**, 123 (1991).
[885] — : Brill-Noether theory for stable vector bundles. Duke Math. J. **62**, 385 (1991).
[886] — : Moduli spaces of (semi)stable vector bundles on tree-like curves. Math. Ann. **290**, 341 (1991).
[887] THADDEUS, M.: Conformal field theory and the cohomology of the moduli space of stable bundles. J. Diff. Geom. **35**, 131 (1992).
[888] — : Stable pairs, linear systems and the Verlinde formula. Submitted to Inv. Math.
[889] THOMASON, R.: Equivariant resolution, linearization, and Hilbert's fourteenth problem over arbitrary base schemes. Adv. Math. **65**, 16 (1987).
[890] TRAUTMANN, G.: Poncelet curves and associated theta characteristics. Expos. Math. **6**, 29 (1988).
[891] — : Orbits that always have affine stable neighbourhoods. Adv. in Math. **91**, 54 (1992).
[892] TSUYUMINE, S.: Thetanullwerte on a moduli space of curves and hyperelliptic loci. Math. Z. **207**, 539 (1991).
[893] TYURIN, A.: The structure of the variety of pairs of commuting pencils of symmetric matrices (in Russian). Izv. Akad. Nauk SSSR Ser. Mat. **46**, 409 (1982).
[894] — : Symplectic structures on the moduli spaces of vector bundles on algebraic surfaces with $p_g > 0$ (in Russian). Izv. Akad. Nauk SSSR Ser. Mat. **52**, 813 (1988); translation in Math. USSR-Izv. **33**, 139 (1989).
[895] UENO, K.: Classification theory of algebraic varieties and compact complex spaces, Lecture Notes in Math. **439**, Springer, 1974.
[896] UHLENBECK, K.: Connections with L^p bounds on curvature. Comm. Math. Phys. **83**, 31 (1982).

[897] — : Removable singularities in Yang-Mills fields. Comm. Math. Phys. **83**, 11 (1981).

[898] — and S.-T. YAU: On the existence of hermitian Yang-Mills connections on stable bundles over compact Kähler manifolds. Comm. Pure Appl. Math. **39**, 257 (1986) and correction **42**, 703 (1989).

[899] VAN DE VEN, A: Twenty years of classifying algebraic vector bundles. Journées de Geometrie Algebrique d'Angers 1979, Sijthoff & Noordhoff (1980).

[900] VERDIER, J.-L.: Instantons. Astérisque 71–72 (1980).

[901] — and J. LE POTIER: Module des fibrés stables sur les courbes algébriques. Progress in Math. **54**, Birkhäuser, 1985.

[902] VERLINDE, E.: Fusion rules and modular transformations in 2d conformal field theory. Nucl. Phys. **B 300**, 360 (1988).

[903] VIEHWEG, E.: Weak positivity and the stability of certain Hilbert points I and II. Inv. Math. **96**, 639 (1989) and **101**, 191 (1990).

[904] VINBERG, E. and V. POPOV: Invariant theory. Encyclopaedia of Math. Sci., Springer, 1992.

[905] VISTOLI, A.: Chow groups of quotient varieties. J. Algebra **107**, 410 (1987).

[906] — : Intersection theory on algebraic stacks and on the moduli spaces. Inv. Math. **97**, 613 (1989).

[907] WALL, T.: Geometric invariant theory of linear systems. Math. Proc. Camb. Phil. Soc. **93**, 57 (1983).

[908] WEHLAU, D.: Some recent results on the Popov conjecture. Group actions and invariant theory (Montreal, 1988) CMS Conf. Proc. **10**, Amer. Math. Soc. (1989).

[909] — : A proof of the Popov conjecture for tori. Proc. Amer. Math. Soc. **114**, 839 (1992).

[910] WEITSMAN, J.: Real polarization of the moduli space of flat connections on a Riemann surface. Comm. Math. Phys. **145**, 425 (1992).

[911] WELTERS, G.: On flexes of the Kummer variety. Nederl. Akad. Wetensch. Proc. Ser. A 84 45, 501 (1983).

[912] — : A characterization of non-hyperelliptic Jacobi varieties. Inv. Math. **74**, 437 (1983).

[913] — : A criterion for Jacobian varieties. Ann. Math. (2), 120, 497 (1984).

[914] WILDBERGER, N.: Convexity and unitary representations of nilpotent Lie groups. Inv. Math. **98**, 281 (1989).

[915] — : The moment map of a Lie group representation. Trans. Amer. Math. Soc. **330**, 257 (1992).

[916] WITTEN, E.: Quantum field theory and the Jones polynomial. Comm. Math. Phys. **121**, 351 (1989).

[917] — : On quantum gauge theories in two dimensions. Comm. Math. Phys. **141**, 153 (1991).

[918] — : Two dimensional gauge theories revisited. IAS preprint IASSNS-HEP-92/15 (1992).

[919] WOLPERT, S.: On the homology of the moduli space of stable curves. Ann. Math. (2) 118, 491 (1983).

[920] — : The geometry of the moduli space of Riemann surfaces. Bull. Amer. Math. Soc. **11**, 189 (1984).

[921] — : On the Weil-Petersson geometry of the moduli space of curves. Amer. J. Math. **107**, 969 (1985).

[922] YAU, S.-T.: On the Ricci curvature of a compact Kähler manifold and the complex Monge-Ampere equation. Comm. Pure Appl. Math. **31**, 339 (1978).

[923] YUKIE, A.: Applications of equivariant Morse stratifications. Harvard Ph.D. thesis, 1986.

[924] ZUCKER, S.: Variation of mixed Hodge structure II. Inv. Math. **80**, 543 (1985).

[925] ZUO, K.: Generic smoothness of the moduli spaces of rank two stable vector bundles over algebraic surfaces. Math. Z. **207**, 629 (1991).

[926] — : Regular 2-forms on the moduli space of rank two stable bundles on an algebraic surface. Duke Math. J. **65**, 45 (1992).

Index of definitions and notations